Safety in Complex Organizations

Managing safety in today's complex organizations is no easy feat. When looking at safety from a cultural perspective, organizations are seen as cultures, with their own internal dynamics and ways of interacting with, and interpreting, the outside world. Looking at organizations from a cultural perspective improves our ability to lead and manage them by increasing our understanding of these internal dynamics shaping their behavior. This book offers a theoretically sound yet practical approach to organizational and cultural issues. It investigates their influence on safety and acknowledges the complexity and dynamic nature of today's organizations.

The book is a comprehensive introduction to a cultural approach to safety. It explains the main principles in thinking about safety from a cultural perspective and provides the reader with tools for monitoring and developing culture for safety in organizations. Methods for developing a cultural approach to safety are introduced, and typical approaches to assessing culture for safety are critically reviewed. It draws on more than two decades of experience that the author has in the safety field as a researcher, consultant, and practitioner and includes practical examples, lessons learned, and research findings from several safety-critical fields, including the nuclear, healthcare, construction, petrochemical, and transportation sectors. The reader will develop a thorough overview towards developing their own cultural approach to safety through overviews of resilience and complexity thinking, organizational psychology, management sciences, leadership research, psychological safety, normalization, drift, decision making, leadership, and communication.

Safety in Complex Organizations: A Cultural Approach is a fascinating read for professionals from any discipline who deal with hazards and safety issues, especially those in the nuclear industry; transportation industries, including aviation, rail, and maritime; petrochemical industries; the food industry; healthcare; mining; construction; and conventional and renewable power industries.

Safety in Complex Organizations

A Cultural Approach

Teemu Reiman

CRC Press
Taylor & Francis Group
Boca Raton London New York

CRC Press is an imprint of the
Taylor & Francis Group, an informa business

Designed cover image: Mielikuvitustoimisto

First edition published 2026
by CRC Press
2385 NW Executive Center Drive, Suite 320, Boca Raton FL 33431

and by CRC Press
4 Park Square, Milton Park, Abingdon, Oxon, OX14 4RN

CRC Press is an imprint of Taylor & Francis Group, LLC

ISBN: 9781032789163 (hbk)
ISBN: 9781032792453 (pbk)
ISBN: 9781003491156 (ebk)

DOI: 10.1201/9781003491156

Typeset in Times
by Deanta Global Publishing Services, Chennai, India

Contents

About the author

Teemu Reiman has more than twenty-five years of experience working in the nuclear, healthcare, construction, and rail sectors, as well as experience in the mining, maritime, aviation, oil and gas, research, defense, and manufacturing industries. Dr Reiman has a doctoral degree in Psychology from the University of Helsinki, Finland, and is an associate professor at Aalto University, Finland. He has worked as a safety culture manager at a Finnish nuclear new-build project and as a senior scientist at the Finnish technical research center VTT. Today, his company Lilikoi (www.lilikoi.fi) offers his clients a deep understanding of organizational behavior, leadership, and culture and provides actionable insights and tailored strategies that help them manage safety.

Preface

Cultural issues in relation to safety and accidents have been considered already since the 1950s, but it was the Chernobyl nuclear accident in 1986 that brought culture into the limelight. Lack of, or deficiencies in "safety culture" were then identified in several other investigations, such as those into the King's Cross underground fire in London in 1987 and the Piper Alpha offshore platform fire in 1988.

Since then, research and practical applications of culture for safety and safety culture have proliferated, even though researchers could not agree on a shared definition of the concept itself. The concept seemed to be something that resonated with both the industry as well as the research community. This sparked my interest in the cultural approach some twenty-five years ago.

I have always considered myself as an applied researcher and a research-oriented practitioner. I have seen how these two worlds of research and practice co-exist but do not always understand, or even appreciate, each other. It is as if they had different cultures … both worlds fascinate me, and I have lived and worked in each.

I believe the concept of culture offers a way not only to understand the differences, but also to reconcile some of them and improve the way we manage safety. However, this requires an elaboration of what the cultural view on safety entails, and how it can contribute both to safety practice and safety science. This is what I attempt to do in this book by asking how the concept can add value to the management of safety in complex organizations.

Safety, or accidents, do not respect academic boundaries or theoretical concepts. I believe that approaching organizations from a cultural perspective can bring a particular view, a diverse way of looking at the challenge of safety, different from the traditional ways of managing it. However, the undefined nature of the concept of culture has also led to simplified conceptualizations and markets for consultants with their silver bullets and miracle cures. Especially the concept of "safety culture" has suffered from this sacralization of culture. Results are often counterproductive.

When clarifying the cultural approach, it needs to be put into perspective with other safety approaches. In 2014 I wrote a critique of the safety culture concept together with my colleague Carl Rollenhagen (Reiman & Rollenhagen, 2014). Our main argument was that safety culture discourse had largely failed to integrate itself with the more technically focused safety approaches and other safety considerations to capture a systems view of organizations and safety. This book is my attempt to respond to this critique and present an approach to safety that is based on a cultural view of the organization.

In this book I aim to balance scientific rigor with a pragmatic approach. My approach is eclectic on purpose. I provide brief introductions to the different safety paradigms, and the interested reader is encouraged to dig deeper. Different paradigms highlight different aspects of safety. My educational background is in organizational psychology, but I integrate findings from engineering, management sciences, sociology, anthropology, social psychology, and cognitive psychology. The reader will be the judge of whether I succeed or not.

This book has several goals. First, I investigate safety and accidents from a cultural perspective. I look at how accidents happen and what meanings individuals and organizations have attributed to them. I will also inspect safety, what it is, and how it has been approached in safety science during the past decades. Second, I wish to convey a view of organizations as cultures, wherein people and technology interact and organize. This view challenges both behaviorist and mechanistic views of the organization. It focuses on interactions, meanings, and preconditions for behavior rather than the actions and behaviors as such. It is more a way of looking at organizations than a particular method of assessing or developing them. However, adopting this way of looking has strong implications for the selection and utilization of safety management methods, as I will attempt to demonstrate. In this view, culture is a lens through which to look at organizations, catching the things that are missed by structural and behavioral ways of looking.

This book illustrates what kind of issues are of interest to practitioners or researchers that approach safety-critical organizations from a cultural perspective. I try to demonstrate how the cultural approach helps in understanding and managing safety related challenges in organizations. This means not only changing culture but also becoming aware of the cultural nuances and using that awareness to tackle concrete daily problems (and changing culture as a side-effect). Finally, I try to elaborate both the content and form of the cultural approach, as well as its specific application to safety.

The approach taken in this book is based on a systems view, moving beyond a mechanistic view of organizations and safety management to a more process-oriented view. In addition, the importance of work context as well as the socially constructed and sometimes dysfunctional sides of organizational reality is acknowledged. The practice and science of safety long ago identified the three elements of importance for safety, namely technology, human behavior, and organization. The particular focus of interventions has varied depending on the method and underlying assumptions. However, what has often been neglected is the interaction between the three elements.

I will attempt to show that progress in safety does not depend only on shifting focus from technology to humans and ultimately to organizations, but also on understanding that these are all part of the same sociotechnical system – and this system has its own peculiar characteristics, including its culture. Only by understanding the interactions and their emergent manifestations as cultural assumptions can we truly understand organizations. And with understanding comes the ability not to control our organizations, but to influence and steer them.

A cultural approach is very much about indirect effects and indirect influences. Culture cannot be changed directly. Even behavior is elusive to direct change efforts. Behavior and culture are emergent and always taking place in an environment where the constraints and possibilities are ever changing yet have a pervasive, but non-deterministic, influence. Understanding these influences and delayed effects is at the root of systems thinking.

Creating a culture for safety is not easy and it does not happen overnight. I do not want to sugar-coat things too much or promise silver bullets and lightning-fast solutions. In fact, if after reading this book, you know better what the problems are, this book has already served its purpose. I do offer solutions and practical tools throughout the book, but all of them require first an understanding of what culture is, and how it affects how we understand and manage safety. I discuss in depth how our understanding of safety is rooted in culture, and how this understanding can sometimes lead us astray in our search for these silver bullets and quick and easy measures.

Culture is a phenomenon that an organization creates for itself and that, once created, affects the organization. The organization itself does not perceive this effect in many ways, because all things that happen according to cultural assumptions are considered "business as usual" by members of the organization. Culture is created, re-created, and reinforced repeatedly as members behave and communicate in ways that seem natural, obvious, and undeniable to them. Culture is learned. We do not have culture when we are born, but we learn and grow into the culture. Culture cannot be directly seen, but its effects are tangible. Sometimes these effects include accidents.

Culture has a purpose – or rather, we learn for a reason. The world is much too complex for any of us to pay attention to all details all the time. We need rules, expectations, guidelines, heuristics, to help us manage our daily lives. However, culture is not the same as behavior. Culture affects behavior, and most behaviors can be considered an artifact of culture – emphasis on the most. Culture concerns what happens before, during, and after behaviors. It is the context in which behavior gains meaning. Culture is beneath behavior, as Alvesson and Sveningsson (2024) succinctly put it. This does not mean that the cultural approach should dismiss behavior but rather that behavior should not be its main focus.

A cultural approach does not mean that we put more focus on people than technology. Rather, it means that we consider all three elements, people, technology, and organization, together. The cultural approach considers how safety is perceived and how structures and interaction support creating and maintaining an adequate view on safety. While culture is inherently subjective and inter-subjective, it has significant implications for safety. The cultural approach begins with an understanding of this inter-subjective reality and then evaluates its impact on various outcomes, including safety. Only by comprehending the current culture and cultural dynamics in general can we identify the most effective strategies to influence it.

I will outline a cultural approach that relies on systems view and clarify the added value of using the concept "culture" when managing safety in complex organizations. I will also illustrate the kinds of questions that are asked and issues that are considered when looking at complex organizations from a cultural perspective and how these may affect safety. An organization is complex when it has many interconnected parts, heterogeneity, and unpredictable interactions. A complex organization cannot be fully understood or managed by looking only at its parts in isolation. A systems view is required to manage safety in complex organizations.

The book is intended for anyone who is genuinely interested in safety at the workplace and wishes to understand the myriad issues affecting it. It is especially useful for managers and leaders of organizations that wish to take safety seriously and contribute to creating a culture for safety in their organizations. The book has been written so that it is approachable for both academics and practitioners. The practical value of scientific theories is explained, and the practical tools and methods are put in a scientific context. Thus, the book also serves as a resource and a handbook for students and academics in the field of safety management, safety culture, and leadership.

THE STRUCTURE OF THE BOOK

The first part of the book, Chapters 1–4, provides the reader with information on culture, complex organizations, human behavior, accidents, and safety.

Chapter 1 deals with the basics of what culture is and what it means to approach safety from a cultural perspective. Basic definitions of culture are provided, and the cultural approach to safety in complex organizations is outlined.

At the end of the chapter, I introduce systems thinking and explain why I consider it a necessary complement to the present cultural theory. Culture distinguishes human systems from most other types of systems.

Chapter 2 deals with the basics of human behavior individually, in groups, and in organizations. Understanding the variability and contextuality of human performance is important to understanding safety in complex organizations. At the end of Chapter 2, the first detailed accident case is presented as we take a look at NASA and the two Space Shuttle accidents, namely *Challenger* and *Columbia*. These accidents illustrate many of the aspects of culture and safety we deal with in this book, such as culture acting simultaneously as constraint and enabler as well as culture as resisting change. We will return to these two accidents multiple times in this book. As they are among the most studied accidents in history, they can teach us a lot about culture and safety.

Chapter 3 looks at the various reasons for accidents from a cultural point of view. The chapter is interested in the explanations that we give for accidents, going full circle from the errors conducted at the front lines to system complexity and finally the strategic mistakes of senior managers. The end of Chapter 3 includes an in-depth look at the classic sociological accident models the Man-Made Disasters model by Barry Turner and the Normal Accidents model by Charles Perrow. Although developed already in the 1970s and 1980s, these models include important findings about culture and safety as well as about the role of power in organizations that are still valid today.

Chapter 4 turns its attention to the question of what safety is. I will argue that safety is a property of the sociotechnical system and that to manage safety, the sociotechnical system needs to be managed. I will also make a differentiation between four types of hazards that contribute to four different safeties: local hazards, core task–related hazards, external hazards, and psychosocial hazards. The chapter also introduces a model of tensions that managing safety in complex organizations has to balance.

The end of Chapter 4 includes the second detailed accident case, a description of BP's Texas City and Deepwater Horizon accidents. These two accidents demonstrate the challenges of defining and measuring safety in complex organizations, along with showing the complexities of learning from experience and organizing for safety.

The second part of the book, Chapters 5–9, deals with various aspects of safety in complex organizations by considering what a good culture for safety looks like on a general level (Chapter 5), how organizing affects safety and culture (Chapter 6), how to lead complex organizations (Chapter 7), and how to organize the safety function and the work of safety professionals (Chapter 8).

Chapter 5 focuses explicitly on the concept of safety culture, which has become a much-used but also a much-critiqued concept in most safety-critical domains. I propose that the concept has some merits, especially as a normative model of issues that a healthy culture for safety should consider. The end of the chapter presents a highly influential organizational safety model with a strong focus on the importance of culture: High Reliability Organizations.

Chapter 6 presents the essential organizational functions and processes needed to manage safety. I consider how the various ways of organizing have an influence on safety and culture, intentionally or unintentionally. The chapter ends with the third detailed accident case, relating to the Fukushima and Onagawa nuclear power stations (NPS). This time we compare two organizations in similar external conditions during the Great East Japan earthquake and tsunami to find out why one of them had a severe accident (Fukushima NPS) and the other did not (Onagawa NPS).

Chapter 7 delves into the challenges of leadership for safety. The model of tensions that was introduced in Chapter 4 is refined into a model of adaptive safety management. The model elaborates eight safety leadership principles. These principles are described in detail along with examples of their implementation in practice. The end of Chapter 7 includes an introduction to two typologies of organizational culture: the Competing Values Framework and the Multi-Focus Model. These models do not have an explicit focus on safety and that makes them especially valuable tools for a cultural approach to safety where safety is considered a property of the entire system.

Chapter 8 looks at safety professionals and safety departments. The same principles as used in Chapter 7 for safety leadership are described in this chapter from the perspective of safety professionals. Safety professionals need to oversee as well as challenge the line organization in safety issues, and thus adaptive safety management requires slightly different principles compared to line organization. Chapter 8 ends with the fourth and final detailed accident case: Boeing 737 MAX crashes, which illustrate the challenges of organizing safety as well as the challenges of a globalized environment.

Chapter 9 illustrates the various methods and tools that can be used to manage safety in complex organizations. The chapter does not aim to teach how these methods are to be used in detail, but it rather gives advice on their use from a cultural perspective. At the end of the chapter, the Design for Integrated Safety Culture (DISC) model is elaborated.

The third part of the book, Chapters 10–12, looks at how to monitor, assess, and develop culture for safety in complex organizations.

Chapter 10 looks at complex organizations and their cultures in the context of global challenges, industry subcultures, and other contextual factors affecting both a current culture and the ideal culture when considered from a safety perspective. The chapter ends by introducing Hofstede's model of six dimensions of national culture and discussing their safety implications.

Chapter 11 is devoted to the assessment and monitoring of organizational culture. At the end of the chapter, the final and most recent theory that helps us in creating a culture for safety is introduced, although at this point it has already been heavily referenced in the previous chapters: Resilience Engineering.

Chapter 12 presents the outline of the cultural approach to safety and summarizes the key messages of the book. The chapter proposes a shift in the perspective from which we approach safety in complex environments. The chapter ends with questions that help readers to get started with creating a culture for safety.

Acknowledgments

The detailed comments and insights from Kenneth Pettersen Gould, Jean-Christophe Le Coze, Saara Reiman, and Kaupo Viitanen have significantly elevated the quality of the final product. I am profoundly grateful to them. Any remaining errors are entirely my own.

My heartfelt thanks go to all my clients and cooperation partners in the various safety-critical domains where I have had the privilege to work. I am also deeply appreciative of the research funding from the nuclear, healthcare, and rail domains over the years, which has enabled me to embark on this journey of lifelong learning.

I have been incredibly fortunate to work with some exceptional colleagues, whose expertise and guidance have profoundly enriched my work. I extend my sincere thanks to Carl Rollenhagen, Caroline Pike, Leena Norros, Pia Oedewald, Vesa Ruuska, Hanna Nokkonen, Pirjo Knuuttila, Lena Kecklund, Jan Skriver, Monica Haage, Egbert Schram, Matti Vartiainen, Heli Talja, Kirsi Pajunen, Erik Hollnagel, Merja Airola, Nadezhda Gotcheva, and Gregory Rolina.

I may have inadvertently overlooked some people due to the limitations of my memory. Apologies for any omission.

I am grateful to the Finnish Work Environment Fund (TSR) for their support in financing the finalization of this book.

The figures in this book, including the cover image, have been beautifully developed by Mielikuvitustoimisto and Anri Veisterä.

Above all, my deepest gratitude goes to Eerika for taking care of everything else.

1 Organizational Culture and Safety

In this chapter, I will present the concept of organizational culture and discuss how it forms through shared learning experiences and interactions. I highlight the importance of understanding both the structural and dynamic aspects of culture, including artifacts, espoused beliefs, and basic assumptions. I will present the basic ideas of the cultural approach to organizations and how it helps to manage safety.

I view culture as both the process and the temporary outcome of people interacting with each other and their environment and trying to make sense (of the environment, of each other, of themselves). I will clarify in this chapter why it is important to separately highlight these different characteristics of culture – culture as a process and an outcome (a structure) – if we want to understand both the individuals' potential to affect culture and culture's effect on each of us in turn. But first, why does culture form in the first place? And what good (or bad) does it do for us?

1.1 CULTURE AS AN EXPLANATION

We are not born with culture, but we grow into it. Metaphorically, culture can be seen as the thread that weaves individuals into groups and groups into organizations, organizations into societies, and societies into humankind.[1] As with all metaphors, this highlights some aspects of culture, while hiding the others. We can also see culture as an iceberg, a lily pond, a river, a brain, a tree, and so on. All these definitions are attempts to make sense of a complex phenomenon. Culture is an abstraction that helps us to understand groups and the reciprocal relation between individuals. Culture is an explanation and a justification borne out of our need to understand, communicate, and belong.

As a holistic, abstract concept, culture is notoriously difficult to define in clear and precise terms. This is not for the lack of trying, both when it comes to concept of organizational culture and safety culture. Ironically, our need to define and make sense of our environments is what creates "culture" in the first place. Culture represents the silent consensus in groups that cooperation and coordination require, including expectations about appropriate behavior and ready-made solutions to various ordinary problems. Culture also represents the values on which behavior is built (Kluckhohn, 1962). Yet, in addition to consensus, culture also incorporates topics that are avoided, disputed, and openly disagreed with.

Most people crave clarity and meaning in their lives. Often a simple story is preferred over a complex truth. We see this in how organizations explain both success and failure, but also in how we make sense of our everyday lives. Culture is a product of our human tendencies. Culture offers clarity; it is the whisper in the wind that reminds us of how things have been in the past, hints at how they are now, and proposes how they should be in future.[2]

Culture comforts. Too much uncertainty and vagueness are distressing. Things must make sense to us. Most of us long for explanations for the unfolding events of our lives. Based on research, we know that people generally abhor randomness and are typically more inclined to associate events with fate or some all-powerful force rather than pure coincidence. For us, any explanation is better than no explanation at all. Our needs for consistency, meaning, and predictability in our lives are the building blocks of culture. Culture is a non-permanent consequence of our joint sense-making as we learn to live and work together. Whether it's national culture, organizational culture, or team culture, culture is always based on identity building, it's always learned and taught, and it brings us a sense of predictability, stability, and continuity in an otherwise ambiguous environment.

DOI: 10.1201/9781003491156-1

Even though culture forms to offer stability and predictability in our lives, it is never static. Our social reality is being continuously created and recreated. This typically happens gradually, so that we don't even notice the change until a certain threshold has been reached.

In addition to craving meaning and clarity, most people have a powerful need to belong. We are inherently social creatures. Being a member of a certain group is an integral part of our identities; it is our *social identity.* In Chapter 2 we will take a closer look at both sensemaking and social identity.

In this book, we are interested in culture, because we are interested in managing complex organizations safely. And to manage them safely, we need to understand how they work, develop, and adapt, and how the people in them behave. Today's organizations are complex, technologically and socially. Simple, linear models can give us a false image of their performance. This is why it is important to specify what we mean by culture and avoid a simplified understanding of it.

1.2 MODELS OF CULTURE

The term "organizational culture" became popularized in the 1980s with bestseller books such as *In Search of Excellence* (Peters & Waterman, 1982) and *Corporate Cultures* (Deal & Kennedy, 1982). Both books emphasized the importance of leaders in creating the right culture for the company to succeed. Among the first scientific publications was Andrew Pettigrew's article where he defined culture as a "system of … publicly and collectively accepted meanings operating for a given group at a given time" (Pettigrew, 1979, p. 574). He emphasized the importance of symbols and language, and their capacity to impel people to action. Ideologies, beliefs, rituals, and myths were other concepts that Pettigrew, borrowing from anthropology, applied to understanding organizations and their cultures. He highlights the importance of understanding the temporal aspect of organizations, and states that a "sound theory must … take into account the history and the future of a system and relate them to the present" (Pettigrew, 1979, p. 570). Linda Smircich (1983) conducted an analysis of the various uses of the concept of culture that is still relevant today. She differentiated between two ways of seeing culture in organizational contexts: viewing organizations as cultures or viewing organizations as having a culture. She proposed the framework as a dichotomy. but in this book, I will propose that both views have their merit and can bring added value to safety.

There is no single generally accepted definition of organizational culture. Generally, cultural approaches are interested in the meanings and beliefs that organizational members give to organizational elements (structures, systems, and tools) and how these meanings influence behavior (Alvesson, 2002; Czarniawska-Joerges, 1992; Reiman & Oedewald, 2007; Weick, 1995).[3]

Geert Hofstede (2011) has described culture as "the collective programming of the mind that distinguishes the members of one group or category of people from others."[4] Culture is a set of unwritten norms that set out what a particular group expects its members to do – and what to believe. Hofstede has identified six key dimensions that help explain differences between culture, especially on a national level. Hofstede's model is elaborated in Chapter 10.

Culture has also been described as "the way we do things around here" and as behavior that takes place when no one is watching.[5] This behavioral view captures some aspects but simultaneously misses deeper dimensions.[6] Probably the best-known theory of organizational culture is American psychologist Edgar Schein's model of organizational culture as semi-unconscious basic assumptions that staff learn together. Schein's (2017, p. 6) definition of culture is as follows:

> The culture of a group can be defined as the accumulated shared learning of that group as it solves its problems of external adaptation and internal integration; which has worked well enough to be considered valid and, therefore, to be taught to new members as the correct way to perceive, think, feel, and behave in relation to those problems. This accumulated learning is a pattern or system of beliefs, values and behavioral norms that come to be taken for granted as basic assumptions and eventually drop out of awareness.

Culture forms and develops as a group interacts with its environment while learning how to interact internally. To understand culture, we thus need to understand both its internal dynamics as well as the environment, or context, in which it operates.

When we consider culture, we can look at it from two perspectives: its structure and its dynamics. Here we will first look at culture as an emergent structure.

1.3 CULTURE AS STRUCTURE

Edgar Schein considered the core of organizational culture to be partially unconscious basic assumptions (Schein, 2017). By *basic assumptions*, Schein means the ways of feeling, perceiving, and thinking that the group has learned when solving problems related to external adaptation and internal integration. These basic assumptions have proven to be so effective in solving new challenges that they are taught to new group members as the right way to perceive, think, act, and feel.

Schein's definition emphasizes the learned and shared nature of culture. Culture is a proven way of acting, thinking, and feeling that is shared by a group or an organization. Culture therefore influences how its members interact with each other, what information and things they consider important, and how they perceive the objectives of the group, as well as how the objectives are pursued. Because assumptions are based on solutions that have been proven to work in some way, they are often taken for granted and not actively questioned. This means that cultural learning is mostly non-intentional: we do not choose to learn culture, nor do we choose exactly what to learn.

A common way to illustrate the levels of culture is to compare culture with an iceberg (IAEA, 2016b). The iceberg models emphasize that most of the culture (aka the iceberg) is hidden. That is, it is outside our daily observations. We only see what is above the surface, the concrete and the observable.

This surface level can be described as the level of *cultural artifacts*. Artifacts include, for example, the language, technology, and products used in the organization. Organization charts and official processes of the organization are also artifacts. They are easily seen, heard, and felt in the organization in question.

Technology and technological solutions can be seen as forming an important part of the culture itself. The visible and structural characteristics of an organization, such as management systems, organizational structures, information systems, and technology are part of the overall culture of the organization[7] – its manifestations.

Management systems are important artifacts of the given organization's culture. This means that the management system – as all artifacts – both manifests culture and influences culture. For this reason, activities that support a healthy culture for safety need to be built into the management system and related controls. The role of management systems is further discussed in Chapter 6.

To really understand culture, we need to inspect the elements "below the waterline" and clarify how artifacts are related to the elements – how they are generated and maintained by the elements and how the artifacts in turn slowly shape the lower levels.

Processes represent the various routines and practices that form in organizations. These can be meaningful for the personnel and manifest organizational values and assumptions. These kinds of routines are sometimes called rituals, which give symbolic, or cultural, meaning to our routines (Anand, 2024). Typically, rituals are intended for the inner world of the organization, not for dealing with the outer world. Routines can also be formed by repetition and habit formation. In this case the original meaning of the routines is typically lost, and the meaning of the routines is retrospectively interpreted – rationalized.

The level below the artifacts is the level of *meanings and beliefs*. Schein (2017) called this level *espoused values*, but the concept *meanings and beliefs* describes better the essential content of this level. It describes how the organization has made sense of various events and how it wants to show itself to others (and to itself). This level of culture includes ideals and thoughts that the organization wishes to convey both internally and externally. These include the espoused values, which are also documented in the organizational artifacts such as the company policy. Strategy is a rationalization of what the organization wishes to achieve (the organizational core task) and how it wants to accomplish this within its perceived environment and other constraints.

The middle-level includes *rationalizations*, which are typically in the form of narratives and stories. Rationalizations are important for safety as the various explanations for both successes and failures of the organization belong to this level of culture. Rationalizations include explanations for non-actions, things that no-one reacted to or in the face of which stayed silent. This is one of the reasons that by choosing not to act, or choosing to stay silent, one also affects the culture. Rationalizations include shared explanations for various safety campaigns and other initiatives from senior management to create a shared "safety culture". It also includes the ready-made solutions and explanations for incidents; for example, what it means if an incident is "caused by human error". These rationalizations are guided by the deeper levels of culture (underlying logics and basic assumptions), but once formed, they can further shape these levels.

Norms indicate how we are supposed to behave in different situations. They provide us with ready-made solutions to different daily problems. Norms can be further categorized into general organizational norms and role-specific norms (Palmer, 2012, p. 135). Role-specific norms provide employees with guidelines on how others in the organization expect them to deal with a range of issues and situations. Norms help us in coordinating activities by anticipating how others will respond. Norms also provide us with ready-made solutions to typical problems and provide guidance on what is acceptable and what is not acceptable in terms of safety and quality of work as well as behavior at work in general. Norms have an effect especially on behavior that can be observed or inferred by others. To have an effect, norms also require that the person identifies with the given group (see Chapter 2 and social identity).

Values represent issues or things that are considered meaningful and important as such. Values are guiding principles and abstract ideals that influence how we interpret different situations and how we behave in them.[8] *Espoused values* are such values that are not based on shared learning in the group but rather represent official goals or ideology. These may or may not be congruent with the other values or the basic assumptions. This means that organizations may say they value certain issues such as safety, collaboration, or equality, but if these are not aligned with the actual practices, they are merely wishful thinking or impression management.

The deepest layer of culture consists of *shared basic assumptions* and *underlying logics* of action. Basic assumptions are shared, pre-conscious, and taken-for-granted beliefs, values, and norms (Schein, 2017, p. 7). *Underlying logics* refer to group's pre-conscious ways of explaining, justifying, and idealizing (White, 2021, p. 61).

Assumptions cannot be observed directly, as they are "beneath the surface" of organizational daily life. Basic assumptions are taken for granted, and it is difficult for members of culture to recognize or verbalize them themselves, without the help of interpretation.[9] To understand culture, it is important to map out these deeper contents. Assumptions affect how people interpret various internal and external signals and events. They create expectations and guide our attention.

The effects of the deepest layer of the culture manifest, among other things, in how risks are assessed and managed, how the current level of safety is assessed and interpreted, and what is considered appropriate and inappropriate behavior. Basic assumptions also help to define formal and informal norms, such as desirable and appropriate behavior in the workplace. Basic assumptions to a large degree define what is normal for a group, and culture thus influences what is considered normal work and when something is amiss (Reiman & Rollenhagen, 2018).

Figure 1.1 illustrates the structural view of an organization and its culture. In the figure, *basic assumptions and underlying logics* denote the "real" culture. The middle level, *meanings and beliefs*, represents the aspirational, rhetorical and often rationalized daily reality in organizations - its climate. The top level, *artifacts and behavior*, represents manifestations of culture but does not constitute culture itself.

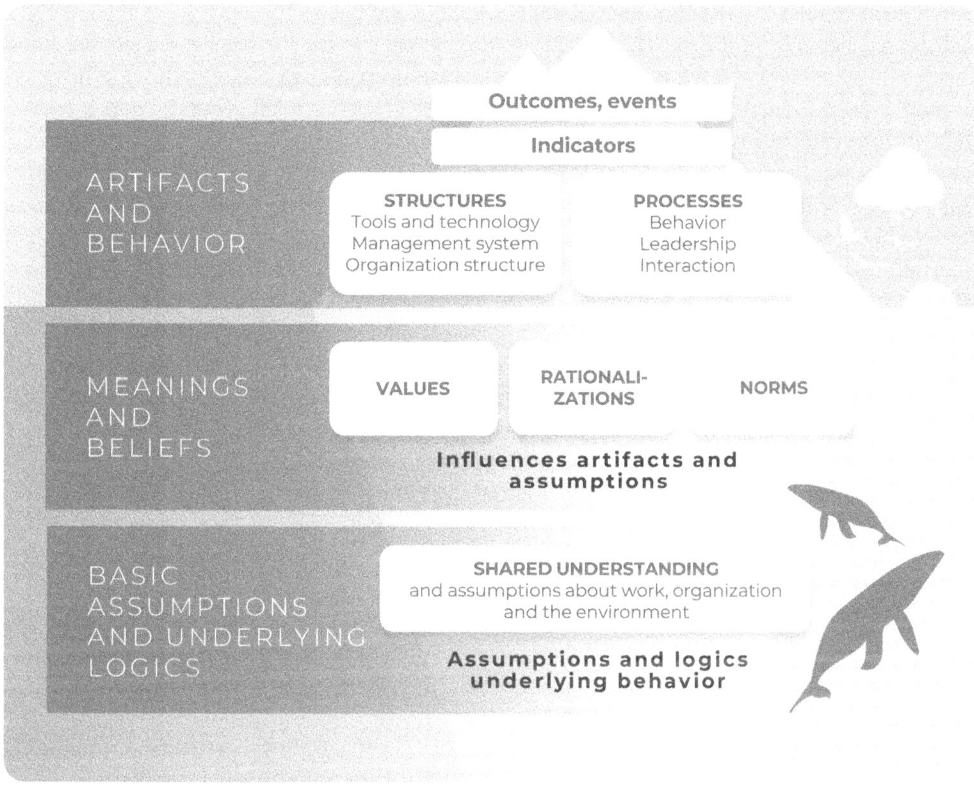

FIGURE 1.1 Organization as an iceberg with cultural assumptions below the surface

In safety-critical areas, the basic assumptions may concern, for example, what is the risk, what are the key hazards threatening the organization's operations, how workplace risks should be talked about, and what is considered a valid warning signal. The assumptions also concern the role of human beings in managing the safety of the organization, the reasons why people make mistakes, and how these mistakes should be addressed (Reiman & Rollenhagen, 2018).

In my assessments and investigations, I have noted some safety-critical assumptions, some of which have also been identified in several notable accident investigations that are illustrated in this book. One such assumption concerns the *current safety of the system*, whether the system is considered safe until proven otherwise (a priori safe system), or whether safety needs to be assured before proceeding (see the *Challenger* and *Columbia* accident cases in Chapter 2). Assumptions concerning the nature of *information*, *uncertainty*, and *control* are important in safety-critical organizations. Due to the presence of hazards, employees' need to be (or feel) in control is typically high. Some organizations seek control by reducing reality into numbers and calculations – by trying to get rid of uncertainty. Others seek control through a more qualitative understanding, dialogue, and employee involvement. These assumptions relate to how the organization defines and deals with *complexity*. Can everything be predicted, or are there always emergent effects and surprises, even in well-functioning organizations? Is the world binary or multi-modal? Can contradictory views on a topic both be true? Are there clear reasons (causes) for events, or can events happen without a (root) cause? What is the role of randomness in success and failure?

Another related assumption concerns the *definition of safety*: is it seen as legal compliance, as a lack of incidents and accidents, as the presence of controls and barriers, or as the more holistic wellbeing of the organization (see Chapter 4)? Furthermore, organizations form assumptions about the typical *causes of events* and failures and about the role of humans in both success and failure.

These in turn are related to assumptions about *leadership* and the best ways of *organizing* for safety (see Chapters 6 and 7). In the nuclear industry, for example, the logics of defense-in-depth strongly affects how work is organized (see Chapter 4 for defense-in-depth and Chapter 10 for industry subcultures). Organizational issues affect interaction, and assumptions form about the appropriate and needed ways of interacting. *Trust* is an interesting concept that many organizations highlight as an espoused value and imperative for safety, but few have a clear shared understanding of what trust means and how it is created, and whether it is always good for safety. Assumptions often form in subgroups about what trust means, what type of issues can be trusted in this organization, and how to earn and lose trust in the organization. All these rest on assumptions about *professionalism*, that is, what counts as professional conduct in the field, the recognized standards of competence, and the legitimate exercise of expertise in defining and pursuing the organizational core task.

Lastly, assumptions also involve what the organization deems *unimportant* for safety. For instance, some organizations have been surprised when I have raised work wellbeing and motivation issues during a safety assessment, seeing these as HR matters rather than safety issues. Likewise, some organizations separate safety from actual work, ignoring how task planning and coordination relate to safety. Assumptions can also be in contradiction with each other. These contradictions are handled situationally and locally.

Safety research and accident investigations indicate that certain assumptions are incorrect or flawed. Nonetheless, since assumptions are derived from learning, they inherently possess a functional aspect. There is a reason why they have formed, and this reason needs to be understood. Artifacts may preserve assumptions, while rationalizations can obscure conflicting or erroneous beliefs. Consequently, directly challenging the validity of these assumptions is likely to provoke significant resistance and denial from the organization.

Basic assumptions affect espoused values, rationalizations, and artifacts. Declared beliefs and values are not always aligned with basic assumptions but may rather embody some kind of ideal state or the organization's perception of itself. The exact meanings of artifacts and the assumptions underlying them can also be difficult for the employees to verbalize and for the cultural analyst to interpret.

Not all behaviors are a result of the deeper levels of culture, and therefore not all behavior can be counted as an artifact of culture. Behavior is also influenced by situational factors, such as time pressure, and individual factors such as personality, emotions, fatigue, and abilities – that is, traditional human factors. These factors need to be considered by the organization when developing its safety management system and culture. For this reason, when mapping culture, the aim is to find behaviors, practices, and routines that the organization repeats in its activities, not just the situational reactions of individuals.

1.3.1 Benefits of the Structural View

The structural view of culture offers several lessons for organizational effectiveness in general and safety in particular:

- Assumptions, norms, and beliefs influence behavior, but never fully predict it.
- The strength of a culture depends on how well in line the artifacts, values, and basic assumptions are.
- Culture is not visible; only its manifestations are. The real culture is under the surface.

Culture is the result of shared learning experiences that influence how a group learns in the future (Schein, 2017). These learning experiences are typically so deeply rooted that they seem obvious and undeniable to the group. It may take a major crisis to overturn these assumptions – and even a crisis is not always enough. Culture is a pervasive force in a group, and all influences from "outside" are interpreted through cultural lenses. For this reason, benchmarking and external reviews are both important and difficult at the same time – the culture (or the people in the given culture) resists and reinterprets everything that comes from outside. Still, a culture always looks different depending on whether you look at it from the inside or from the outside, and often both perspectives are needed to truly understand it. We will return to assessment of organizational culture in Chapter 11.

Schein (2017) identifies the following characteristics of culture: "structural stability", "depth", "breadth", and "patterning or integration". *Structural stability* implies that culture is not only shared but also defines the group's identity and survives even when some members of the organization depart. *Depth* refers to the fact that culture denotes the deepest and often unconscious aspects of a group. One cannot control a culture. The essence of a culture is invisible and intangible, yet it is tangible in its manifestations. Once a culture forms, it covers the group's functioning in its entirety (*breadth*), including issues related both to safety and efficiency. Finally, *patterning or integration* means that culture ties various elements, such as rituals, climate, values, and behavior, together into a coherent and sensible whole (Weick, 1995).

1.4 CULTURE AS PROCESS

To understand culture and its significance in organizations, we need to view culture as both a process and an outcome. When we look at culture as an outcome, we focus on the structural features of culture, its constituent components, its depth, and its stability. This was the focus in the previous section. This view of culture highlights the visible and invisible aspects of culture as well as qualitatively different components, ranging from physical artifacts to the "real" culture of shared basic assumptions.

When seeing culture as a process, the focus is on interaction and the reciprocal way culture is being created and recreated by our everyday encounters at work. This view is needed to complement the structural view by highlighting the self-referential nature of culture: we create culture for us, but once created it "takes on a life of its own" and affects how we interact. For this reason, it is a simplification to say that culture drives performance (or safety), since success or failure drive culture as much as the other way around (cf. White, 2021).

Culture is something that an organization creates for itself and that, once created, influences the organization. The organization generally doesn't notice this effect, as its members view everything that aligns with their cultural assumptions as routine operations ("business as usual"). Culture is created, recreated, and strengthened repeatedly, as members behave and communicate in ways that seem natural, normal, and undeniable to them.

In order to look at the dynamics of culture, it is helpful to differentiate between culture, structures, and practices in the organization (Antonsen, 2009c; Guldenmund, 2000) (Figure 1.2).

Structures and practices are two main elements of an organization as a culture. An important process facilitating these two is the process of sensemaking, indicating how activities and structures both influence and are influenced by each other. In essence, culture is this process of collective sensemaking. Practices include individual behaviors and more collective routines and habits,[10] as well as various daily local adaptations and communication. Structures are both physical and interpersonal; not only do technology, tools, organizational structure, rules, and formal roles belong to this category but also norms, shared meanings, and underlying logics. Seeing organization as a culture thus focuses on the interactions between practices and structures and the ongoing sensemaking process.

Structures store culture, but as with all things stored, culture can also become sour with age. *Organizational structures* refer to formal arrangements, division of power and responsibilities, management system, tools, and physical facilities and technology. *Affordances* are characteristics or properties of an object that suggest how it can be used (Gibson, 1979). For example, organizational structure and the physical workspace provide us certain "natural" ways of interacting, communicating, and sharing information and hamper other ways. The quickest way to change the climate and

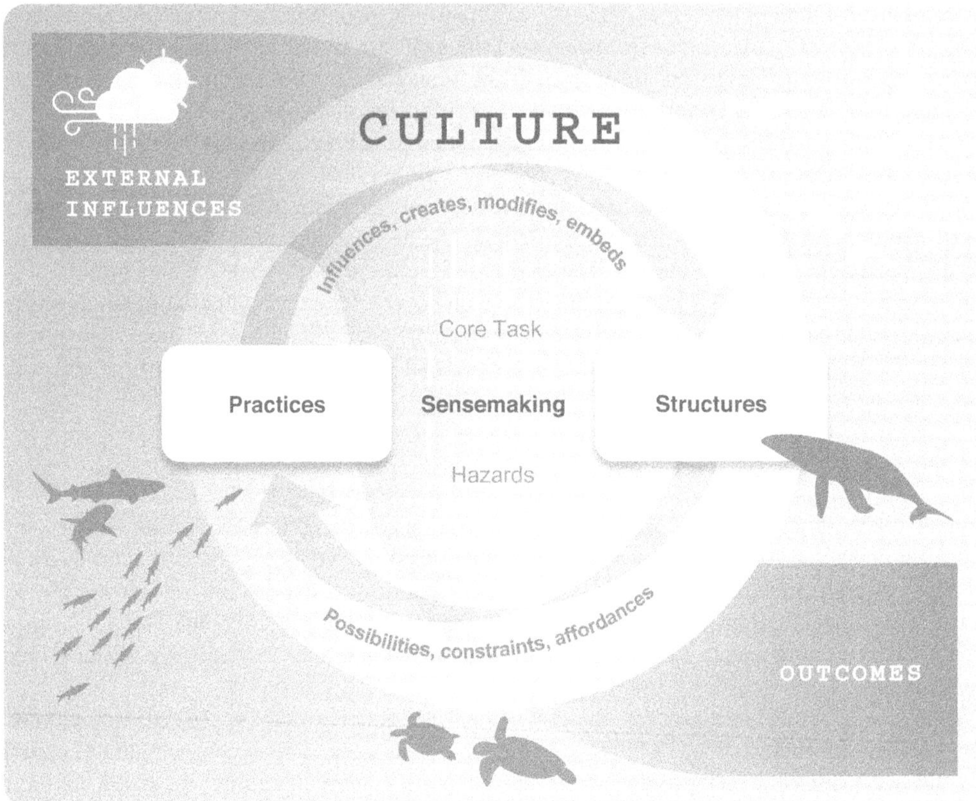

FIGURE 1.2 Process model of culture – organization as a culture

ways of interacting in an office environment is to change where people sit[11] – this is not always for the better, though, as many open-office experiments in the early 2000s demonstrated.

Structures help us cope with the complexity of our modern sociotechnical systems by providing us with ready-made guidelines and solutions. In daily work, we do not need to analyze in detail all situations and develop from scratch the needed measures, since these issues are already embedded in the structures. This improves efficiency, but it obviously has a downside in terms of reduced mindfulness and readiness to act in case of new and unanticipated contingencies. Structures simultaneously constrain and facilitate action. They do not directly determine behavior. Structures represent the history of their creation and utilization. The meaning of structures is built and rebuilt in social interaction.

One of the central social processes in any organization is *sensemaking*. Sensemaking refers to the process whereby the so-called external reality is enacted by the personnel, where people create the "reality" they later perceive and interpret (Weick, 1995; Weick et al., 2005). Sensemaking is a process of active agents together structuring the unknown to be able to act. Organizational reality is an ongoing accomplishment that emerges from efforts to create order and make a retrospective sense of what occurs.[12] In Figure 1.1, meanings and beliefs denote the collective end results of sensemaking. Sensemaking as a social process is elaborated in Chapter 2.

Despite the inherently unified and self-reinforcing nature of culture, it gradually and continuously changes. This change is usually so slow that it is not easy to notice from the inside. At the same time, it is very difficult to intentionally get culture to change quickly. Usually, this requires some kind of crisis or calamity – and often even that is not enough, as members of a culture are very adept at explaining events to themselves in a way that does not require a change in the deep elements of the culture (Baumard & Starbuck, 2005). However, with systematic and persistent work, culture can be developed. Further, external attempts to change the culture are often met with resistance, or the ideas and methods are interpreted within the cultural frames and transformed into an acceptable form.

Culture forms via two processes of internalization, one of which is active and the other one passive. They are intertwined in reality but distinct in form. In the active process, people make sense of events, interact, learn, and create meaning together. In the passive process, different structural forces such as affordances, power, resources, repetition, and so on influence activity and interaction without conscious sensemaking.

The core task of the organization and the technology that it uses to fulfil the core task have a significant impact on the formation and development of culture (Reiman & Oedewald, 2007). The influence of the core task and technology on culture is elaborated in Chapter 10.

One can never understand culture based on documents and records alone. These can be a starting point for cultural analysis, but they do not offer a view of how the culture forms and develops over time, or how the people in it make sense of various issues, including the records and the documents. Sometimes the reality described in the documents is more wishful thinking and rationalization than actual practice. Moreover, the documents do not reveal how people make sense of events, and how they in their daily interactions create and recreate cultural meanings, beliefs, and assumptions. Documents provide a static view of an inherently dynamic phenomenon.

The tragedy of the so-called safety culture approach (and many other safety management approaches) is that it too often seeks to implement a unified mechanistic approach to an organic system without considering what is already there, and how the system reacts when a new "program" is being pushed by the management. This widens the gap between the formal system (the safety culture program and the safety management system) and the actual culture – the meanings, beliefs, values, and patterns of action in the organization. These approaches thus miss the one thing that they advocate to consider: culture (see Chapter 5).

The dynamic aspects of culture also have several consequences for organizational effectiveness and safety:

- A culture resists quick changes because the structures that store culture are stable.
- Despite this inherent resistance, a culture changes gradually via its internal dynamics.
- Every action influences a culture either by maintaining the existing structural features and the symbolic meanings or slowly changing them. It is important to note here that a visible non-action (choosing not to say or do anything in a particular situation) is at least as effective in terms of influencing a culture than is visible action.
- A culture forms as the group adapts to external environment and learns to work together. Sometimes these two aspects of the culture do not align. The ways of integrating internally may lead to some external features being overemphasized and some being neglected. Similarly, ways of responding to external pressures may lead to internal disputes and disparities in power, with some groups perceived to be more important to the organizational core task.
- A culture is always a product of history. This means that a culture represents the world as it was, not as it is. Sometimes the solutions of the past become outdated as external or internal environment changes. New technologies or old technology change risks, as do changes in competition, contractor networks, the workforce, and so on.

1.5 CULTURE AS MEANING

The cultural approach shows an interest in things that the organization does not consider relevant, as well as things that are perceived as laborious, distressing, and threatening to wellbeing. Only risks that are considered – that is, that are perceived as risks – can be managed. If everyone agrees on something, but they are wrong, then consensus and team spirit become a risk. Weick (1998, p. 74), building on Turner's (1978) ideas (see Chapter 3), emphasizes that organizations and their activities are strongly defined by what they ignore and how consistent this inattention is within the organization. The inclusion of neglected and distressing issues in cultural analysis is important especially when we are interested in hazards that, if neglected, may have fatal consequences for employees or society at large.

We can also feel that some aspects of the current culture are stressful, even distressing, but at the same time feel that there is nothing we can do – that we are in the minority and have no power to affect the majority opinion. However, one of the typical fallacies in groups is the fallacy of collective ignorance (also called pluralistic ignorance, see Chapter 2) – we just do not know what the majority opinion is without asking everyone and not only listening to those that eagerly, and without being asked, let their opinion be known. Many people publicly support opinions that they consider to be in line with the majority opinion, even if they privately disagree with them. One of the many paradoxes of culture is that we often are willing to tolerate its negative sides in exchange for certainty – any culture is better than no culture at all!

Cultures are typically built around some form of consensus. Even on the internet, people gather around and protect (in different discussion platforms and chats) ideas and meanings that resemble those they already have rather than seek to change their ideas or construct new meanings. The company they seek on the internet reinforces their old ideas and prejudices. This happens in organizations too. People will readily choose a good story over a truth that is bad or too complex. For those of us working for safety, this unfortunately means that even good stories can prove fatal if they are not based on adequate understanding of hazards and risks.

The culture of an organization is rarely, if ever, completely uniform. Subcultures can typically be found in any organization of significant size. A member of an organization may also simultaneously belong to several cultural groupings. Subcultures are typically formed based on, among other things, age, job function, professional education, work experience, physical location, or nationality (Schein, 2017). Schein noted that

> when sales and research units or engineering and manufacturing units have trouble talking to each other and collaborating effectively, the problem can be viewed essentially as an intercultural one.
>
> **(Schein, 1985, p. 39)**

Subcultures may differ from each other, among other things, in how they perceive the core task of the organization and their own, as well as others' role in fulfilling it. They may also share different norms and beliefs about the work and its risks. Subcultures in themselves are not inherently negative from the point of view of safety, for example, as they bring the necessary diversity to views that, without internal variation, easily become too narrow (the so-called groupthink phenomenon). Different professional groups can also develop subcultures related to the type of work (professional identity) that transcend a specific organizational boundary (medical doctors, pilots, maintenance technicians, etc.). Thus, any organization can be considered a melting pot of multiple subcultures interacting simultaneously (Reiman & Rollenhagen, 2018).

1.5.1 Culture, Leadership, and Power

Schein (2017) considers leadership essential for the creation of culture. He calls this the leader's most important task. He has elaborated primary embedding mechanisms as well as secondary articulation and reinforcement mechanisms to depict how leaders create and change culture. Primary embedding mechanisms include issues leaders pay attention to, measure, and control on a regular basis; how they react to critical incidents and organizational crises; and how they allocate resources, rewards and status, as well as how they recruit, select, promote, and terminate employees. Secondary mechanisms include structural and artifact-level (Figure 1.1) issues such as organizational design and structure, physical spaces, and formal statements as well as rites, rituals, and stories. If the secondary mechanisms are in line with the primary embedding mechanisms, they can support the creation of culture, but on their own they are not enough for culture creation. We will discuss leadership on multiple occasions throughout the book. Chapter 7 is devoted to leadership issues.

Leadership is a way of using power in an organization, a "soft" and informal way, but one that is not at all less efficient than authority-based power use. Leadership is a way of influencing in a manner that gets people to act or to change their behavior *willingly*, whereas other ways of using power may be more coercive. Both power and leadership are relational, dependent on situations, and value neutral, meaning they can be used for various outcomes.

Culture and power are closely associated in organizational contexts (Antonsen, 2009c). Power, leadership, and culture are intertwined, requiring acknowledgement and consideration of all three when addressing one. They all involve social interaction and meaning, with both structural and relational characteristics. Pfeffer (1981) emphasized that power is a structural phenomenon, created by the division of labor and departmentation that characterizes the specific organization or set of organizations being investigated. Power is wielded by altering either the physical or social reality of the person(s) subjected to power use. Altering physical reality means rewarding, making decisions, organizing, or doing other acts with consequences. Altering social reality refers to how people perceive the significance and meaning of various issues and events. It can also mean the power user's perceived potential to alter physical reality. At the same time, structures embed power and can help to maintain power relations. Organizational culture can thus be seen as a conduit of power.

Power means the ability to make others act in a certain way or believe in something. In an organizational context, power has been defined as the capability of one organization member to direct the behavior of others, or to bring about the outcomes the power wielder desires (Pfeffer, 1981). Power is a relationship between the power holder(s) and a specific group. Several different types and sources of power have been identified in the literature. It is possible to differentiate between coercive (perceived ability to mediate consequences or punishments), reward (perceived ability to mediate rewards), legitimate (perceived right to prescribe behavior due to a formal position or role), expert (perceived special knowledge or skills), referent (identification with or admiration of the other), and information power (having access to valuable information). Another type of power is resource-dependent power, with resources ranging from money to information to prestige.

Conceptualizations of power originally assumed that power is something that an individual holds (Fleming & Spicer, 2014), yet they are all linked to culture, either directly (e.g., expert power) or indirectly (e.g., coercion focusing on behavior). Fleming and Spicer (2014) identify four faces of power (coercion, manipulation, domination, and subjectification). The first two are episodic exercises of power, whereas the two latter are systemic. Episodic exercises refer to direct and explicit actions, relying on identifiable acts that shape the behavior of others. Systemic power refers to more subtle exercises of power relying on institutional, ideological, and discursive resources to influence organizational activity.

Coercion refers to direct mobilization of power via, for example, formal position, personality, or access to valuable resources. Manipulation, in turn, attempts to ensure that action and discussion occur within accepted boundaries. It is more covert than coercion; it aims to influence what issues are on the table and what issues are left out. In domination, actors establish influence through the construction of ideological values. Culture is one mechanism of ideological domination. A key issue in domination is to shape people's perceptions so that they accept their role, "either because they can see or imagine no alternative to it, or because they see it as natural and unchangeable" (Lukes, 2005, p. 24). Finally, subjectification seeks to determine an actor's very sense of self, including emotions and identity (Fleming & Spicer, 2014).

Ideological power means the ability to shape, influence, and control the beliefs, values, norms, and perceptions of a group. It is a form of power that operates through ideas, ideologies, and narratives, rather than through physical force or coercion. This type of power can influence what people think is "normal", "right", or "acceptable", shaping their understanding of the world, their behavior, and their choices. Ideological power is the ability to shape how people think and what they believe, often without them realizing they are being influenced. This type of power can refer to assigning cause and consequences after an accident, to legitimating a certain explanation of the event over other potential explanations (Dekker & Nyce, 2014; Sagan, 1993). Power and culture together explain why human errors are still such a prevalent causal attribution after accidents (ibid.).

In organizations, formal authority embeds reward and legitimate (partly coercive as well) power into specified, and typically documented, roles in the organization structure. But even this most formal type of power is deeply cultural, as the behavior of the role holder is subject to expectations (how the people occupying the role should behave) as well as interpretation and rationalization by the personnel (what the recent behavior meant, whether they issued rewards or sanctions, and whether we should comply). Obedience to authority also differs on average between types of organizations (Palmer, 2012, pp. 179–180).

As power and culture are intertwined, using power always influences culture. For example, change resistance during reorganizations partly stems from fear of losing power, and using power to change an organization has an influence on the cultural power dynamics. Power also affects information flows, organizational silence, rationalizations of various events, and so on. Power issues affect what and how the organization learns. This will be elaborated in Chapter 3.

1.6 CULTURE AND INDIVIDUALS

It is important to emphasize that organizational culture is not an individual trait, but rather an organizational collective property. Culture is what individuals share with each other regarding meanings, beliefs, and assumptions. Thus, individuals do not have a culture separated from the social context. We also need to separate individual behavior and cultural practices. Culture affects behavior, but so do many other things, such as personality, external factors, and mood. Alvesson and Sveningsson (2024, p. 45) describe this well:

> Culture is expressed in behavior and can be seen in actions, events and other material aspects but does not refer to these exterior elements per se but to the meaning and beliefs these have for people. Culture is thus behind and beneath behavior.

It is worth avoiding overemphasizing the shared nature of culture. Culture is what is familiar to its representatives. However, this does not mean that all cultural artifacts are interpreted in the same way. Each of us makes our own interpretations based on our own preferences, values, perceptions, and the countless identities to which we all belong. This may create a false impression that the meaning of these things is shared. In fact, we may only share the *familiarity* of these things. This is particularly typical of safety-related issues, which are often repeated in organizations through communication, in meetings, posters, safety walks, and so on. Repetition is often thought to ensure that the message gets through (aka is understood), but in reality, it only guarantees that the message becomes familiar. Thus, we can simultaneously share culturally the familiarity of a safety message (and bemoan "it's these empty slogans again") and interpret the content of the message individually – make sense of its meaning. Culture always influences these interpretations and in due time can align them. This means we can actively aim to promote the spread of certain meanings within that culture.

Still, it needs to be remembered that every individual experiences culture from their own perspective. A cultural approach, as I see it, reminds us to pay attention to subjective issues such as meanings and beliefs, whether these are shared, individual, or something between. It emphasizes interaction and the ongoing construction of meanings, routines, and ways of working, in addition to more stable and shared outcomes such as basic assumptions. Instead of promoting and advocating homogeneity, the cultural approach should facilitate and acknowledge both the diversity of views in organizations and sometimes their integration and homogeneity. Rather than treating everyone as homogenous and expecting them to behave similarly, the cultural approach focuses on the context of behavior. Human behavior is always contextual, and by analyzing and better understanding the context, we gain a deeper understanding of the mechanisms shaping individuals' thoughts, feelings, and actions.

A paradoxical aspect of our uniqueness in organizational environments lies in our deep social nature, as we all belong to numerous groups and share cultural and social identities with them. Family, friends, hobbies, and political or religious ties all shape our beliefs and attitudes. While organizations are a key source of social identity, we constantly navigate the many identities and affiliations within us when interpreting the world.

There is no denying the potential influence of key individuals on the culture of the organization. As discussed in the previous section, power and leadership are deeply intertwined, and people with power and good leadership skills are able to change cultures. We will look more closely at individual issues and the socialization of individuals into groups and cultures in Chapter 2, the role of powerful individuals in Chapter 3, and leadership in Chapter 7.

1.7 A CULTURAL APPROACH TO SAFETY

A cultural approach to safety considers culture, structure, and interaction in complex sociotechnical systems. It considers how safety is perceived and how structures and interaction support creating and maintaining an adequate view on safety. The cultural approach to safety recognizes that culture both creates and solves issues, reveals and hides risks, and contains both shared and conflicting aspects. Although culture is subjective and inter-subjective, it has real safety implications. A cultural approach starts by understanding this inter-subjective reality and then assessing its effects. Only after grasping the current culture can we find the best ways to influence it for better safety outcomes. A culture should not be an explanation but a description of and an approach to viewing the organization (cf. Hopkins, 2018).

The cultural approach should avoid overemphasizing the mindful and meaningful aspects of culture. All organizations have mindless routines and patterns of behavior that are not actively analyzed, but rather just executed out of tradition. Many of the meanings that are embedded in the tools and structures are embedded so deeply that they lie outside daily awareness. A particular culture includes at the same time highly meaningful and mindless elements – and employees may not be aware of how meaningful things are before something is changed or lost, or before a major contingency forces the meanings to surface.

The cultural approach provides insights into the underlying dynamics behind an organization's daily life and routines. A cultural perspective can help us understand how practices develop or what needs to be improved for future safety. The cultural viewpoint highlights that employees' subjective experiences impact the organization's objective outcomes, including safety. Additionally, the cultural approach examines how the organization manages the whole and identifies areas that might be neglected. It considers interaction and relationships, including communication within organizations, acknowledging their importance for safety. Furthermore, the cultural approach recognizes the role of artifacts such as management systems, official procedures, plans, tools, and technology in shaping interactions and influencing sensemaking within organizations.

The cultural approach aims to examine the organization as a whole, taking into account the influences that gradually shape it and guide its developmental trajectory. This approach to safety considers both informal elements of the organization and formal ones, such as tools and technology. It focuses on the meanings attributed to technology and how personnel interpret their tasks and responsibilities and collaborate to manage perceived risks. The cultural approach explores what constitutes safety and how it is achieved within the specific organizational context.

A culture forms based on a shared interpretation of history, not always objective facts. A good story is almost always better than a boring, or too complex, truth. History lives in the culture, but always in some altered form. It is socially constructed phenomenon, continually produced and reproduced through the interaction of individuals within the context of existing social structures (Antonsen & Almklov, 2020).

A cultural approach is interested in the meanings and beliefs which members of the organization assign to organizational behavior and in how these assigned meanings influence the ways in which the members behave. The social world is constructed by people and reproduced by networks of symbols and meanings shared by people, allowing us to collaborate and communicate with each other. Even technology acquires meanings from its designers and users that influence its subsequent exploitation (and subsequent reinterpretation of these meanings).

1.8 SUMMARY

The chapter explored the concept of culture within organizational contexts. In essence, a culture is an abstraction that helps us understand groups and the reciprocal relation between individuals and groups. It is seen as both a process and a temporary outcome of people interacting with each other and their environment. It is often used as an explanation, and groups form cultures to explain, clarify, and predict. However, for safety purposes, it is better used as a description and a viewpoint to widen, instead of limiting, how we see safety in complex organizations.

Organizational culture refers to the way the members of the work community have learned to relate to their work and interact with each other. Culture thus arises from shared learning experiences that influence how the group learns in the future (Schein, 2017). Real culture is not what is written in policies or official values, but what is believed in and what is not, what is *considered important and what is not, what is done in everyday life.* Even though culture cannot be seen directly, its effects and (temporary) outcomes are tangible. Like the tip of an iceberg above water level, only a small part of an organization's culture can be easily detected. Most cultural elements – such as attitudes, values, beliefs, and deep-seated assumptions shared by employees – are not visible or otherwise directly perceptible. They form part of the iceberg below the water level. They are not tangible and are often even unconscious, but they influence behavior and affect concrete things "above the surface".

A culture defines what is considered normal work, how it should be carried out, what the potential warning signals are, and how to act in abnormal situations. Cultural norms define the correct ways to behave in risky situations and the correct ways to talk about safety, risks, or uncertainty. This influences the perception of risks and hazards, as well as the feeling of individual responsibility. The cultural standards and norms create an environment of collective responsibility, where the individual feels their main responsibility to be not one of making decisions but one of conforming to the collective norms. In Chapter 3, we look at how this may in some cases lead to rationalizing unethical behavior as cultural compliance.

A culture develops and changes gradually. Every action and non-action by organizational members influence the culture. Leaders need to bear in mind that a culture must be cultivated actively. However, individual behavior as such is not purely cultural. Many situational and individual-related factors affect behavior. Leadership is essential for creating and changing culture. Leaders influence culture through what they pay attention to, and how they react to crises, allocate resources, and model behavior. Consistency in leadership behavior significantly impacts culture.

All organizations develop a culture. A culture forms as personnel seek meaning, clarity, and belonging. It is based on identity building and is learned and taught, providing predictability and stability in an otherwise ambiguous environment.

Culture and power are closely linked in organizational contexts. Power can influence culture by shaping behavior, information flows, and organizational dynamics. Leadership is a way of wielding power to influence culture positively or negatively.

Organizational culture is a collective property, not an individual trait. While individuals interpret cultural artifacts differently, a culture influences these interpretations and can align them over time. Key individuals can significantly impact a culture through leadership and power. Individuals are accountable for their own behavior, but that always takes place in a context wherein the individual may be unaware of all influencing factors and feel some influencing factors as coercive.

The cultural approach to safety considers culture, structure, and practices in complex socio-technical systems. It emphasizes understanding the current culture to influence it for better safety outcomes and recognizes both mindful and mindless elements within culture.

Reflection Questions

How do you define culture in your organization?

How does your organization discuss hazards and risks? How do you keep the discussion alive and constructive?

What are the situations in your organization where you can best influence the culture?

How do your employees rationalize safety efforts?

Who has power over safety issues in your organization?

How do you gain power and influence in your organization?

What are the issues the members of your organization disagree about most? What are the conflicting priorities? Are these explicitly discussed or hidden?

How does *your* behavior maintain or change current culture? If you want to change the culture, what do you need to change in yourself?

What is the role of the safety department in your organization? How aware of their own culture and its influence on interaction with the line organization are the members of the safety department?

SYSTEM THEORIES AND CULTURE

What is a System?

A system is comprised of interconnected or related activities or entities, such as hardware, software, buildings, spaces, communities, and people, with a shared purpose. It includes links between the entities that can involve state, form, function, and causation. The system changes and modifies its state and interactions based on circumstances and events. It is conceptualized within a boundary and has inputs and outputs with many connections, and the whole system is usually more than the sum of its parts (Wilson, 2014).

There are different types of systems, ranging from simple to chaotic. Between these two extremes are complicated and complex systems. Typically, all systems where humans are involved are complex systems. In simple systems, the relation between cause and effect is linear and direct. The system has a limited number of components that interact in clear and predictable ways. Complicated systems have many interconnected parts, but they operate according to predictable rules and relationships. These systems are still predictable if every component's function and interaction with other components is known. Chaotic systems have no perceivable cause-effect relations (Kurtz & Snowden, 2003).

Complex systems have many interacting parts, like complicated systems, but the interactions are dynamic, adaptive, and often non-linear. This means that small changes in one part can lead to significant and unpredictable effects on the whole system. Outcomes in complex systems are not fully predictable because of emergent properties (new behaviors that arise from interactions among the system's parts) and feedback loops. In complex systems, attention needs to be placed on the *interconnections* between the components instead of, or in addition to, individual components. The relationships among the components are *non-linear* in nature. This means that inputs are not necessarily proportional to outputs. Small actions can lead to big effects and big actions can lead to small effects on the system level.

Early Systems Theories

Biologist Ludwig von Bertalanffy introduced general systems theory (GST) in 1937. His thesis was that a new type of entities, systems, had entered scientific discourse. According to him, this challenged the scientific methods of that time, as in systems, due to their non-linear interactions, the whole is more than the sum of its parts. Bertalanffy developed the theory based on his insights from biology but deduced that the same principles apply to all systems, whether physical, biological, or social (Jackson, 2024). Interactions, non-linearity, emergence, and adaptation were some of the early concepts that are still in use. General systems theory makes a distinction between open, closed, and isolated systems. Open systems exchange energy, matter, people, and/or information with the external environment, whereas closed systems only exchange energy. Isolated systems do not exchange anything.

Operations research developed during the Second World War. Norbert Wiener introduced cybernetics in 1948 as the science of control and communication in animals and machines. Concepts that were introduced and elaborated within cybernetics include circular causality, feedback, and information. Information in particular had a crucial role in cybernetics. The introduction of computers led to system dynamics and formal modeling of systems. System dynamics was pioneered at MIT in the 1950s by Jay Forrester.

The notion of requisite variety developed by Ashby has important implications for safety. Ashby (1956) argued that for a system to effectively control or adapt to its environment, it must be able to respond to the variety (complexity) of inputs it might encounter. This means that the variety of possible responses that a system can generate must match or exceed the variety of disturbances or changes it faces from its environment. The system must have an equal or greater variety of responses (options for control) to maintain effective functioning or adaptation. If the system doesn't have enough variety, it won't be able to adapt properly to changes or disruptions. The basic idea is captured in the title of a paper by Conant and Ashby (1970): "Every good regulator of a system must be a model of that system."

Deming, who popularized the PDCA cycle (Plan, Do, Check, Act) in the 1950s, was influenced by early system theories, mostly systems engineering and its application to statistical quality control.

Katz and Kahn (1966) applied the concept of open systems to the organization and popularized systems theory in organization studies. The theories of system dynamics (SD) developed by Forrester, the viable systems model by Beer (1984), and soft systems approaches by, for example, Ackoff (1979) and Checkland (1981) were attempts to modify the systems approach to be more applicable to social systems.

The Tavistock Institute of Human Relations, founded in London in 1947, combined systems thinking with psychoanalysis, social psychology, and group dynamics to understand human behavior in complex systems such as organizations and societies. They viewed organizations as sociotechnical systems (Trist & Bamforth, 1951). They showed how changes in either technology or people had an effect on the other systems. The technical and social systems had to be jointly optimized to achieve the primary task of the group or organization (Jackson, 2019).

Tavistock, together with Kurt Lewin, pioneered action research, a method whereby researchers do not remain independent observers, but collaborate with organizations to diagnose issues and implement real-time changes. The Tavistock Institute studied many issues still relevant today, such as culture, power, leadership, decision-making, involvement, and social change.

Peter Senge in his book *The Fifth Discipline* popularized systems thinking for managers, especially its system dynamics tradition (Senge, 1990). The fifth discipline, according to Senge, was systems thinking, tying together the four other disciplines (personal mastery, mental models, shared vision, and team learning) that are needed for organizational development and learning. Senge made many important practical observations about organizations from a systems perspective, including challenges connected with feedback, non-linearity, and applying simplified solutions to complex problems. One interesting observation concerns blame: Senge emphasized that systems thinking shows there is no outside element to blame for system problems, but rather "you and the cause of your problems are part of a single system" (Senge, 1990, p. 67). This insight is important both for learning in complex systems (Chapters 3 and 6) as well as for changing those systems (Chapters 7 and 12).

The various open systems theories and especially their applications to complex systems laid the groundwork for the development of the theory of complex adaptive systems. Notable early figures were Ilya Prigogine and Murray Gell-Man as well as John Holland and Stuart Kauffman at the Santa Fe Institute.

Complex Adaptive Systems

A principal feature of a complex adaptive system is its ability to adapt. Chapman (2004) notes that institutions and organizations have internal processes that allow them to survive changes in the environment in which they operate. Changes in the environment will reflect on the organization, and it may be forced to react (e.g. cut costs, reorganize), but it will remain recognizably the same institution. Its internal organization, core values, and culture are a source of its ability to adapt – its resilience (Chapman, 2004).

The history of a complex system influences its present behavior (e.g. by forming a culture), and these systems are sensitive to initial conditions. Additionally, no single member of the system has complete awareness of the system's overall behavior, leading individuals to respond based only on local knowledge.

Outcomes in systems emerge from a complex network of causal interactions instead of a single causal factor. Dekker (2005) describes the systems approach as seeing "socio-technical complexity not as constituted of parts and their interactions, but as a web of dynamic, evolving relationships and transactions". The systems approach thus necessitates taking a holistic view of the topic of interest and defining its relations to the other neighboring topics.

According to Cilliers (1998), a complex system has many elements that interact in a rich manner. This means that any element affects and is affected by several others. These interactions are non-linear in nature, meaning that small changes in inputs can cause significant changes in outputs. It also means that any interaction can feed back to itself immediately or after a delay. Many effects in complex systems are delayed, making it more difficult to pinpoint which interactions have been influential in creating them. Nevertheless, interactions take place primarily, but not exclusively, with immediate – or local – neighbors. The locality of interactions combined with complexity means that the elements, for example, individuals working in an organization, may be ignorant of the behavior of the system as a whole. This includes ignorance concerning all the actual and potential effects of one's own work and working practices on the organization.

An important characteristic of a complex system is that the overall behavior of the system cannot be predicted by the behavior of the individual elements (Cilliers, 1998). This challenges the typical scientific method of reduction, where variables are inspected one at a time, while controlling for the other variables. Many psychological experiments suffer from the fact that they have been conducted in laboratory settings, without the "intervening variables" present in everyday work life.

In complex systems, everything affects everything else in a non-linear and often delayed manner. This means that within complex adaptive systems, the pursuit of any single quantified target is likely to distort the operation of the system and thereby reduce its overall effectiveness. Chapman (2004) notes that one of the significant dangers of specifying targets and simple measures of performance in complex systems is that it results in sub-optimization. He notes that

> a specific target can encapsulate only one element of a complex organisation, and its dominance is likely to undermine other aspects of the organisation that are crucial to its general and long-term effectiveness.
>
> **(Chapman, 2004, p. 58)**

Complex systems self-organize. *Self-organization* denotes the spontaneous emergence of new structures and new forms of behavior in the system (Prigogine & Stengers, 1984). Self-organization is the collective result of local yet non-linear interactions among the system components. The phenomenon of self-organization means that control in complex adaptive systems is always distributed rather than centralized, since self-organization takes place without centralized control or external influence. Self-organization often produces emergent phenomena. *Emergence* denotes the arising of global characteristics of the system (in an organizational context, these characteristics refer to, e.g., practices, structures, processes, and cultural assumptions) from characteristics of agents and their relationships, without being reducible to these characteristics (Reiman et al., 2015). Yet these patterns have an effect on the individual agents. The irreducible nature of emergent properties means that the properties of the whole are distinctly different from the properties of the parts. The elements or agents in a complex adaptive system may not be that complex themselves, but as they interact with many other elements, they produce emergent properties for the system, the higher order properties which make the system what it is (Cilliers, 2010). Social identity and culture are such emergent phenomena (McMillan, 2008, p. 63).

Complex systems evolve in time (Cilliers, 1998). They are partly constituted by their history, and they are irreversible, meaning that they cannot be rewind back to their earlier form and state. Actions are thus irreversible,[13] and the past helps to shape present behavior. History dependence also means that solutions can seldom be copied from one system to another; what works in one organization cannot be replicated in another organization, since they each have their own distinct histories (McMillan, 2008, p. 112).

A complex adaptive system is in a continuous process of flux and change. Change in these systems is a natural tendency, not something initiated by an outside force (McMillan, 2008). These systems have several inherent tensions that need to be balanced in one way or another. Chapman (2004, p. 60) notes three tensions:

- Adaptability (or flexibility) versus adaptation (or efficiency);
- Change and innovation versus stability;
- Centralized control versus decentralisation and autonomy.

A complex adaptive system exists within its environment, but it is also part of that environment. Environmental changes require a change in the system. However, since the system is part of its environment, change in a system changes its environment, creating a process of mutual change and evolution. This idea of nested systems means that we can look at systems at different levels, from individual and group to organizations, industries, and societies.

Table 1.1. summarizes the characteristics of complex adaptive systems.

Because of the above-mentioned features, complex adaptive systems cannot be controlled in the traditional meaning of the word *control*, that is, exercising restraints and direction over the outcomes of the system (Reiman et al., 2015). There is always unpredictability in the outcomes of a complex adaptive system. Thus, management needs to focus on creating *potential*

TABLE 1.1
Characteristics of complex adaptive systems (Cilliers, 1998; Meadows, 2008, Kurtz & Snowden, 2003; Reiman et al., 2015)

Characteristic	Description
Rich interactions	A complex system has a large number of elements that interact in a rich manner. This means that any element affects and is affected by many others.
Non-linearity	Effect is rarely proportional to cause. Small changes in local conditions can have major effects on the overall system.
Locality of interactions	Components of the system typically interact most with other local components. However, in human systems, "components" can choose to interact with the more remote others.
Self-organizing	Self-organization denotes the emergence of new structures, new patterns, and new forms of behavior in the system as a consequence of agent interaction and connections.
Emergence	Emergence refers to phenomena and behaviors that rise from the interactions of the system's components, which cannot be predicted by studying the individual components in isolation.
History-dependent	Many actions are irreversible, and the system cannot be rewound back to its earlier form and state.
Adaptive	Organizations and people learn new ways of responding.
Changing	Complex adaptive systems are not stable, but they are constantly changing.
Feedback loops with time delays	Feedback in complex systems is delayed.
Coevolution	A complex adaptive system exists within its environment, but it is also part of that environment.

in the organization, and in identifying and influencing patterns in the organization (Carrillo, 2020). The behavior of complex adaptive systems is impossible to predict in detail, and the actors in those systems are always acting under uncertainty (cf. Norros, 2004).

The complexity of the system implies that safety cannot be comprehended or managed by examining its individual components in isolation. Instead, safety should be regarded as an emergent property of the sociotechnical system. Consequently, any attempts to understand or control systems by breaking them down into separate elements are inherently limited. Therefore, effectively managing safety requires addressing the entire complex adaptive system.

The modern world has been described with the acronym VUCA, referring to volatility, uncertainty, complexity, and ambiguity. Volatility describes the rapid and unpredictable nature of change. Changes can be more unpredictable, extensive, and occur more frequently than in a less complex world. An example of an extensive change is the COVID-19 pandemic. The increased use of artificial intelligence is another example.

Uncertainty characterizes the unpredictability of events and issues. During the early months of the pandemic, little was known about the virus and how to guard against it. Climate change is also increasing the uncertainty concerning external events. Inside organizations, new technologies combined with old, aging technologies create new hazards.

Complexity denotes the intertwined forces and issues, making cause–effect relationships unclear. Digital instrumentation and control (I&C) and the use of artificial intelligence are examples of helpful tools that at the same time increase the complexity of our systems. New technology combined with the old also increases complexity. Adding redundant and diverse technical or organizational safety systems also adds to the system complexity.

Ambiguity points to unclear realities and potential misunderstandings stemming from mixed messages. Wherever there are groups of people working together, there is ambiguity due to communication issues but also different ways to make sense of reality, different values and goals, and different ways to react to each other. These features also mean that effective learning is more difficult in the VUCA environment. Feedback is delayed, and the relationship between cause and effect is difficult, if not impossible, to decipher. Furthermore, we are only able to get feedback on the actions that we take, not on the myriad possibilities we omit. We learn by simplifying and assuming (see Chapter 2 and 3) in an environment that never stays the same but rather constantly evolves. Thus, even if we manage to learn something about the system in question, we need to constantly update that view since the system does not remain the same.

What Did We Learn about Safety in Complex Organizations?

A systems approach highlights that all things are interconnected; we always need to choose the level of analysis – where to draw the system boundaries, what to leave out. This challenges the idea of modeling and predicting system behavior in exact terms; all models are incomplete, but they may still be useful.

Systems theories remind us that to understand a system, we need both a microscope and a "macroscope". We need to pay attention to interactions and feedback loops and look at the whole. Reductionist approaches are insufficient with complex systems. Interactions, interrelations, and influences are more important than linear cause-effect mechanisms.

In a complex system no one can understand all the interconnections and all the consequences of actions. This impossibility of understanding applies to both managers and experts. Humility instead of hubris is needed to manage a complex system. Managers as well as experts need to admit that they depend on other people. There is a practical need for deference to expertise (see HRO in Chapter 5).

Technology, management systems, personnel, and organizational structures are all part of the system, and they cannot be looked at in isolation. Various human and organizational factors approaches need to be complemented with a view of how the factors interact and what emergent properties this interaction creates. As the focus of attention is on interactions and emergent properties, safety assurance should also look at the whole system instead of each component individually.

The features of complex systems may be positive or negative in terms of their safety effects. Negative effects are considered in safety science with terms such as drift and normalization (see Chapter 3). Positive effects include the ability of the system to adapt to various disturbances and the ability to learn and develop (see Chapter 4).

Since complex adaptive systems evolve over time through learning, feedback, and adaptation, today's strengths can be tomorrow's weaknesses, and vice versa. Their inherent change also puts an emphasis on systems of learning (see Chapter 6).

Management and control of complex adaptive systems means creating the potential for the system to self-organize in effective ways. Instead of episodic planned change interventions (cf. Weick & Quinn, 1999), complex systems require adaptive management where the system is guided, nurtured, and influenced rather than controlled. Changing a complex system entails establishing the conditions for the system to adapt and self-organize rather than defining the ideal state in advance and implementing it.

Finally, complex adaptive systems are resilient rather than robust. We will return to resilience in Chapter 11.

NOTES

1. I will not take a stance here on whether other animals create cultures, and if they do, how they differ from human cultures.
2. At the same time, the invention of the term *culture* represents our tendency to simplify and categorize. For this reason, many of our approaches to culture in fact offer us only consolation, not understanding – consolation in the form of an easy explanation, a simplification, or a soothing story. This problem plagues in particular the concept of safety culture, an issue that will be discussed multiple times in this book.
3. For the history of the concept and different definitions and operationalizations of organizational culture, see, for example, (Alvesson, 2002; Martin, 2002; Pettigrew, 1979).
4. Hofstede's model is focused on national culture, but the theory applies also to organizational culture. Hofstede sees organizational culture as the way in which members of the organization relate to each other, their work, and the outside world in comparison to other organizations. This definition is in line with Schein's definition that is utilized in this book.
5. Herb Kelleher, the co-founder and former CEO of Southwest Airlines, has said that "culture is what people do when no one is looking".
6. Hopkins (2018) notes that the emphasis on practices does not exclude the importance of norms and values. I agree, but the formulation "the way we do things around here" may lead people to consider mostly behavioral issues on an individual level. For this reason I prefer defining culture as referring to the underlying assumptions, and treating practices as manifestations of culture and behavior as only partly determined by culture.
7. To be specific, in the structural view technology and other structures represent the surface manifestations or artifacts of the deeper organizational culture. This culture consists of the shared basic assumptions. As many of these assumptions are embedded in the technology and other structures, the culture of an organization de facto includes technology and structures.
8. There are several different definitions for values, and theories of culture differ in how they see values. For example, Schein (2017) considers two types of values in the organization: espoused values that can be discussed and questioned, and assumptions that are shared values that have proven useful and are taken for granted. Hofstede et al. (1990) in turn treat values as individual traits affected mostly by demographics: nationality, age, education, and gender. Hofstede's model is discussed in more detail in Chapter 10.
9. These assumptions are only partly in verbal form. They are implicit knowledge, which means that they are more like propensities to act (think, feel, and behave) in certain ways than verbalized beliefs. Thus, when we translate them into language, we always interpret the culture, and risk losing some of the nuances. This also means that the assumptions are not unconscious as such, but rather outside of daily reflection – pre-reflective (Leisti & Poskiparta, 2022).
10. There is a fine line between practices and structures when it comes to routines. The existing, culturally embedded routine as such can be seen as a structure, but its myriad instantiations in daily life belong to practices.
11. For this reason the COVID-19 pandemic and increased teleworking was, and still is, a major cultural change.
12. Weick et al. (2005, p. 409) define sensemaking as follows: "Viewed as a significant process of organizing, sensemaking unfolds as a sequence in which people concerned with identity in the social context of other actors engage ongoing circumstances from which they extract cues and make plausible sense retrospectively, while enacting more or less order into those ongoing circumstances."
13. This does not mean that the effects cannot in principle be counteracted by other actions. It means that the particular action always leaves a lasting impression on the system.

2 Individuals and Groups in Organizations

Chapter 2 presents the main issues of individual and group performance in organizational contexts from a cultural perspective. The aim is to demonstrate both variability and commonality of performance at individual, group, and, subsequently, organizational levels. I will describe individuals as balancing between autonomy and belonging, individuality and conformity. I emphasize that human behavior needs to be understood in context, but that context never fully determines our behavior. Humans continually make sense of their environment in order to act in it. This sensemaking is not driven by objective facts but rather plausibility and our extraordinary capabilities to create patterns and order to (act in) a complex world. A culture is a result of our collective sensemaking, but in order to understand why cultures form, we need to understand human behavior, individually, in groups, and in organizations. The chapter concludes by presenting the Space Shuttles *Challenger* and *Columbia* accidents in detail. These accidents show the challenges of managing safety in complex organizations and the interrelations between various organizational levels, from individual and groups to the entire organization.

2.1 HUMAN PERFORMANCE AND INDIVIDUAL VARIABILITY

2.1.1 Individual Performance in Context

Human behavior is always contextual. We act based on affordances provided by the environment (in terms of cues for action embedded in tools, procedures, and technology) and expectations and norms provided by significant others (the peer group). Human behavior needs to be understood in the context of people attempting (together) to make sense of themselves and their environment, and act based on perpetually incomplete information while relying on social conventions, affordances provided by the environment, and the available cognitive heuristics. Human behavior is dictated by the local rationality principle: people (usually) do what makes sense given the situational indications, operational pressures, and organizational norms existing at the time (Dekker, 2011).

Attitudes are evaluations of people or objects as positive or negative. They specify how we feel about the people or objects in question and influence how we behave. Attitudes orient us. What happens after that depends on many factors, and thus there is debate in research on the matter of how much attitudes actually affect behavior. Personality affects attitudes to some degree but in no way determines them. *Personality* refers to the enduring patterns of thoughts, feelings, and behaviors that define an individual. Personality shapes how a person interacts with the world and approaches different situations. According to the so-called Big Five personality theory, personality can be described with the help of the following five traits: 1) openness to experience, 2) conscientiousness, 3) extraversion, 4) agreeableness, and 5) neuroticism[1] (Goldberg, 1992).

Openness to experience refers to a general appreciation of variety of experiences. People who are open to experience are intellectually curious, open to emotion, sensitive to beauty, and willing to try new things. *Conscientiousness* refers to a tendency to act dutifully and strive for achievement against measures or outside expectations. High conscientiousness indicates a preference for planned rather than spontaneous behavior. *Extraversion* refers to a person's breadth of activities and engagement with the external world. Extroverts tend to be enthusiastic and action oriented. *Agreeableness* means a general concern for social harmony. Agreeable individuals are generally considerate, kind, generous, trusting and trustworthy, helpful, and willing to compromise their interests with

DOI: 10.1201/9781003491156-2

others. Agreeable individuals are more probably transformational rather than transactional leaders (see Chapter 7). *Neuroticism* refers to a tendency to have strong negative emotions, such as anger, depression, or anxiety.

No strong links can be made between personality and job performance. Conscientiousness is the dimension with the highest positive correlation to work performance (Salgado et al., 2020). However, different tasks and environments interact differently with various personalities, so no general conclusions can be made. The person–job fit is a more important factor to consider when organizations recruit and promote people. For example, there is some evidence that openness to experience is important for performance in highly complex tasks but not for simple tasks (Salgado et al., 2020).

Personality traits have been studied also from a safety perspective. Salgado (2002) studied the effects of the Big Five personality characteristics on counterproductive behaviors such as absenteeism, accidents, deviant behaviors, and turnover. His results indicated that conscientiousness was negatively associated with deviant behaviors and turnover, and extroversion, openness, agreeableness, and emotional stability were positively associated with the turnover criterion. However, none of the Big Five personality measures were found to be predictors of absenteeism or accidents.

Some general observations about personality and safety can be made. Conscientiousness is associated with rule-following behavior. Individuals who are conscientious are more likely to adhere to safety protocols in the workplace, meticulously following guidelines to ensure safe environments. Additionally, individuals with high levels of agreeableness are more inclined to consider others' safety concerns and contribute to a cooperative and safe work environment.

The relationship between extraversion and openness to experience and safety is complex. Extroverted individuals, who tend to be more socially engaged, may have a positive influence on others' adherence to safety practices. However, they might also distract or pressure others into engaging in riskier behaviors. Regarding openness to experience, while open-mindedness can foster innovation and improvements in safety, it may also result in an overemphasis on novel or untested methods that may not be safe without thorough evaluation.

Organizations should not seek to standardize the personality of the people they recruit. Different personalities bring needed diversity into the workplace and help to avoid, for example, groupthink (see later in this chapter). Certain key positions such as process operators or airline pilots may require less variance in personality profiles.

During the early years of safety science (Swuste et al., 2021), there were attempts to tie personality and accidents together with the concept of "accident proneness". This theory has been debunked, but it still lives as a complementary approach to blaming individuals for human errors, as we will see in the next chapter.

Emotions are deeply involved in everything we do, from decisions to our daily routines. Contrary to some popular beliefs, emotions do not hinder rational decision-making. In fact, we always rely on emotions when making decisions of any kind. People with impaired capacity to feel emotions also have impaired ability to make decisions.

Emotions influence work performance significantly and this influence can be positive or negative depending on emotion and the situation. Positive emotions are associated with motivation, engagement and collaboration whereas negative emotions are associated with conflict and dysfunctional team dynamics. Emotions are also associated with many of the heuristics and they are heavily involved in risk perception (see below).

2.1.2 Bounded Rationality and Decision Heuristics

It was realized already in the 1950s that humans are inherently incapable of processing all the information in their environment. March and Simon termed this "bounded rationality" (March & Simon, 1958; Simon, 1957). According to them, people make decisions based on partial and inaccurate information, and on incomplete and flawed analysis – not a step-by-step analysis. Despite (or maybe at least partly because of) these "flaws", humans excel in making sense of ambiguous situations and acting under uncertainty. Our decision-making is based on recognition of patterns and looking for "good enough" – solutions instead of perfect decisions. Simon (1957) labeled this human tendency "satisficing", a combination of satisfy and suffice. Instead of analytic processing of information, we tend to rely on heuristics and intuition for guidance.

As culture is for groups, intuition is for individuals a result of learning and experience. Despite allowing us to cope with the vast amount of information in today's workplace, intuition sometimes makes us vulnerable to certain biases.

Barry Turner (1978) utilized the concept of bounded rationality in his "Man-Made Disasters" theory to explore the nature of unintended consequences of decision making in organizations (see the end of Chapter 3).

Tversky and Kahneman introduced the topic of heuristics and biases originally in their 1974 article in the journal *Science* titled "Judgment Under Uncertainty: Heuristics and Biases"[2] (Tversky & Kahneman, 1974). Since then, numerous heuristics, or rules of thumb, have been identified in the literature. I will briefly summarize the most important ones from the perspective of culture and safety. It should be noted that these heuristics are inherent features of how humans process information. As such, they are not reasons for events or accidents. They are issues to consider when designing our sociotechnical systems for humans. In fact, saying that an accident was "caused" by human error just means that any human could have done the same thing. Consequently, it makes no sense to pursue the human element in the accident in isolation from context, culture and technology, and the myriad interactions in the system prior to, and during, the event.

2.1.2.1 Heuristics Affect How We Pay Attention to Things

Anchoring refers to a phenomenon whereby an individual's judgments are influenced by a reference point – by an "anchor". People start reasoning with a number they know and adjust it in the direction they consider appropriate. This anchor can be totally irrelevant to the problem at hand, but it still influences the decision. Anchoring may affect how we interpret the various safety indicators and changes in them, or the risks of various activities.

Confirmation bias means that we naturally look for information that confirms our preconceptions. It means that we see what we are prepared to see (Moskowitz, 2024, p. 147). This happens by looking for and putting more weight on information that is consistent with, and reinforces, our existing beliefs. Expectations create orderliness and predictability and offer guides for performance and interpretation, but they can also undermine reliable, resilient performance because they encourage confirmation seeking, reliance on existing categories, and simplification (Weick & Sutcliffe, 2007, p. 41).

Especially under pressure, people are more likely to search for confirming information and ignore information that is inconsistent with their expectations. Confirmation bias is also strengthened in groups that lack psychological safety and trust so that diverse opinions and worries are not disclosed. Moskowitz (2024, p. 178) summarizes that when information is ambiguous, we have a tendency to see it in a way that confirms what we already believe. As discussed in Chapter 1, one characteristic of complex systems is that information is ambiguous.

The bias also contributes to what Taleb (2012) refers to as "mistaking evidence of no harm for no evidence of harm". The absence of evidence on risks is interpreted as evidence of absence of risks. This leads to us assuming that things are safe unless we are confronted with evidence indicating otherwise. Confirmation bias is the reason why norms that emphasize how good the group is, or how safe activities are, can be so dangerous. The accident example at the end of this chapter illustrates the dangers of forming an assumption that things are safe until proven otherwise.

In research, we try to actively fight against confirmation bias with the research design and hypothesis setting. When setting a hypothesis, we try to look for evidence that disconfirms the hypothesis rather than for evidence to confirm it. Furthermore, if such evidence is not found, the hypothesis is still not confirmed but rather it remains valid (until more studies may prove it wrong or validate it further). Of course, all this is mostly research-as-imagined; research as done in practice is very variable indeed.

2.1.2.2 Heuristics Affect What We Remember

Our memories are not exact copies of events as they have transpired. Once information has been committed to memory, it can be altered. Our recollection of memories can be manipulated, and even entire sets of events can be confabulated. Confirmation bias (see above) affects also what we remember. People will scan their memories in quite a similar manner to scanning the environment, and look for confirming evidence, or in this case confirming memories (Moskowitz, 2024, p. 154). In practice, this means that we can remember events very differently based on our expectations, for example, how people typically perform or how accidents typically happen.

Priming occurs when an individual's exposure to a certain stimulus influences their response to a subsequent prompt, without conscious guidance or intention. In priming, accessible information shapes how ambiguous information is interpreted (Moskowitz, 2024, p. 270). Priming explains why after safety training, people are (for a short while) perceptive of those risks that the training concerned – sometimes even overemphasizing those. Various slogans and safety posters also try to build on this heuristic (see Chapter 9).

Framing influences what we recall. The 1974 study "Reconstruction of Automobile Destruction" by Elizabeth Loftus and John Palmer is a landmark experiment in cognitive psychology, focusing on the malleability of human memory. It explores how language used in questioning can influence a witness's memory of an event, specifically in the context of automobile accidents. The researchers conducted two experiments in which participants watched videos of car accidents and were then asked to describe what they saw. The critical manipulation was the verb used in the question about the speed of the vehicles. Some participants were asked, "How fast were the cars going when they 'smashed' into each other?" while others were asked using verbs like "hit", "collided", "bumped", or "contacted". The wording of the question significantly influenced participants' speed estimates. For example, participants asked about the cars "smashing" into each other reported higher speeds than those asked about "hitting". In a follow-up experiment, participants were asked a week later whether they had seen broken glass in the video, even though there was none. Those who had been asked the "smashed" question were more likely to report having seen broken glass, suggesting that the wording of questions can distort not just speed estimates but also the memory of other details. Loftus and Palmer (1974) concluded that memory is not a static record of an event but can be reconstructed and influenced by post-event information, including the way questions are framed. Framing happens in, for example, incident investigations when we ask, "why did you do that?" instead of "what happened?" (Busch, 2016, p. 22). We will return to potential biases in incident investigations in Chapters 6 and 9.

Availability heuristic refers to a mental shortcut that occurs when people make judgments about the probability of events based on how easy it is to think of examples. A typical example is our tendency to overestimate the number of homicides and underestimate the number of suicides due to the former being reported in the news media more widely than the latter. Another example is overestimating the risks of nuclear power plants due to high profile accidents such as Chernobyl and Fukushima. Availability bias also explains why people who have personally experienced, or witnessed, an accident are typically more prone to take appropriate safety measures. Availability heuristic stems from our need to assign an explanation to why some information is readily available in the mind (Moskowitz, 2024, p. 135). "If it is readily available, it must be important", says our internal logic.

2.1.2.3 Heuristics Affect How We Attribute Causality and How We Learn

Individuals typically attribute successes to their own merit but failures to the presence of external forces. This kind of thinking is good for our self-esteem, as we can take credit for successes and explain away failures. Similar to how we view our own success, we tend to blame someone else's mistake on the characteristics of that person (laziness, indifference, lack of ability), instead of the situation or work conditions.

A bias called *fundamental attribution error* (also called *correspondence bias*) refers to the tendency for people to overemphasize dispositional, or personality-based, explanations for behaviors observed in others while underemphasizing situational explanations (Fiske & Taylor, 2008). People have a general tendency to perceive others as having quite stable traits and see their behavior as less dependent on context than their own (Fiske & Taylor, 2008). Thus, the errors and mistakes of others are perceived as being due to stable traits ("bad apples") rather than contextual reasons. Furthermore, these errors are typically attributed to the actions of people closest to the hazard source, such as operators or maintenance technicians. *Self-serving bias* refers to the tendency to attribute our own, or our own reference group's (see *social identity*), successes to stable traits and failures to situational factors (Moskowitz, 2024, p. 323).

The *just world fallacy* is another, connected, cognitive bias. This bias makes us believe that good things happen to good people, and bad things happen to bad people. Thus, a person involved in an accident must have done something to deserve it, for example, they made an error or conducted a violation – that is, if the world would be just. This bias hinders our ability to see systemic influences on the successes and failures of people. It also helps us to maintain the belief that we are safe, that these negative events could never happen to me (because I am a good person). This is related to yet another bias, called the *illusion of uniqueness.* We all have this illusion, and paradoxically this makes us especially prone to manipulation and group influence.

Hindsight bias refers to our tendency to exaggerate what could have been anticipated in advance (Fischhoff, 1975). After an event has happened, it seems much more plausible and simpler. In hindsight, we can see the overwhelming evidence pointing to the outcome and wonder why people did not see that in time – why they acted as (stupidly) they did. It is very difficult for people to ignore the knowledge of an actual outcome to generate unbiased inferences about what could or should have happened (Fiske & Taylor, 2008). Finding out that an outcome has occurred increases its perceived likelihood. For this reason, we easily blame front-line workers for missing the obvious (in hindsight) danger signs or for making the obvious (again, with the benefit of hindsight) wrong decisions. We easily think that "I would have perceived the danger" or "I would have made the right decision", because we know what the danger was and what the right decision would have been. People are, however, unaware of the effect that outcome knowledge has on their perceptions (Fischhoff, 1975).

Outcome bias shares many similarities with the hindsight bias. It refers to our tendency to evaluate decisions based on their outcomes so that decisions leading to good outcomes are judged as good, and vice versa (Baron & Hershey, 1988). Because of the outcome bias we tend to think that major accidents have to have major errors behind them, and minor accidents are caused by minor errors. Likewise, actions that cause harm are judged as more unethical than actions that do not cause harm, irrespective of the underlying intentions of the actor or other contextual factors (Bazerman & Tenbrunsel, 2011, p. 95).

Representativeness refers to our tendency to categorize and form stereotypes. It means assessing similarity of objects and organizing them based around the category prototype.

2.1.2.4 Heuristics Affect How We Behave

Self-fulfilling prophecy resembles confirmation bias (see above), but in this bias the perceiver's own behavior is shaped by their impression or expectation of the other person's behavior, and this behavior draws out expectations confirming behavior from the other person. Thus, the perceiver produces the behavior in the other person. An example is trust, and especially distrust. A person who does not trust others treats them with suspicion, does not share information, and actively avoids involving others. When the others then do not in turn share information with the given person or do not involve the person in their circle, this is considered proof that it was the right call not to trust them.

Overconfidence and the illusion of superiority refer to our tendency to overestimate one's own capability to make correct decisions. People tend to overrate their abilities and skills as decision makers. It also means that people overestimate their personal immunity to harm, which may facilitate excess risk taking, especially if the actualized risk (accidents) are not easily recalled into memory (see *availability bias*). Most of us see ourselves as better than the average person (which is impossible, especially if we are talking about median scores and not average scores). People reason about their abilities in a self-motivated way to improve their self-esteem (Moskowitz, 2024, p. 172). Leaders often overestimate their effect on their subordinates' performance, or their effect on a culture.

Furthermore, the Dunning-Kruger effect refers to the fact that we are mostly ignorant about our level of ignorance and that even people with an actual low ability on a task will overestimate their ability – claiming higher ability than they actually possess (Dunning, 2011; Kruger & Dunning, 1999).

Loss aversion (Tversky & Kahneman, 1991) and *status quo* (Samuelson & Zeckhauser, 1988) biases refer to our tendency to avoid losses and stick to what we have. The pain of losing something is psychologically more intense than the pleasure of gaining something of identical value (often measured in monetary terms in experiments). In practice this means that we tend to be more concerned with risks associated with a potential change rather than the risks of not changing.

Omission bias (Baron, 2008) is related to loss aversion. We tend to underestimate the risks associated with omitting something, when compared to risks of doing something. This leads us to prefer omission (inaction) over commission (action) in uncertain situations. Omission bias is also used to refer to people's tendency to evaluate harm done through omission as less morally wrong and less blameworthy than commission when there is harm (Yeung et al., 2022).

Decision-making in dynamic situations is further hampered by "*misperceptions of feedback*", which refers to human tendencies to adopt an event-based view of causality that underappreciates time delays between action and response and is insensitive to non-linearities (Diehl & Sterman, 1995).

Our skills in subjectively simplifying complexity sometimes work against us objectively, as we will see in Chapter 3, when we examine how accidents in complex organizations happen.

2.1.3 Risk Perception

Risk perception refers to the subjective judgments people make about the characteristics and severity of various hazards. Like decision making, it has an intuitive and analytical component. Emotions are heavily involved in risk perception. Negative emotions increase perceived risks, and positive emotions decrease them. Confounding this causality, perception of high risk can cause stress and negative emotions. Loss aversion bias (see above) is partly related to our fear of loss and the negative feelings associated with loss. The *affect heuristic* (Slovic et al., 2005) refers to the tendency to rely on immediate emotional reactions to judge the risks associated with a particular decision, rather than evaluating the facts rationally.

We can differentiate between perceived, or subjective, risk and objective risk, even though the distinction can be questioned, as perceiving the risk in one way can change our behavior and thus affect the so-called objective risk. However, with the term *subjective risk* we want to emphasize that people perceive risks differently, and people perceive different risks differently. The *social amplification* or attenuation of risk (Kasperson et al., 1988) depends on several factors, such as the voluntariness of the potentially risky endeavor, its familiarity, and the "dread factor" (uncontrollable catastrophes). The following summary discusses the factors that influence how risky an issue is perceived to be (Pidgeon et al., 1992, 2003).

Risks are emphasized and potentially overestimated if:

- The issue/situation is unique or very rare (e.g. a nuclear accident or a meteor strike);
- The issue is rare but has major consequences (disasters in general and nuclear accidents and meteor strikes in particular);
- The matter/situation is out of one's own control or decided by someone else;
- The matter is new or there is no personal experience of the matter (fear of the unknown);

- The issue or situation is psychologically scary;
- The benefits of the issue are difficult to perceive, or the benefits go to someone else;
- The risk is caused by humans and not by nature;
- Other members of the group/organization express that they consider the issue as risky;
- It is easy to imagine the thing/situation as dangerous, for example a recent plane crash that has been shown on TV;
- The risk assessor experiences negative emotions caused by something other than the risk in question, such as anxiety, fearfulness, or depression.

The risks are typically underestimated if:

- The situation/issue is "normal" and repetitive (e.g., smoking, driving) – it has not been a danger before;
- The matter/situation has been decided by oneself, for example, due to interest in it;
- The issue is a personal risk and not a risk to others;
- Other members of the group/organization do not express that they consider it a risk;
- The potential reward for the risk taker is big;
- The situation or issue requires *not* doing something instead of doing something;
- The risk assessor perceives the risk to be so big that they resort to some psychological coping mechanisms, such as denial;
- The issue happens in future and not immediately (e.g. climate change);
- The risk assessor is a sensation-seeker (Horvath & Zuckerman, 1993).

Ultimately, risk perception depends on competence and understanding of the hazards as well as the local affordances. Local affordances mean how the environment, tools, and technology highlight and potentially hide different hazards.

2.1.4 Rationality and Decision-Making

Daniel Kahneman (2011) has differentiated between two modes of thought: System 1 is fast, instinctive and emotional. Intuition and gut feelings are often associated with thinking according to System 1. System 2 is slower, more deliberative, and more logical. It requires conscious attention and effort. System 1 corresponds to *intuition* and System 2 to a *rational* and analytic thought process.

Even though these are presented as separate systems, both intuitive and rational thought are intertwined, and both systems are needed for adequate decision-making. Often the rational mind comes into play after the decision has been made, to rationalize it. According to the latest research, our thinking is predominantly intuitive, with reflective and analytical thinking only coming to play in special circumstances. Analytical thinking often follows our decisions rather than preceding them, and therefore does not increase our rationality (Leisti & Poskiparta, 2022, p. 123).

Analytical thinking involves examining one's impressions and experiences rather than solely relying on objective facts. People often rationalize their decisions after making them. These rationalizations are typically perceived as reasons for the initial decision, rather than being recognized as retrospective justifications, thus maintaining the appearance of rational decision-making. This also affects how we make sense of events retrospectively.

Analytical thinking is a slow and cognitively more demanding process when compared to intuitive thinking. People use experience to overcome some of the limitations of analytic thinking, such as its slowness and cognitive load, as well as difficulties with uncertain or ambiguous situations. *Naturalistic decision making* (NDM) is a research tradition that studies how individuals make decisions in real life (Klein, 2008). Field research has been conducted in the NDM tradition to discover the strategies people such as firefighters, military commanders, nurses, design engineers, and pilots use in decision making. NDM investigates the types of decisions made in their work and how they utilise their experience to manage challenging conditions such as time pressure, uncertainty, unclear goals, high stakes, organizational constraints, and team coordination requirements.

The NDM approach contrasts with heuristics and decision paradigms (e.g., Kahneman, 2011) that employ predefined tasks assigned to inexperienced subjects under controlled laboratory conditions. Most of the biases presented in this chapter have been discovered in these laboratory experiments. The recognition-primed decision (RPD) model was one of the main outcomes of their empirical research. RPD relies on two main processes: pattern matching and mental simulation. According to Klein (2008), the RPD model is a blend of intuition and analysis. The pattern matching is the intuitive part (corresponding to Kahneman's System 1), and the mental simulation is the conscious, deliberate, and analytical part (corresponding to System 2). Klein argues that recognition-primed decision-making is more common than pure analytical decision-making. Recognition-based decision making means that the expert searches for similar conditions in previous cases while being aware of the differences between those earlier cases and present situation (Klein, 2008; Klein et al., 1993).

Heuristics and intuition are not in contradiction. Kahneman and Klein (2009, p. 518) note that

> the intuitive judgments of some professionals are impressively skilled, while the judgments of other professionals are remarkably flawed.

They note that an important distinction is whether the intuitive decision-making is based on experience and manifests skill or whether it is based on general heuristics without specific experience (Kahneman & Klein, 2009, p. 519).

In addition to recognition-primed decision-making, people rely on metaphors and analogies to make sense of complex environments. Creating metaphors is a fundamental mechanism of the mind. It allows us to use what we know about our physical and social experience to provide understanding of countless other subjects (Lakoff & Johnson, 1980). When we speak about a "healthy" culture, or a culture as a "family" or an "iceberg", we are using a metaphor to understand culture (cf. White, 2017). The same applies to understanding of organizations as machines, brains, and so on (Morgan, 2006). When relying on analogies, we seek patterns in our environment and identify similarities between the current problem and an analogous situation from the past. This is what happened at Three Mile Island (see below) when the control room operators misdiagnosed the situation (they recognized a wrong pattern) as a problem that they had been trained to deal with.

An unfortunate side-effect of our active search for patterns is that we tend to see them also in places where they do not exist. One example of this is confusing correlation with causation. If B happens right after A, we tend to assume that A caused B. Various conspiracy theories are also partly due to our extraordinary capacity to discover patterns.

The confidence we have in our decisions isn't strongly linked to the difficulty of the issue at hand or even to our competence in the issue. Additionally, we're not very proficient at evaluating our own decisions. Due to confirmation bias, we may get more certain when we try to analytically reflect on our decision, but in reality, we are just coming up with further proof for sentiments that we already have made up (Leisti & Poskiparta, 2022, p. 181). If we are not aware of our biases and assumptions, gathering more information may increase our certainty in the decision, but it does not increase the quality of the decision, as we gather information that supports our initial viewpoint and disregard evidence that would disprove it. Strong and cohesive cultures can similarly influence certainty and enhance our inherent tendencies to ignore opposing evidence. This phenomenon, groupthink, is examined later in this chapter.

Decision-making heuristics are social properties in the same way that social properties are information-processing properties (Leisti & Poskiparta, 2022). People should not be treated as detached from their social environment. Human cognition does not occur only in the brain but rather it operates in dynamic interaction with the environment, including the technological and social environments (White, 2017, p. 48).

The mind is distinctly different from a computer in that it does not operate in isolation from the environment but is in reality embodied (White, 2017). This means that the work we do, the technology we use, and the people and environment we interact with all structure our cognition. This is one more reason not to focus too much on individual decisions and their outcomes when considering a culture, but more on the interaction of work, technology, people, and the environment.

Finnish psychologists Leisti and Poskiparta (2022, p. 93) note that to understand the human factors involved in decision making, we need to consider, in addition to the processes in the brain, the individual history of the decision maker; the social, institutional, and cultural context; and even the bodily state of the decision maker (e.g. fatigue, hunger, and mood). When we make sense of issues and events, we do not do so in isolation, but we build on our professional experience and expertise, our interactions with significant others, and the various cues offered by the environment.

The features of human performance described here help to make sense of why it is so difficult for us to learn from accidents and why we still so readily deem them to be caused by individuals failing to act safely (see Chapter 3). The various heuristics and biases also partly explain why we sometimes have unrealistic expectations about our own performance. Bazerman and Tenbrunsel (2011, p. 37) noted that "most of us dramatically underestimate the degree to which our behavior is affected by incentives and other situational factors". The real or imagined presence of others is one such strong situational factor.

2.2 MAKING SENSE IN SOCIAL SETTINGS

As we established in the previous section, thinking and reasoning are deeply social and cultural. Mercier and Sperber, in their article "Why do humans reason?" (2011) have proposed that the primary function of reasoning is argumentative. It is to devise and evaluate arguments intended to persuade, rather than a means to improve knowledge and make better decisions. This view on reasoning explains many of the biases presented in this book. If the reason why we reason is to find arguments to support and communicate our views to others, it makes sense to focus on confirming evidence and disregard disconfirming information. I doubt it is as simple as this, as surely we are able to use our reasoning for better decision making too. Moskowitz (2024), citing one of the early thinkers in psychology, William James, notes that "thinking is for action". A key requirement is that we need to know what to expect from an object or a person (Moskowitz, 2024, p. 9).

Developmentally, thinking predates language. This also means that the deepest cultural assumptions, especially when it comes to national cultures, are only partly linguistic. White (2021, p. 10): "language and culture are parallel systems ... with overlaps and touch points that reflect, support, or at times serve as adaptations to each other". Language should be seen as one of the tools of thinking and sensemaking, but not the only one.

Our cognition, with all its heuristics and stereotypical ways of handling information, is built to predict what will happen in our (local) environment. This does not mean that the brain is a passive recipient of information. Rather we take part in continuous and active sensemaking as we interact with our environment and the people and objects in it. This sensemaking aims to enact us meaningful courses of action.

Scott Snook (2000) provides an excellent analysis of the friendly- fire incident over Iraq in April 1994 where two F-15 fighter pilots shot down two friendly Black Hawk helicopters, killing 26 people. In his analysis, he questions whether the F-15 pilots "decided" to shoot and whether asking that question leads anywhere in the analysis. He points out that it takes us to decision-making, and especially wrong decisions: "why did they make the wrong decision [to shoot two friendly helicopters]". When the problem is approached from a decision-making perspective, it falls squarely onto the shoulders of the decision maker and away from the context and situational influences. Snook notes that framing the issue as a question of meaning, as "good people struggling to make sense" rather than as "bad people making poor decisions" allows better understanding of the event (Snook, 2000, p. 207).

Weick et al. (2005, p. 409) define *sensemaking* as follows:

> Viewed as a significant process of organizing, sensemaking unfolds as a sequence in which people concerned with identity in the social context of other actors engage ongoing circumstances from which they extract cues and make plausible sense retrospectively, while enacting more or less order into those ongoing circumstances.

Sensemaking happens when we categorize and put into words the flow of organizational circumstances. In this cultural process of sensemaking, people define what are the problems that need to be solved, and how to solve them. Once the personnel have named the problems, they already have predefined ideas on the right solutions (cf. Weick, 1995).

Sensemaking means an attempt to tie actions and beliefs more closely together. Decisions are often justified by emphasizing some "facts" over others, by reconstructing the pre-decision-making history. Sensemaking is driven by plausibility and coherence rather than accuracy or facts as such. People do not make sense of events only once but rather engage in a continual revision of their understanding based on subsequent events (historical revision) and the interpretation of others (social influence) (Weick, 1995; Weick et al., 2005).

Three Mile Island Nuclear Accident[3]

On 28 March 1979, an accident occurred at Unit Two (TMI-2) of the Three Mile Island Nuclear Power Plant in Harrisburg, Pennsylvania, resulting in the meltdown of part of the reactor core and the release of small amounts of radioactive material into the environment. No one died or was injured because of the accident, but the public image of nuclear power as a safe form of energy suffered a severe blow.[4] *TMI-2 was a pressurized water type nuclear power plant (PWR). In a PWR, water from the primary circuit is supplied to steam generators, which produce steam in the secondary circuit to rotate the turbine. PWR has two closed circuits, and the water in the primary circuit is prevented from boiling by high pressure. The water in the primary circuit is radioactive, whereas the water in the secondary circuit is not. The TMI-2 reactor was new, and at the time of the accident it had only been at full capacity for about 40 days.*

The event began when polisher pumps in the turbine side tripped, causing the turbine to trip. Left without heat removal, the temperature of the reactor core and the pressure in the primary circuit began to rise rapidly. This led to an automatic shutdown of the reactor, that is, control rods that effectively absorb neutrons landed in the reactor core, stopping the nuclear reaction. A relief valve (PORV) opened as planned to relieve excess pressure from the primary circuit. The valve was supposed to close but it stuck open, allowing large amounts of coolant to escape. Due to factors such as control room design, the operators failed to detect that the valve was open.

Initial events escalated quickly, with the leak occurring 15 seconds after the feedwater pumps stopped, leading to the water in the reactor boiling within minutes. The operators conducted many suboptimal activities that made a serious situation worse. They failed to detect that they were in the loss-of-coolant accident situation. The operators misdiagnosed the situation as it progressed.

About 15 minutes after the start of the event, the rupture plate of the water tank into which the water from the primary circuit had discharged broke due to high pressure. Radioactive water was discharged directly into the reactor containment for the next couple of hours. Two minutes after the initial pump failure, automation turned on two high pressure injection (HPI) emergency cooling pumps to add water to the primary system.

Operators noticed that the pressure in the primary circuit continued to rise and the water level in the pressurizer increased – at least according to their indications. At this point – about an hour after the start of the event – the operators made a mistake that significantly aggravated the situation. They turned off the emergency cooling pumps because they were afraid that the pressurizer would fill with water. This had been warned in the training, as it would make it more difficult to manage the plant ("reactor goes solid"). If the emergency cooling pumps had not been shut down, severe damage to the reactor would probably have been avoided. In reality, the water in the reactor was boiling, and the water level decreased as it formed into steam and escaped through the relief valve. About an hour and a half after the start of the event, the main circulation pumps began to vibrate (cavitate) as steam passed through them instead of water. The operators feared that the pumps would break and stopped them. The flow of water stopped, and the formation of steam accelerated.

Due to the shutdown of the main circulation pumps, the reactor core was exposed and overheated. About two and a half hours after the accident, the fuel cladding began to melt, and part of the upper part of the reactor core collapsed into the lower and middle parts. At the same time, the PORV valve, which was left open, was finally noticed and the pipeline was closed. Half an hour after this, the operators realized that the fuel was damaged. They declared a plant emergency and turned on one of the main circulation pumps, as well as the stopped emergency cooling system, to get water into the reactor. However, two-thirds of the reactor core had already been exposed. The exposed core had heated to a temperature of more than 1000 C due to residual heat, and part of the fuel had melted. Through an open valve, the water from the primary circuit had been discharged into the water tank, and when its rupture plate broke (about 15 minutes from the start of the chain of events), due to high pressure, radioactive water had been able to discharge directly into the reactor containment.

In retrospect, the symptoms of containment leakage were clear. However, operators missed the jammed valve due to a poorly designed monitoring panel that showed command status rather than actual valve position. It took a new operator, who came to shift over two hours later, to identify the open valve.

The control panel was misleading, indicating the valve was closed when it was only commanded to close. With hundreds of poorly arranged alarms simultaneously going off, and many meters off-scale, operators failed to understand the situation correctly. Operators noticed the exhaust tube had a high temperature reading but wrongly assumed that it was caused by a chronically leaking valve – an issue they had had no time to repair in late 1978 as they hurried the plant into production. The accident investigation highlighted substantial training and competence deficiencies at the plant alongside issues in emergency preparedness and organizational functioning, including, for example, lack of independent safety assessments of line operating procedures and lack of learning from experience. After the event, the Institute of Nuclear Power Operations (INPO) was established to set industry standards and share experiences in the United States.

Operators at TMI had to make sense of an unfolding and complex situation. The initial situation did not make sense to them until they found a pattern: that there is too much water in the pressurizer and this needs to be avoided. This misdiagnosis guided their actions until a new crew arrived and was able to discover the open PORV valve. Instead of continuous decision-making, work involves continuous sense-making grounded in how we see ourselves as professionals and members of the work community.

2.3 SOCIAL NEEDS AND SOCIAL IDENTITY

Moskowitz (2024) summarizes that people have *social needs* in addition to biological needs. Three main social needs have been identified: 1) a need for social connection, or affiliation needs; 2) a need for positive self-regard, or self-esteem needs; and 3) a need to have understanding and meaning that allow us to prepare for action, or epistemic needs.

Epistemic needs refer to our need to identifyl/label, understand, and be able to make predictions about people and objects. These needs arise out of the fact that people do not wish to feel as if they are subject to random and haphazard forces that are beyond their control. We want to see and feel the world as controllable and predictable. The unfortunate effect of this is that people tend to blame the victims when something bad happens to them, either intentionally or by accident (Moskowitz, 2024, pp. 26–28). This tendency is captured by the various biases described above, such as the fundamental attribution error (see also Chapter 3 and why we fail to learn).

Self-esteem needs refer to our need to experience a positive sense of esteem and identity (Moskowitz, 2024, p. 28). This may lead to the biases mentioned earlier, such as confirmation bias, as we try to maintain a positive view of ourselves or defensiveness when presented with negative information about ourselves. The way we make sense of our qualities and actions is filtered through the lens of a need to enhance and protect self-esteem (Moskowitz, 2024, p. 29). This is one more reason why we may fail to learn from incidents and accidents (Chapter 3), as we tend to protect our self-esteem by attributing failures to external factors and successes to internal factors (see *fundamental attribution error*).

A strong source of positive identity comes from the groups with which we are associated (Turner et al., 1979). We tend to seek groups that would facilitate our social identity, which brings us to the third social need. *Affiliation needs* refer to our need to affiliate with other people, to feel approved of, and to belong to something larger than the self (Moskowitz, 2024, p. 29).

A key psychological process underpinning group behavior is a shift in self-perception from a personal level in which the self is defined by a unique personal identity ("me") to a group level in which the self is defined by a social identity that is shared with the other members of a psychological ingroup (Haslam et al., 2022). The more the members see themselves as part of that group, the more they are motivated to conform to the group's values and beliefs, and the more likely they are to see the group norms and goals as a point of reference for their own behavior (Turner, 1991). Individual identification and contextual salience are important concepts that affect the strength of social identity at any given moment. Social identity is central to the formation of a culture, as it provides the psychological platform for mutual influence that helps to transform individual views, opinions, and attitudes into shared values and beliefs (Haslam et al., 2020).

Social identity refers to a sense of belonging to a certain organization, profession, or group in the organization and the differentiations made between the ingroup and other groups (Haslam, 2004). Social identity differentiates one group from another. It clarifies the group's characteristics, activities, membership, and expected behavior. Healthy social identity also increases engagement and motivation and binds the group together so that everyone supports each other.

Social identity emerges by categorizing the world into "we" and "them", shaping one's perception of "who we are". Furthermore, we tend to favor our own group over others and exaggerate the differences. This *ingroup favoritism* (or ingroup bias) means we show preference to the members of our own group over the members of other groups.

Research demonstrates that even artificially created groups in laboratory settings quickly develop a rudimentary social identity, favoring their own members over other similarly artificial groups (Tajfel, 1978). Social identity enables cooperative and purposeful behavior aimed at common objectives. It also enhances human interaction and communication, collective action, and going beyond official duties (Haslam et al., 2020). Social identity plays a crucial role in fostering understanding and predictability in an individual's world (cf. Weick, 1995).

We tend to define our group in relation to other groups. Often this is done with negations: "we are not like those X companies", or "we are different from our competitor Y". The challenge when defining ourselves in this way is that it still leaves open the question "what are we?" And as we tend to have a stereotypical and overly narrow view of other groups than our own, our definition of what we are not also is stereotypical and narrow. Definition by negation is not an optimal starting point for developing a healthy culture.

Social identity formation has also implications for learning. As mentioned previously, we tend to attribute successes and failures differently depending on whether we are judging our own or someone else's behavior. However, once we form a social identity with a group, we judge its behavior as we would our own. Pettigrew (1979) labeled this, in the spirit of the "fundamental attribution error" (see above), the *ultimate attribution error* – our preference to using traits instead of situational or contextual factors to describe negative behavior of outgroups and the positive behavior of the ingroup (Moskowitz, 2024, p. 325).

In practice, this means that when the ingroup makes a mistake, people typically attribute it to situational causes such as bad luck, poor information from others, weather, and so on, but when others (the outgroup) fail, this is attributed to their competence, wrong attitude, or other more or less stable trait. An organizational challenge is to organize activities in such a manner that the negative effects of heuristics are minimized (see Chapter 6).

Individuals outside one's own group are also considered less diverse and more homogeneous than members of one's own group: "they" are all the same, but "we" all have our own individual differences (Haslam et al., 2020), as long as these are still in line with the group's goals and values. However, in conflict situations, people may see their own group as more homogeneous than it really is, forming stereotypes about both groups (Haslam et al., 2020). This assumes interchangeability among members in goals, values, and norms, fostering trust within the group to act cohesively. Simultaneously, there's an expectation that the opposing group will behave typically for their group. These biases can pose risks to group dynamics.

While a group can offer support and is a significant source of identity for an individual, it can also treat its members quite harshly. The black sheep effect implies that "pleasant" ingroup members are judged more favorably than "pleasant" outgroup members, whereas "negative" ingroup members receive harsher evaluations compared to "negative" outgroup members (Marques & Paez, 1994). This indicates that there is substantial social pressure within the group for individuals to act "pleasantly".

Generally, individuals derive social identity from multiple sources. People belong to multiple cultures and carry multiple identities simultaneously. The relative strength of these identities changes according to the context. When at work, we may identify with the company or our occupational identity (civil engineer, nurse, I&C technician, psychologist, etc.), the particular office or function (emergency care, R&D department, operations, etc.), or our age group (or gender). Leaving the office, other identities typically become more salient as we go home, spend time with friends or with our hobbies and spare-time activities. We can also identify with our national culture. At a workplace with multiple nationalities, subcultures can form based on the national cultures and already existing social identities.

Different social identities are salient depending on the situation and context. A safety professional can have a hobby that involves taking risks not accepted at work, and this does not make him a bad safety professional.[5]

2.4 NORMS AND SOCIAL CONFORMITY

Social influence, by, for example, peer pressure, is a type of power (see Chapter 1). Social influence operates through norms, expectations, and the desire for acceptance within a group. People are influenced by the pressure to conform to what is seen as "normal" or "appropriate" in a given social context. Groups establish norms to guide their actions, dictating appropriate behavior for varying situations and roles. These norms streamline, regulate, and predict social interactions, thus unifying group activities (Hogg & Abrams, 1988).

Norms develop about, for example, the dress code, the appropriate and non-appropriate ways to behave, how to communicate, authority relations, proper ways of working, fair work rate, how to interpret and apply formal rules, and so on. New members infer the group's actual norms by observing members' behaviors. Certain members, often called opinion or informal leaders, have a greater influence on the formation of norms despite lacking official status. Essentially, leadership is about shaping the group's norms and culture, as elaborated in Chapter 7 (compare with Schein, 1985).

Group norms often persist independently, even when most members think they are wrong or immoral, as individuals hesitate to challenge what they assume is majority opinion. Some organizations struggle to discard harmful attitudes because members bond over complaining about them, making it hard to initiate real change. In order to change these norms, the change should be carried out in a manner acceptable to these norms, but the ways accepted by the norms do not necessarily lead to real change (see e.g., Weeks, 2004). The higher the social cohesion of the group, the higher the level of conformity to the group norms.

Pressure to conform, or conformity bias, refers to situations where people are more likely to follow the orders of higher authorities without questioning the validity or morality of the action involved. Conformity stems from the importance of social identity to self-esteem. People conform to the group's views to avoid dissent (Moskowitz, 2024, p. 177).

Asch (1956) conducted his famous conformity studies in the 1950s where he demonstrated that subjects in an experiment consistently gave wrong answers about the length of a line when others in the group (who were actors) all gave the same wrong answer. The experiment demonstrated how the group pressure shaped the individuals' perceptions and public response. An interesting and often downplayed finding in Asch's studies was that a single ally – another person who answered contrary to the popular but outright wrong opinion – allowed the subject to maintain their correct view and avoid succumbing to the group pressure. Sometimes it takes just one person to enable others to stand their ground, or change the world.

Groupthink refers to a phenomenon where a highly cohesive group develops informal norms and intragroup social dynamics that aim to preserve internal harmony and team relationships. Alternative viewpoints are dismissed, and conflict avoided. This is especially prone to happen under stress (Janis, 1982). These pressures to censor own's own views can cause groups to confirm their initial preferences and forego the search for potentially disconfirming evidence. Cohesive groups that work under stress (and under perceived external threat) can hold an inflated sense of their superiority (moral and otherwise) and underappreciate outsiders. They can also hold a stereotypical view of others external to them. Groupthink leads us to overemphasize success factors and avoid conflicts.

Groupthink can be best avoided by having diverse individuals working together in teams where there is adequate psychological safety and trust, allowing all opinions to be heard. Unfortunately, given a choice, the majority of us tend to surround ourselves with people who are similar to us. Similarity increases cohesion but narrows the possibilities for innovation and novelty. Inclusive leadership is needed, one that emphasizes that it is safe to disagree in this group (Carrillo, 2020; Edmondson, 2018).

Pluralistic ignorance refers to the fact that we do not know what other people are really thinking. This explains why some routines and practices may persist even though the majority of the employees despise them. Likewise, it may explain why organizations get rid of practices that people actually enjoy. In pluralistic ignorance, people in a group misperceive the group's norms because members of the group act inconsistently with their private beliefs because of a social pressure they believe exists (Moskowitz, 2024, p. 176).

The *Abilene paradox* (Harvey, 1974), originating in a story based on the author's real experience, illustrates pluralistic ignorance. The story takes place one hot July afternoon in Coleman, Texas, when a dust storm was blowing through the house. Despite the heat, a family (a husband and his wife were visiting her parents) was enjoying a fan-cooled porch with cold lemonade and a game of dominoes. Suddenly, the father-in-law suggested driving 53 miles to Abilene for dinner. Although the husband, the storyteller of the story, doubted the idea, everyone else seemed enthusiastic. According to his wife, the trip "sounds like a good idea". The husband notes that "since my own preferences were obviously out of step with the rest I replied, 'Sounds good to me,' and added, 'I just hope your mother wants to go.'" She did. The trip in an unairconditioned 1958 Buick was as uncomfortable as expected, the heat was brutal, and the cafeteria food wasn't worth it.

But the story is not over. Four hours and 106 miles later, they returned to Coleman, exhausted. They sat silently in front of the fan until the husband broke the silence, saying, "It was a great trip, wasn't it?" No one responded initially. Then the mother-in-law irritably admitted she hadn't enjoyed it and had only gone because she was pressured. The husband was shocked and claimed that it was actually him that went to satisfy everyone else. He blamed them – "You're the culprits". The wife cut in and denied blame, arguing that she just went along to be sociable and keep everyone happy. Her argument: "I would have had to be crazy to want to go out in heat like that". Her father cut in, stating he never wanted to go either and preferred staying home. They realized they had taken the arduous trip to Abilene despite none of them wanting to go, which made the situation perplexing and nonsensical (Harvey, 1974).

Harvey (1974) summarizes the lesson to organizations:

> Organizations frequently take actions in contradiction to what they really want to do and therefore defeat the very purposes they are trying to achieve. It also deals with a major corollary of the paradox, which is that the inability to manage agreement is a major source of organization dysfunction.

In organizations, both conflict and agreement need to be managed, as both can be either healthy or unhealthy in terms of the organizational core task.

In social interactions, we develop stereotypes about others. Stereotypes simplify and speed up actions by setting expectations based on assumptions. A stereotype is basically a knowledge structure made up of "learned expectations regarding the way things usually go" (Strauss & Quinn, 1997, p. 49), and as mentioned above and discussed later in Chapters 3 and 7, we are typically not very objective when it comes to learning.

Stereotypes lead us to treat individuals as representatives of their group, affecting our perceptions and behaviors. Stereotypes also influence how others behave towards us through self-fulfilling prophecies. For example, believing that a group misbehaves causes us to treat them poorly, prompting negative reactions that confirm our stereotype. Conversely, positive expectations foster kinder treatment and friendlier responses. Typically, stereotypes are negative because threatening information is more memorable. Studies have shown that even when people are told equal amounts of positive and negative details about certain others, they recall more negative information and form negative impressions (Fiske & Taylor, 2017). This is called the *negativity bias* (see Vaish et al., 2008).

Haslam et al. note that groupthink-like processes take place also in successful decisions. Similar to human errors and decision-making biases, group-level phenomena stem from similar sources, contributing to both successes and failures. Haslam et al. (2022) specify that the strength of the group's social identity determines whether the group is able to act as a group, while the content of the group norms determines the nature of the influence that the cohesiveness has on the group's behavior. Norms that emphasize consensus can contribute to groupthink in cohesive groups, whereas if norms support critical thinking, highly cohesive groups will more likely perform well (Haslam et al., 2022).

2.5 COLLABORATION, COMMUNICATION, AND TEAM-WORK

Work requires us to collaborate and interact with each other in a joint purpose of accomplishing something that no one of us could accomplish on our own. Effective teamwork and effective teams are cornerstones of any effective organization. Plenty of research has been conducted with the aim of identifying the traits and features of high-performing teams. Similarly, knowledge of poorly performing, or dysfunctional, teams has increased.

A team is "a group of people who are interdependent with respect to information, resources, and skills and who seek to combine their efforts to achieve a common goal" (Thompson, 2018, p. 4). Not all groups are teams. To function as a team, groups need a shared purpose; structure, including clear roles and responsibilities; and adequate interaction to collaborate, form norms, and find a solution to issues concerning internal integration as well as their core task (cf. Schein, 1985). In essence, all teams develop a culture, or a subculture, of their own, within the constraints and possibilities created by the wider organizational culture.

Communication is a key issue in safety. In aviation, in the past many problems have occurred because the co-pilot did not speak up about their concerns to the pilot, who was regarded as having higher authority. For example, the collision at Tenerife between the Pan-Am and KLM planes in 1977 was partly attributed to communication challenges between the pilot and the co-pilot (Weick, 1990).

Similar problems have been encountered in many other domains, for example in surgical teams and intensive care units in healthcare (Edmondson, 1996, 2003). Thus, information sharing and open communication are highly important for effective teams. Especially when teams make safety-critical decisions, it is important that all members of the team consider their input as welcome and valuable. Further, they need to understand that it is OK and necessary to challenge opinions, present critique, and, through dialogue, aim at jointly created solutions. Team leaders need to realize their effect on the team. If they present their views too strongly and too early, it may limit the decision-making process. Similarly, if they criticize those who speak up, they contribute to organizational silence. Leaders need to instead reduce power-based barriers to speaking up (Edmondson, 2003) and listen to their employees.

Communication is typically mentioned both in accident investigations as one of the contributing factors as well as in various safety culture models as an essential aspect of good culture for safety. It is also one of the most critiqued aspects in any workplace – a frequent finding in all assessments that I have carried out in many different safety-critical domains. However, communication is a concept that is almost as vague as culture or leadership. What does it mean to communicate? What needs to be communicated and how? Who communicates, and with whom? Furthermore, a critical success factor of communication that is often forgotten is listening (receiving and understanding the communicated message). We will speak about the importance of listening widely in this book, including the next section on trust and psychological safety.

Communication is about transmission of information as well as creation of meaning (sense). It is an essential tool for sensemaking (see above). Key characteristics of good communication are transparency, clarity, dialogue, and sensitivity to the audience.

Transparency means that information is shared openly, and that there are no hidden agendas in the reasons for sharing information, or in the reasons for sharing some type of information while not sharing other information. One reason why lack of communication is a typical critique in employee surveys is that the staff *think* the management knows more but have decided not to tell the employees. Transparency should also extend to issues that the management does not know, or issues that include uncertainties.

Clarity refers to the fact that messages should be clear, straightforward, and to the point. There should be a reason for communicating, and a clear message is able to get this reason across. Jargon and ambiguous statements should be avoided. However, if there is uncertainty about the contents of the message, this needs to be openly admitted rather than hiding under vague statements.

Dialogue refers to the two-way nature of any communication. It always involves not just speaking but also listening. Encouraging active listening fosters a culture of respect and understanding, allowing employees to feel heard and valued. Sensitivity to the audience means recognizing and respecting diverse backgrounds and perspectives and shaping the message so that it is not considered inappropriate or offensive. It also means tailoring messages to the audience's needs and a level of understanding. In addition to these characteristics, communication should be timely (e.g. prompt communication of changes or important decisions) and conducted via appropriate media.

When leaders communicate, consistency in communication helps reinforce messages and reduces confusion. It's important that managers at different organizational levels convey aligned messages. The role of leadership and communication is discussed in Chapter 7.

Conflict management and the ability to have constructive conflicts are also important for team effectiveness. Teams without any conflict may not be very effective in the long run, as they are prone to groupthink. However, conflict that is focused on people instead of tasks is destructive to team spirit and team effectiveness.

In aviation, *crew resource management* (CRM) was introduced after the Tenerife accident to improve the non-technical skills of pilots. It has been applied to healthcare and nuclear settings also. CRM skills include teamwork, leadership, situation awareness, decision making, communication, and personal limitations (Flin & Martin, 2001; Salas et al., 2006).

O'Connor et al. (2008) studied the team skills required by nuclear power plant operations team members with a critical incident technique. They proposed a taxonomy of team skills consisting of five categories:

1. Building situation awareness, consisting of developing understanding, anticipation, maintaining overview, and performance monitoring;
2. Team focused decision making, consisting of analytical decision making, procedure following, intuitive decision making, and initiative;
3. Communication, consisting of assertiveness and information exchange;
4. Co-ordination, consisting of adaptability, supporting behavior, and team workload management;
5. Collaboration, consisting of leadership, co-operation, and followership.

Team performance has been studied a lot in safety-critical environments such as control rooms in nuclear power or chemical plants, air traffic control and aircrews in aviation, and critical care and surgery teams in healthcare (Künzle et al., 2010b).

Similar to individuals and organizations, adaptive capacity is one of the key elements of high-performance teams (Salas et al., 2008). High-performance teams also have a clear purpose, to which all members are committed and which guides decision-making in the team. The team also needs diverse enough competence to carry out the task as well as a climate where the competence of each team member can be adequately utilized.

Some years ago, Google conducted internal research to find out what sets apart their most effective teams from the others.[6] The researchers interviewed more than 200 teams across all areas of the business, from engineering to product management and sales. They identified five key dynamics that set effective teams apart from others:

1. *Psychological safety*: Can we take risks on this team without feeling insecure or embarrassed?
2. *Dependability*: Can we count on each other to do high quality work on time?
3. *Structure and clarity*: Are goals, roles, and execution plans on our team clear?
4. *Meaning of work*: Are we working on something that is personally important for each of us?
5. *Impact of work*: Do we fundamentally believe that the work we're doing matters?

There are several different types of teams. Some teams function together over an extended period, whereas others are formed to perform a specific set of activities. For example, medical practitioners form a team during each surgery in healthcare, and pilots and aircrew form a team during each flight in aviation. For these ad-hoc teams, clarity in communication and coordination is of utmost importance, as members may have limited experience of each other's personalities and working styles. Standard operating procedures and exact language (e.g. two- or three-way communication) are typically used (Wiggins, 2022).

In addition, *situation awareness*, meaning an awareness of the big picture and what is happening, is important for team effectiveness in complex safety-critical organizations. This situation awareness does not necessarily need to be an individual activity or an individual state, but leadership is in an important role in creating preconditions for a situation awareness to be created and maintained (see also *sensitivity to operations* and HRO in Chapter 5). Situation awareness also includes understanding the *intentions* of other actors. Shared culture helps in creating this type of understanding as shared values, system goals, and decision-making premises.

2.6 TRUST AND PSYCHOLOGICAL SAFETY

Trust and psychological safety are both important concepts for group performance. They are interrelated but distinct phenomena. *Trust* is about the beliefs one person has about another. It involves the expectation that others will act in a reliable, honest, and competent manner and consider your interests while doing so (Mayer et al., 1995). Psychological safety is collective and focuses on group norms and dynamics (Edmondson, 2018).

Psychological safety refers to the shared belief that there will be no punishment or humiliation in the team or work community for speaking up with ideas, questions, concerns, or mistakes (Edmondson, 2018). This means that the team is safe for interpersonal risk-taking. The concept of psychological safety has its roots in Amy Edmondson's studies of intensive care unit functioning (Edmondson, 1996, 2003). She found that psychological safety was important for team learning, both when introducing new technology and new treatment practices, and when dealing with mistakes and near-misses. Psychological safety varies within the same organization between different teams. Supervisors have a great impact on the psychological safety of their work community. If the supervisor presents their own views too strongly and too early, it can limit the decision making process by excluding possible dissenting opinions (Edmondson, 2003).

Psychological safety has mostly been studied on operational teams, but one can hypothesize that it is as important, if not even more so, on the level of senior management. They should be able to share their doubts and concerns about the direction of the organization, or about their level of understanding concerning the workplace reality.

Carrillo (2020) has noted that the key to building psychological safety is for the supervisors to build genuine inclusive relationships with their subordinates. The threat of exclusion or rejection is what makes people conform (e.g., groupthink) or stay silent (Carrillo, 2020). The more complex the work is, the more important psychological safety is for the group performance in the long term.

Noort et al. (2021) studied speaking up in aviation based on cockpit voice recordings of conversations that took place before accidents. They noted that the problem often lies in listening, not in speaking up. In the studied incidents, the co-pilots regularly spoke up about their concerns before the accident, but the captain did not listen.

Psychological safety is a property of a group. This means that to ensure open reporting, in addition to psychological safety, an organization needs trust and processes aligned with the principles of just culture (see Chapter 5) so that people have the confidence that they will be treated fairly, also outside their own reference group. High psychological safety in one's own reference group can also be due to a belief that whatever is said in the group will stay in the group. Cohesive teams are known to protect their members and cover them in case of mistakes or even mental problems.

Psychological safety is also dependent on the *salience of social identity* (cf. Fransen et al., 2020). If people identify with, and hence care about, a given group of people, they are much more likely to speak out when something threatens the group (Packer, 2009). Steffens et al. (2017), in their extensive analysis of literature, showed that social identification (with both workgroup and/or organization) is associated with increased well-being at the workplace. Furthermore, this positive effect was stronger when the social identification was shared by other members of the group. The findings highlight the fact that "individuals' stress and wellbeing at work are grounded not just in individuals' personal capabilities and shortcomings but also in groups and organizations as a whole" (Steffens et al., 2017, pp. 22–23).

Weiss et al. showed that the salience of the social identity affects the way a leader can encourage employees to speak up. When the team members shared the same professional group membership (shared social identity) as the leader, implicit inclusive leader language (for example, speaking about "we" and "us") promoted speaking up. In turn, when the team members belonged to a different professional group (had a different social identity) than the leader, explicit inclusive leader language (invitations and appreciations) was more effective (Weiss et al., 2018).

All in all, leadership has a major influence on psychological safety. Interestingly, an authentic leadership style that contributes to psychological safety has been found in some studies to lead to lower perceived safety risks (Nielsen et al., 2013). This indicates yet again the complex relationship between psychological and physical safety, as lowered perceived risk is positive for wellbeing as long as it is based on increased control of those risks rather than merely on a feeling of being taken care of. We will investigate this influence in more detail in Chapter 7, which is devoted to leadership and leaders' role in managing safety in complex organizations.

Trust represents a positive assumption about the motives and intentions of another party. More specifically, trust refers to the willingness of a party to be vulnerable to the actions of another party – their willingness to take risks (Mayer et al., 1995). This is an important distinction that has safety consequences: trust increases risk taking. This does not mean risk taking in its traditional safety management meaning, that someone would break safety rules or act recklessly. It means our ordinary behavior at the workplace: we trust that our subordinates carry out their tasks as we have instructed, we trust that our supervisors share important safety-related information with us, we trust that most employees try their best, and so on. In essence, trust is a similar heuristic (or bias) as other information processes heuristics presented in this Chapter (cf. Fiske & Taylor, 2017, p. 98). And as with all heuristics, trust works more often than not, but in some cases may lead us astray.

McEvily et al. (2003) develop the notion of *trust as an organizing principle*. By an organizing principle, they mean the logic by which work is coordinated and information is gathered, disseminated, and processed within and between organizations. They give three examples of organizing principles: market, hierarchy, and clan (Ouchi, 1980). Trust is most closely related to the clan principle. We will return to these different types of cultures in Chapter 7.

The main point is that trust and psychological safety can both have positive as well as negative consequences for safety at an organizational level. As with everything else, these need to be considered on a system level. They are connected to the cultural assumptions discussed in Chapter 1 and the joint learning experiences of the group: What issues are we allowed to raise into discussion and how? Should we keep silent on certain topics? Can anyone speak? What are the personal consequences for me if I "cry wolf" and nothing bad happens? Both trust and psychological safety are thus connected to risk, and both are related to hazard risks as well as to interpersonal risks, risks to our identity. Finally, it needs to be always remembered that we are all individuals and what feels safe for one person may not feel safe for another. Leaders need to be mindful of variability at individual, group, and organizational levels (see Chapter 7).

Organizational commitment refers to a desire, a need, and/or a sense of obligation an individual feels toward the organization (Meyer & Allen, 1991). Different components of commitment have been identified in research. A normative commitment reflects a feeling of obligation to continue employment. An affective commitment refers to employee's emotional attachment to, identification with, and involvement in the organization. This type of commitment is mostly associated with psychological safety and trust. A continuance commitment in turn refers to an awareness of the costs associated with leaving the organization due to e.g. rewards or lack of alternatives, and it is the type of commitment least likely to correlate positively with performance. An individual can experience all three forms of commitment to varying degrees, and different antecedents affect these commitment types.

2.7 SUMMARY

To understand safety in complex organizations, we first need to understand those who create it, humans, and how they behave individually and in groups. Safety always starts and ends with people and how they interact together and how they form more or less shared cultures. Culture is learned, and to understand this we need to understand how humans and groups learn.

Individuals are accountable for their own behavior, but the behavior always takes place in a context where the individual may be unaware of all influencing factors and feel some influencing factors as constraining, or even coercive. This context is never same for different individuals, as the individual's personality, experience, and professional background are among the multitude of factors affecting what aspects of their identity are salient at any given time. Many actions and decisions are made intuitively, guided by contextual cues. Human action is guided by local rationality and satisficing rather than analysis and search for absolute truths. This reflects also to how cultures form, learn, and develop both at group level and organization level.

We tend to interact with people who share our interests and exert a positive effect on our self-esteem. This has its pros and cons in terms of creating a healthy organizational culture. Sometimes we need to go outside our comfort zone in order to develop. This may require interacting with people we do not regularly interact with, risking exposure to new ideas or new meanings attributed to old and well-known concepts. It may also require questioning our habits and routines that bring a feeling of security, but not always physical or objective safety.

Communication, trust, and psychological safety are closely interlinked: communication shapes and is shaped by trust and psychological safety. And as in complex systems everything is connected, communication in turn can further increase or decrease both trust and psychological safety.

It is important to note that the decision-making heuristics/biases are not only characteristics of the individual but are also features of the entire social system. Similarly, social identity, culture, and norms determine what is considered a bias or wrong decision in the first place. In an organization that emphasizes efficiency, circumventing the rules to speed up work can be a sign of motivation and commitment, not a violation. Individual performance is never culture-free but it is based on the human tendencies illustrated in this chapter.

Reflection Questions

Do you consider yourself less susceptible to the biases described in this chapter than the average worker?

How does one show commitment to the various groups in your organization (e.g., obeying rules, disobeying rules from the headquarters, working long hours, producing good quality)?

What kind of a person is a stereotypical member of your organization? What is the meaning of safety to this person?

Do you recognize any of the decision-making heuristics in your own behavior and your own decisions?

When you speak about "we" at work, do you mean your own team, your department, the entire organization, or some other group? Does the meaning of "we" change when you interact with different groups inside and outside your organization?

Is the safety organization an insider ("we") or an outsider ("them") in your organization?

What is the level of psychological safety in your team? How about other teams in your organization?

Do employees trust management in your organization? How about management, do they trust the employees?

SPACE SHUTTLES *CHALLENGER* AND *COLUMBIA*[7]

NASA's[8] Space Transportation System, first envisioned in 1968–1969 – before Project Apollo landed on the moon – was supposed to provide the United States an easy and convenient access to space. The original purpose of the system was twofold: to reduce the cost of spaceflight by utilizing reusable spacecraft; and to support ambitious follow-on programs including permanent orbiting space stations around the Earth and Moon, and a human landing mission to Mars. The Shuttle was designed to ferry cargo as well as passengers.

The Space Transportation System was marketed to the US Congress as a "safe, cost-effective and routine access to space". NASA had settled on the basic layout of the Space Shuttle in 1972. It would consist of two solid rocket boosters and three main engines burning hydrogen and oxygen for the eight-minute flight to orbit. The fuel and the oxidizer for the main engines were to be stored in an external fuel tank attached to the orbiter. The orbiter would be the manned and winged shuttlecraft that would return to earth.

The *Challenger* Space Shuttle Accident

The accident was the result of a rubber O-ring seal failure between adjacent sections of the solid rocket boosters (SRB).[9] Unprecedented cold weather on the day of the launch reduced the rubber's resiliency, which, combined with the faulty design of the booster joints, allowed hot combustion gases from the burning rocket to escape. The flames and hot gases burned through the metal supports holding the rocket in position. When the rocket assembly came loose, it ruptured the side of the external fuel tanks, allowing liquid hydrogen and oxygen to mix and explode. The Shuttle exploded 73 seconds after launch. However, even if the final explosion happened in seconds, the accident took years to unfold.

The prime contractor for the design and manufacture of the SRB motor segments was Morton Thiokol Incorporated (MTI). The weather on launch day was exceptionally cold (36 F, ≈ 2oC), 15 degrees lower than that measured for the next coldest previous launch. The durability of the O-rings had not been tested at such temperatures, and worries about the ability of the O-rings to seal effectively were expressed. Thus, before the launch, a teleconference was held between NASA and MTI, where the Morton Thiokol engineers recommended postponing the launch due to uncertainties about the O-rings. NASA strongly challenged MTIs position, and a thirty-minute caucus was held. During the caucus, MTI's management reversed the engineers' original decision and recommended launch to NASA. Bob Lund, vice-president of engineering at MTI, was famously told to "take off your engineering hat and put on your management hat" by a senior executive at MTI. Allan McDonald, who acted as the director of the Solid Rocket Motor Project at Morton Thiokol at the time of the *Challenger* accident, describes his shock in participating in the teleconference and his surprise at how harshly the NASA representatives were speaking:

> *I had heard [NASA manager for the SRB project] talk that way before, but it was always in situations where he was challenging our rationale on why it was safe to fly. This was the first time that NASA personnel ever challenged a recommendation that was made that said it was unsafe to fly. The flight readiness review process was always structured around the contractors having to prove beyond a shadow of a doubt that their hardware was safe to fly ... in the past, data that was inconclusive or not quantifiable was automatically considered as being unsatisfactory to support a recommendation to launch.*
>
> **(McDonald & Hansen, 2009, p. 106)**

McDonald was participating in the teleconference at the NASA end at the launch site, so he did not participate in the caucus and was as surprised as NASA when MTI reversed its position and recommended the launch. Next morning, the Space Shuttle *Challenger* left for its final voyage.

The management at NASA was under a lot of pressure to make the launch. The launch had already been postponed for four days due to poor weather and a technical fault (unrelated to the field joint). In addition, NASA was behind the planned launch schedule (twelve planned flights in 1986) as *Columbia*'s launch had been delayed multiple times due to various reasons – having a domino effect also on *Challenger*'s schedule.

Adding to the pressure, there was a civilian onboard as part of the "Teacher in Space Program", and her entire class had arrived to see the launch and hear her lecture from space. President Reagan was supposed to address the nation in his state of the nation speech the next day, and the successful launch of *Challenger* with the first ever teacher in space was supposedly high on his priorities of things he wanted to say. Postponing the launch would have ruined that opportunity. There had also been repeated mockery on the television news of previous scrubs, and the NASA management was probably not keen on being a topic of yet another TV joke.

The "Teacher in Space" flight needed to be completed to reserve time for prepping *Challenger* to flight status again for next scheduled launch in May. The next flight would be impossible to postpone because it needed to be launched during a specific time window.[10]

Morton Thiokol in turn was at the same time negotiating a new contract with NASA, and its management was in a precarious position of not wanting to upset their most important customer (McDonald & Hansen, 2009).

History of O-ring Problems

Challenges with O-ring erosion were not new. There was a well-documented history of problems with the rocket booster design, including the integrity of the rubber O-ring joints. Instances of erosion and blow-by had been documented by the Morton Thiokol engineers for years.

Design changes to the joints were first proposed already in 1978 during the testing phase, and small fixes to the original design were made along the way.

When the NASA team observed O-ring damage after the second shuttle flight, they made changes to the assembly process and continued with future flights. Over time, and with 14 of the previous 24 shuttle flights prior to *Challenger* showing O-ring damage without causing incident to the flight, the managers normalized the deviation, so that it became acceptable and non-deviant to them.[11] Engineers at Morton Thiokol were worried about the issue and a task force was gathered to inspect it, but it did not have sufficient mandate to implement any immediate solutions. A redesign of the solid rocket boosters had also been proposed by Morton Thiokol, but NASA did not approve it before the *Challenger* accident.

The *Columbia* Space Shuttle Accident

The loss of Space Shuttle *Columbia* was a result of damage sustained 81 seconds after launch, when a piece of foam insulation, the size of a small briefcase, broke off the Space Shuttle external tank (the main propellant tank) under the extreme launch forces. The debris struck the front edge of the left wing of the manned orbiter, damaging its thermal protection system (the reinforced carbon-carbon (RCC) panel on the leading edge[12]). The mission continued but ended tragically sixteen days later when the orbiter was re-entering the earth's atmosphere. During re-entry, the damaged area allowed hot gases generated by air resistance against the vehicle to penetrate and destroy the internal wing structure, which caused the in-flight breakup of the orbiter.

NASA's original shuttle design specifications stated that the external tank was not to shed foam or other debris. Potential strikes upon the shuttle were safety issues that needed to be resolved before a launch was cleared. Gradually, engineers came to see the foam shedding and debris strikes as inevitable and irresolvable, with the rationale that they were either not a threat to safety or were an acceptable risk, merely a maintenance issue. Most shuttle launches recorded such foam strikes and thermal tile scarring. The tolerance for foam strikes increased with each successful launch. This normalization of deviance paralleled what had happened with O-ring corrosion some years before.

Immediately after launch, the cameras at the launch site were reviewed to assess the extent of damage. Some engineers were concerned about the potential damage caused by the debris, but they had challenges in getting their voices heard. Due to previous budget cuts, the video quality was not adequate to determine the damage. A request for a spacewalk to inspect the damage was denied due to time and resource constraints. A Debris Assessment Team was established, but its work was hampered by bureaucratic issues and unclear mandate. Three different groups in NASA requested additional photographic images from non-NASA sources to get a clearer picture of the potential damage, but all requests were denied by the Mission Management Team.

The *Columbia* Accident Investigation Board (CAIB) noted several missed opportunities during the problem analysis while *Columbia* was in orbit. They note that leadership challenges cast barriers that kept at bay both engineering concerns and dissenting views, and helped create "blind spots" that prevented them from seeing the danger the foam strike posed (CAIB, 2003, p. 170). CAIB summarized that among the organizational contributors of the *Columbia* accident were normalization of deviance in accepting the debris strikes; denial of vulnerability; ignoring opinions of safety staff; lack of consistent, structured approaches for identifying hazards and assessing risks; and worker intimidation, contributing to communication breakdown.

Similar to the *Challenger* accident, the Mission Management Team at NASA took safety as an assumption and required quantitative proof for risks. Dekker (2019, p. 377) asks how different things would have been if the Shuttle management had instead asked the engineers to "prove that the *Columbia* has not been harmed". This would have been impossible without the additional imagery they were denied – imagery that would have revealed the damage.

Common Factors in Both Accidents

In the case of *Challenger*, the technical failure was erosion of seals (O-rings) between segments of the solid rocket booster after launch. In the *Columbia* Space Shuttle accident, the physical cause of the accident was a breach in the thermal protection system on the leading edge of the left wing, caused by a piece of insulating foam that struck the wing immediately after launch. In both cases, with the benefit of hindsight, multiple signals of danger were normalized. The O-ring erosion took place regularly, but it never compromised the success of the launches until the launch of the *Challenger* Space Shuttle in January 1986. Foam strikes happened regularly during launches, but before the launch of the *Columbia* Space Shuttle in February 2003, they had not caused any major damage. CAIB concluded that both accidents were "failures of foresight", and that their similarity demonstrated that NASA had not been successful in fixing the causes of the institutional failure responsible for the *Challenger* accident (CAIB, 2003, p. 195).

NASA operated under immense pressure to launch that originated from the promises made when the Space Transportation System was first introduced in the 1970s. It was marketed as a routine bus-like transportation to space with one flight every week. That schedule proved impossible to achieve. The highest rate ever achieved by the shuttle program was nine flights during the entire year 1985. The launch pressures were increased when President Reagan in 1984 announced the intent to build the International Space Station (ISS), which required an operational space shuttle for deliveries of cargo and personnel (Farjoun, 2005).

At the same time, the fleet was aging. Maintenance costs for shuttles were rising, in sharp contrast with demands to cut costs. Shuttles were planned to be replaced by newer vehicles, but in the 1990s, the planned date for replacing the Shuttle shifted from 2006 to 2012 and then to 2015 or later. Aging and increasingly expensive shuttles were expected to continue safe performance for some 10 years past what had been originally planned (CAIB, 2003).

NASA, based on, for example, an examination of the contractual arrangements and opportunities for privatization by the Kraft Committee in 1995, believed the Shuttle program to be well-run and the Shuttle a mature and reliable system. This led NASA to shift responsibilities increasingly to contractors and reduce its own direct involvement in ensuring safe Shuttle operations, instead monitoring contractor performance from a more detached position (CAIB, 2003, p. 118).

After the *Challenger* accident, resources to safety were increased, but then gradually taken away during the 1990s. Different projects at NASA bought safety assurance from a separate function – projects decided how much "safety" they wanted. In 1992, a new philosophy of Faster, Better, Cheaper (FBC) was introduced, partly due to new budget cuts by the government. This approach included major cuts in safety programs and personnel[13] and the downsizing of the safety function. There was a perception in NASA that the agency had overreacted to the Rogers Commission recommendations by introducing many layers of safety inspections in launch preparation that had created a bloated and costly safety program (Farjoun, 2005, p. 30).

During the *Columbia* investigation, it became evident that the safety improvements made after the *Challenger* accident had not been effective. Farjoun (2005, p. 27) writes that after the *Challenger* accident, "the [Space Shuttle] program went into an increasingly ritualistic set of reforms that temporarily increased safety consciousness as the system geared up for a return to flight program that would have a scope far beyond what was advised by commissions and panels". CAIB (2003, p. 101, see also Evans, 2021, p. 158) concluded similarly that even though NASA underwent many management reforms in the wake of the *Challenger* accident and appointed new directors at the Johnson, Marshall, and Kennedy Centers, the agency's powerful human space flight culture remained intact, as did many institutional practices. This culture acted over time to resist externally imposed change. By the eve of the *Columbia* accident, institutional practices that were in effect at the time of the *Challenger* accident had returned. These practices included inadequate concern over deviations from expected performance, a silent safety program, and schedule pressure (CAIB, 2003, p. 101).

The silent safety program refers to the fact that the position, and authority, of safety and oversight in the NASA organization was weak during the *Challenger* accident, and it was weak again during the *Columbia* disaster (CAIB, 2003). The *Columbia* Accident Investigation Board was surprised about the status of safety at NASA, noting that personnel had been eliminated and careers in safety had lost organizational prestige (CAIB, 2003, p. 181). Safety assurance focused on identifying new risks, not on monitoring the old ones. O-ring failures and foam strikes were both categorized as "acceptable risks", but as operating experience accumulated, risk estimations were not updated. Both O-ring erosion and foam strikes were "in-family" issues, which meant that risk was acceptable but also that problems were to be expected.

Launch pressures combined with budget cuts were an influencing factor in both accidents. NASA was in a position where it was not able to set its own launch schedule. It had several customers who each had their say in the schedule.

These two accidents tell little about culture at NASA as of today. Instead, they illuminate the ways any organization can drift into accident, and the challenges that any organization faces when trying to identify the vulnerabilities in its operations, or when trying to change its culture. The lessons from these two accidents are profound, and we will return to them multiple times in this book.

What Did We Learn about Safety in Complex Organizations?

The stories of *Challenger* and *Columbia* demonstrate that in complex systems, accidents can happen without anyone intentionally breaking the rules. Vaughan (1996, pp. 409–410) summarizes:

> The explanation of *Challenger* launch is a story of how people who worked together developed patterns that blinded them to the consequences of their actions. It is not only about the development of norms but also about the incremental expansion of normative boundaries: how small

> changes—new behaviours that were slight deviations from the normal course of events—gradually became the norm, providing a basis for accepting additional deviance. No rules were violated; there was no intent to do harm. Yet harm was done.

In both accidents, the organization gradually drifted to a state in which it no longer operated safely. Earlier danger signals had become part of "normal" work, and they were no longer adequately considered. Similar drift took place within the safety function, which did not have an adequate involvement or influence on decision making (Monteiro et al., 2020a). Safety functions will be inspected in Chapter 8.

Many scholars have gone on to elaborate the conditions that contribute to drift and normalization (Dekker, 2011; Rasmussen, 1997; Snook, 2000), including scarce resources, competition, and the overdesign of the system. Higginbotham (2024, p. 96) notes how from the early design and engineering phase, NASA's plans were based on the overoptimistic expectation that no failures or setbacks would occur.

The accidents indicate the learning potential of analysis combined with theoretical understanding of safety science. Vaughan (1996), a sociologist, analyzed the *Challenger* accident from a cultural perspective and challenged as well as complemented the original investigation. The *Columbia* Accident Investigation Board included academics and explicitly relied on, for example, High Reliability Organization theory (see Chapter 5) and Normal Accidents Theory (see Chapter 3).

Different systems are exposed to different risks. Pariès (2021) reminds us that the level of safety achieved by a system is defined by its exposure to risk, much more than its ability to manage it:

> NASA invented most current risk management methods, and continues to apply them more professionally than the airlines that adopted them. Yet the space shuttle program has a per-flight accident rate that is nearly 100,000 times higher than that of the best airlines! The reason is obvious: it operates at the limits of scientific and technical knowledge, with extreme levels of energy and uncertainty—in other words, levels of risk—that cannot be compared to the day-to-day experience of airline companies.

The example of NASA also demonstrates the difficulties of changing culture, especially if the operating environment and its pressures remain the same.

Finally, the two accidents demonstrate how important it is to assume our systems have risks and try to prove them safe rather than assume they are safe unless proven otherwise (see the biases in Chapter 2 and assumptions as core of culture in Chapter 1). The accidents also demonstrate how important information easily gets lost in complex organizational arrangements, especially if the organizational mindset does not encourage curiosity and the challenging of assumptions (see *organizational learning* in Chapter 6). Finally, they show the importance of deferring to expertise in uncertain situations (see HRO model at the end of Chapter 5) and creating a climate where everyone feels safe to speak up on concerns (see Chapter 2 on psychological safety and Chapter 7 on leadership).

NOTES

1. The Big Five model has been criticized and some of the personality dimensions of specific interest to safety are lacking from that model. These include sensation seeking and the so-called "Dark Triad" of Machiavellianism and the more clinical dimensions of narcissism and psychopathy (Paulhus & Williams, 2002), dimensions that are relevant from safety and accident point of view (see Chapter 3). An interesting alternative approach to the Big Five is the HEXACO model, which includes an "Honesty–Humility" dimension (Lee & Ashton, 2005). This dimension is related to the Dark Triad, and it is also important to leadership (Javalagi et al., 2024).

2. Note on terminology: *Heuristics* and *biases* have been used quite interchangeably ever since. I prefer the term *heuristic* since it implies these are tendencies, not faults. They only become liabilities in complex environments where we should be more accurate than is (or has been) typically required from our cognition. Since some of the heuristics have established names (e.g. confirmation bias and hindsight bias), I will use both terms when discussing the heuristics and the biases they may cause.
3. The accident description is based on the following sources: Filburn and Bullard (2016), Kemeny (1979) and Mosey (2006).
4. The accident was classified as class 5 on the INES scale, which describes hazards and accident events in the nuclear industry. The scale is a seven-step scale. Class 5 stands for "environmental hazard accident".
5. This also means that the programs of 24-hour safety culture that require the same safety mindset while at home and work, while misunderstanding the concept of safety culture (Hopkins, 2018), also misunderstand how people think and behave.
6. https://www.thinkwithgoogle.com/intl/en-emea/future-of-marketing/management-and-culture/five-dynamics-effective-team/ See also: https://www.linkedin.com/pulse/five-keys-successful-google-team-laszlo-bock/
7. Unless otherwise noted, description of the accidents is based on the Presidential Commission report for *Challenger* (1986) and the CAIB report (2003) for *Columbia*.
8. NASA comprises 10 separate centers that serve the different formal missions of the agency. In this analysis, we are exclusively concerned with NASA's human spaceflight program and the centers that serve this program.
9. These rubber O-rings were used to prevent hot gases and combustion products from escaping through the gaps at the joints during ignition and launch. When the O-rings got cold, the rubber hardened and could not expand as quickly. The O-rings were protected by an insulating putty that was meant to block burning gas from reaching them in the first place. However, there were small imperfections in the putty allowing gases to pass. The shuttle had several O-rings, made of a rubber compound, which were used to seal the SRB field joints. The SRBs were composed of seven individually manufactured steel segments. There are two types of joints to hold the segments together: the permanent "factory joints" were sealed at the Morton Thiokol factory in Utah; the temporary "field joints" were sealed before each flight at the Kennedy Space Center in Florida (Feynman, 1988).
10. The planet Jupiter was about to make a close flyby of earth, and a space probe called Ulysses had to be in orbit to make the rendezvous with the magnetic effects of that flyby.
11. There were two mechanisms that threatened the integrity of the O-rings: erosion and blowby. These were differently affected by temperature and manifested differently in the primary and the secondary O-rings, which complicated analyses. The most probable immediate cause of the *Challenger* accident was a blowby of hot gases through both primary and secondary O-rings.
12. After each flight, the insulation tiles on the Space Shuttle had to be inspected and often replaced or repaired by hand. The manufacture, installation, and repair of the tiles was one of the most labor-intensive aspects of the Shuttle program (Higginbotham, 2024, p. 100).
13. During the 1990s, NASA reduced its workforce by 25%.

3 Why Accidents Happen – A Cultural View

In this chapter, I look at the various reasons and rationalizations for accidents. I will start with the "traditional" and still widely used rationalization of accidents caused by human error or wrongdoing by individuals or groups. After discussing these explanations and showing their limits, I will tackle the more systemic yet not so comforting explanations for accidents. Namely, we must look beyond individual errors or intentional wrongdoing and acknowledge the complexity of the system and the challenges it places on organizing and understanding the workplace reality. Furthermore, relying on outdated assumptions about safety, problems of learning in complex environments and gradual change in the system make organizations prone to accidents. Finally, the role of individuals is most apparent at the top, where organizations decide on strategy-level issues that may, intentionally or unintentionally, downplay safety. The chapter concludes by presenting two classic accident theories, Man-Made Disasters and Normal Accidents, that have heavily influenced the modern view on safety in complex organizations.

3.1 HUMAN ERRORS

It is often argued that over 80% of all accidents are caused by human errors or unsafe acts. The argument is old, but for several reasons (see e.g. the biases in Chapter 2) it has resonated with people in charge of organizations as well as with many safety professionals. According to James Reason, *human error* is a general term that encompasses all those events in which a planned mental or physical chain of actions fails in such a way that the failure cannot be attributed to some external factor (Reason, 1990, p. 9). Thus, error is a human action (or lack of action) that does not lead to the outcome that the person intended, but to something worse, *and* this is not due to external circumstances. Errors are different from violations, which are intentional deviations from rules.

Human errors were established in the Human Factors area after the 1979 nuclear power plant accident on Three Mile Island. Reason (1990) divided human error into three categories: skill-based, rule-based, knowledge-based errors. The division is based on Jens Rasmussen's model of the three categories of activity (Rasmussen & Jensen, 1974). Knowledge-based errors consist of mistakes in decision-making. They happen in unknown situations when we need to collect the information and decide without any rules or routines to help us make sense of the situation. Rule-based mistakes can be divided into abusing a good rule and using the wrong rule, and they take place in familiar settings. Skill-based errors include perceptual errors, memory lapses, and slips. These generally take place when we perform familiar tasks requiring little conscious attention.

A widely adopted approach to mitigating human error is the implementation of safety barriers across the various levels of the sociotechnical system. The purpose of these barriers is to prevent various errors and malfunctions or to reduce their consequences. According to Reason (1997, p. 17), an accident typically develops in such a way that hidden faults or weaknesses in the system and errors in immediate operation together render the safety barriers developed to protect the system ineffective. In this way, events proceed step by step towards an undesirable outcome. According to Reason, both latent weaknesses and active errors can occur in any part of the sociotechnical system (human, technology, organization) and at any level of the organization. Similarly, safety barriers can exist in all these areas. Reason's well-known model of the origin of accidents is the so-called Swiss Cheese model, where an accident occurs when the "cheese holes"

DOI: 10.1201/9781003491156-3

in different protective barriers are positioned so that the hazard passes them all simultaneously (Reason, 2000).

Reason (1990) introduced the idea of latent conditions, or latent errors. These are errors that have not been discovered and that remain in the system. They make the system more prone to fail. These conditions can weaken the system's defenses, hide some hazards, create gaps, and increase opacity and hide information. These conditions make active failures more probable and severe in their consequences. They can also promote violations (e.g. time pressure, conflicting or poor procedures). Latent errors have been discovered in many accidents and incidents, with some of the errors originating in the design and construction phases of the particular sociotechnical system, in some cases years before the accident.

Typical latent conditions that have been identified in accident investigations and other safety research are (DOE, 2009a; Patankar & Taylor, 2004; Reason & Hobbs, 2003):

- Deficiencies in planning and organizing work;
- Lack of competence and training;
- Workload and stress;
- Deviation from rules or procedures;
- Poor communication;
- Poor teamwork;
- Ineffective leadership and management;
- Inadequate rules or procedures;
- Inadequate interface design;
- Inadequate tools or equipment;
- Wear and tear on technology, technical failures.

In addition to latent conditions, there are several *error precursors*, which are unfavorable conditions embedded in the work or the work context that create mismatches between a task and the individual (DOE, 2009a). These precursors have been classified as follows:

1) Task demands (time pressure; high workload; simultaneous actions; repetitive actions/monotony; irreversible actions; interpretation requirements; unclear goals, roles, or responsibilities; lack of or unclear standards);
2) Individual capabilities (unfamiliarity with task/first time/rare task);
3) Work environment (lighting, tight spaces, poor interface, etc.);
4) Human nature/capabilities (DOE, 2009a).

One of the commonly identified error precursors is *interruptions* in work, either momentary interruptions or when the work needs to be suspended overnight, or sometimes for longer periods. Among the challenges associated with suspending work are the potential changes in the system that happen before the work is restarted (e.g. an isolated piece of equipment is needed and power put back on, or other conditions change) while the suspension of the work may cause challenges to other parties (who may depend on the job to be completed, or in case of communication failure, may assume the work has been completed). There are many examples of incidents and accidents where temporary suspension of work has played a contributing role, ranging from electrical maintenance works in nuclear power plants (TUKES, 2004) to offshore platform maintenance; Piper Alpha, see Cullen (1990).

Reason and Hobbs (Reason & Hobbs, 2003, p. 105) note that

> a breakdown in coordination is one of the most common circumstances leading to an incident. In many cases, coordination breaks down when people make unspoken assumptions about a job, without actually communicating with each other to confirm the situation.

Another challenge for *coordination and cooperation* emerges from the different cultures of the professional groups in safety-critical organizations (such as maintenance and operations), which affects how they understand each other even in cases where they do communicate with each other. Finally, communication can fail also due to problems in understanding the message due to various individual and environmental factors such as noise or problems in articulation (see Chapter 2).

Breakdowns in *communications* have been observed in many industrial accidents as well as patient safety incidents (Peadon et al., 2020). The 2005 Texas City oil refinery explosion was affected by communication problems between shifts (see Chapter 4). The launch of the *Challenger* Space Shuttle was largely due to a communication breakdown (see Chapter 2). The challenges of collaboration and information flow in organizations are both horizontal (between shifts, departments, or employees) and vertical (subordinate to the manager). *Shift changes* have been identified as a risk for information and communication lapses in many safety-critical contexts.

Many accidents have shown how leaders mistook "no news" for "good news". For example, in the capsizing of the *Herald of Free Enterprise* shortly after leaving the Belgian port of Zeebrugge in 1987, the person in charge of closing the bow doors fell asleep. Common practice was that the captain was notified if there were problems in closing the bow doors, but when the captain heard nothing from the sleeping crewman, he assumed the doors were closed.[1]

Most of the major accidents covered in detail in this book include *design-related contributing factors*. The Space Shuttle solid rocket boosters had multiple design issues, and a redesign was already on table (see Chapter 2). The original design for the solid rocket boosters was taken from the Titan III missiles used by the US Air Force, and as Higginbotham (2024, p. 165) notes, "in creating a man-rated fail-safe joint they had also modified it so extensively that what they produced was, in effect, a quite new and experimental design". Furthermore, the BP Deepwater Horizon's blowout preventer suffered from several design issues (Hopkins, 2012, p. 69; see Chapter 4). In the Chernobyl nuclear accident (see Chapter 5), design errors were not known by the operators (e.g. the positive void coefficient in the reactor and the design error in the control rods) and thus the flawed decisions they made in the control room were premised on wrong understanding of the system.

Design issues at Boeing 737 MAX are the crux of the entire story, as the decision to redesign an existing airframe due to financial and schedule reasons, combined with a promise made to the customers that the plane would be similar to fly as the previous version, led Boeing to introduce software that acted behind the pilots' backs without them even knowing that such software existed (see Chapter 8).

Taylor (2007) notes that design error plays a major role in process plant accidents. According to his study, current design review methods are able to discover and remove between 80% and 95% of the errors made, but still between 20% and 50% of the accidents and incidents that happen in chemical process plants have design issues as contributing factors. Similarly, Kinnersley and Roelen (2007) found that about half of events in aviation, rail, chemical and nuclear industries have design errors as contributing factors.

Moura et al. (2016) reviewed 238 accidents based on investigation reports in various high-risk industries such as oil & gas, food industry, aviation and nuclear. They noted that design failures were detected in 157 of the 238 accidents reviewed. They note that "design failures particularly appear to trigger failures in the human capacity to interpret system status (wrong reasoning and faulty diagnosis), enable potential observation misses and cause some execution errors (sequence, timing and type)" (Moura et al., 2016, p. 44).

Accident models that focus on errors provide tools to identify and control human errors and error precursors. One key problem with error-driven models is the assumption that safety equals error avoidance. People are seen as a threat to safety, because people may take measures that differ from what is planned or expected. Various protective walls must be erected against human activity. On the other hand, these are being erected by other fallible humans. It has also been criticized that different protection mechanisms increase the complexity of the system and thus increase the likelihood of human error and the severity of the consequences of errors (Perrow, 1999).

Everyone makes errors. Experts make different types of errors than novices, but they still make them. Making errors and correcting behavior based on feedback is a key mechanism of learning. Most errors are spotted immediately by the person who made them – sometimes even a bit before physically conducting them! However, we do not spot all errors, especially if they have no immediate negative consequences. We are most likely to make mistakes when we are tired or distraught. Or when there is too much information available. And when there are ambiguous signals about the needed actions, or distractions in the flow of work. In short, when the working conditions do not support us adequately, or when our ability or willingness to perform has been compromised. The challenge is that in many complex environments, errors do not have immediate effect, or they only have an effect in combination with other errors. Thus, sometimes what is an error can only be judged afterwards.

The idea that human errors directly cause accidents is based on a rather simplistic linear model of accident causality (Hollnagel, 2004). It may apply to some occupational accidents and local hazards (see Chapter 4) in systems with very little formal safety management, but in today's complex organizations it is very rare to find an accident with no wider systemic influences.

It has been increasingly recognized that traditional approaches to safety management, which strive to eliminate adverse outcomes and errors in full, fall short in dealing with the complexities of modern systems. Newer views on safety have considered errors as symptoms of the underlying issues within the system, and not the root causes of adverse outcomes (Dekker, 2005). Recently the entire concept of human error has been questioned, especially by the Resilience Engineering paradigm (Dekker, 2005; Hollnagel, 2014; Woods et al., 2010). Resilience Engineering will be described in more detail at the end of Chapter 11.

I do not think it makes sense to abandon the concept of human error, but I do concur that the concept is insufficient in itself to explain accidents in complex organizations. Error needs to be viewed in context, and as a label given to failed actions with the benefit of hindsight. Furthermore, safety management and safety management systems need to take into account the fact that human performance is variable, and humans are sometimes inattentive, careless and even reckless. Performance variability is a must in any complex environment, but it is also a source of errors. Accidents in complex systems do not need to be caused by errors, however, but rather unexpected interactions. From a cultural point of view, the most interesting aspect of human errors is why they are so tempting for us to use as reasons for accidents (see the biases in Chapter 2).

3.2 INDIVIDUAL AND COLLECTIVE WRONGDOING

In addition to errors, violations of rules and other intentional wrongdoings as considered among the contributors to accidents. Similar to human errors, the explanation quickly turns out to be a simplification of reality. Often increased production pressures or reduction in the workforce underlie employees' so-called rule breaking or various shortcuts (Singer, 2022, p. 32). Singer (2022) shows in her book "There Are No Accidents" how various industries actively campaigned to attribute accidents to human errors or "accident-prone" individuals, instead of their products or production facilities[2]. She claims that "accidents are the predictable result of unequal power in every form – physical and systemic" (Singer, 2022, p. 13). Reason (2016, p. 33) notes that clumsy procedures are among the main causes of violations.

Rule violations and shortcuts are often taken with the best of intentions – to meet the organizational goals for efficiency. Accident analyses highlight the challenge of **balancing thoroughness and efficiency**. Hollnagel (2009a) refers to this as the ETTO principle (efficiency-thoroughness trade-off). He suggests that most organizations do not prioritize thoroughness over efficiency, at least not until a hazard has been detected or an incident occurs. Emphasizing efficiency is usually accepted and even preferred unless it leads to negative outcomes. Hollnagel also notes that focusing on efficiency isn't inherently wrong, as lower efficiency can negatively impact safety. People adapt their behavior to manage multiple goals and optimize efforts. The ETTO principle highlights the inherent tension between the desire for quick, efficient operations and the need for careful, thorough approaches, especially in high-stakes environments.

Drugs and alcohol inhibit performance and can contribute to accidents. In traffic, both have a significant relation to road accidents. One of the most famous accidents typically attributed to alcohol is the Exxon Valdez tanker oil spill in Alaska 1989[3]*. The oil supertanker struck a reef during a voyage from Alaska to California. The captain of the ship had been drinking the night before the accident and was probably intoxicated during the accident. However, the captain was not at helm during the accident – an inexperienced third mate was helming the ship at the time of the accident (NTSB, 1990). Several systemic issues contributed to the accident, including undermanning, broken radar that was expensive to fix, changes in the route due to icebergs in the area. Furthermore, Exxon Valdez did not have a double hull. Finally, the company and the crew knew the captain had an alcohol problem, but they did not take action but rather looked the other way (Barondes, 2014). Yes, there was a wrongdoing, but it was not as simple as one drunken captain, even though he was made a scapegoat, the villain of the story.*

Sanne noted in his study that the railway technicians had a complicated relationship to risk as their job of protecting the public from risk necessarily exposed them to occupational hazards. In addition, technicians had to balance rail safety against other goals, such as punctuality, time limits, economic constraints, and their own safety (2008, pp. 645–646). He notes that

> 'interdependence in a team and fatal consequences imply that to be a responsible railway technician, one must be able to take care of oneself and ensure team safety, irrespective of formal roles and responsibilities' (Sanne, 2008, p. 652).

On the other hand, the technicians' occupational responsibility also entails responsibility towards passengers. An attitude or collective aim of 'making it work' existed in the technicians' community, and it biased their actions, encouraging them to do what is needed under given circumstances, even if this entails 'rule bending and risk taking' (Sanne, 2008, p. 653). Sanne concludes his analysis by noting that technicians take occupational risks to compensate for inadequate planning, time or resources: matters that should have been handled by the corporation. Thus, risk taking is an expression of responsibility, skill and control, and it is not conceived as 'risk taking' for the sake of thrills or adventure. It is not considered wrong. It is considered necessary.

Another type of wrongdoing concerns **extended suicide**. Aviation industry woke up to this threat in 2015 when Germanwings flight from Barcelona to Dusseldorf crashed into French Alps with 150 people on board. The crash was deliberately caused by the co-pilot, who suffered from depression and suicidal tendencies. His doctor had declared him unfit to work, but he did not disclose this information at work. The incident led to the revisal of the approach to mental health issues at many safety-critical organizations. An organizational conundrum is whether we allow mentally destabilized pilots to fly (if properly treated) or categorically deny that, and "force" the pilots to hide their mental health issues. Mental health problems are increasing in the western world, and this creates a challenge to develop cultures where we are able to constructively and proactively deal with mental health and see mental health as an integral part of a healthy organizational culture (see Chapter 10).

Donald Palmer argues in his book "Normal Organizational Wrongdoing" that even the most ethical, law-abiding, and socially responsible organizational participants are at risk of engaging in wrongdoing" (Palmer, 2012, p. 9). He (2012, pp. 66–68) provides examples of how cultural assumptions and values can foster wrongdoing in organizations. His examples include bid rigging, price-fixing agreements between competitors, illegal sales practices and fraud. He also shows how this wrongdoing was considered normal by the employees, and that they felt their employer expected that of them. He also shows how wrongful organizational practices of the senior management can breed more wrongful practices elsewhere in the organization.

Another perspective is provided by Max Bazerman (2022) who argues, based on Greene (2013) that our loyalty to our own "tribe" (in-group, see Chapter 2 on social identity) makes us often defer to authority and stick to following the rules, even if it makes us complicit in wrongdoing. The strength of loyalty is that it is not based on rational decision making but rather an intuitive one (see decision making in Chapter 2): if our group did it, it must be good, done for a purpose, and deliberate (cf. Bazerman, 2022, p. 104). Bazerman gives examples of the Volkswagen emission fraud and General Motors faulty ignition switch scandals in the 2010s as well as the Boeing 737 MAX accidents (see Chapter 8). In all these cases, loyalty to the company led people to act in ways that were wrong and ultimately harmed both outsiders as well as the company itself (financially at least).

Palmer notes that organizational cultures can develop practices and artifacts that convey acceptance and even **appreciation of actions that break rules**, especially if these actions are risky or innovative. Practices and artifacts that emphasize achieving ends without providing guidance on the means to achieve those ends are a source of risk according to Palmer. Cultures can also facilitate organizational wrongdoing by exhibiting artifacts and practices that convey norms, beliefs, values and assumptions that stipulate extenuating circumstances under which behavior that is understood to be wrongful in general, can be considered acceptable. Building upon the work of Ashforth, Anand and Joshi (2004), Palmer elaborates six "techniques of neutralization", that members of the culture can use to conceptualize questionable behavior or block any feelings of guilt:

- Denial of responsibility by referring to conditions that are out of the control of the person requiring them to.
- Denial of injury by emphasizing that no harm was done.
- Denial of victim by e.g. considering that "stupid people demand to be exploited".
- Social weighting, by considering an action acceptable if it is no worse than what others have done, or if the legitimacy of the entity labeling the act as wrong can be questioned (e.g. the authorities).
- Appeal to higher loyalties, by considering action acceptable if it achieves a higher moral goal or a sacred principle, such as belief in the free market in the case of Enron.
- Balancing the ledger by emphasizing that good deeds will offset the wrongful acts.

Leaders are in a key role in creating cultures that either endorse or condemn wrongdoing (see below, and Chapter 7).

Palmer also notes how wrongdoing may develop step by step due to a human tendency to compare our prospective behavior with recent behavior. If prospective behavior does not differ much from recent behavior, people tend to conclude that it is a sound course of action. In this manner, recently enacted behaviors establish new benchmarks against which subsequent behaviors are evaluated (Palmer, 2012, p. 105). Max Bazerman (2022, p. 168) observes that individuals are less likely to perceive, and more willing to accept unethical behavior when it develops incrementally rather than switching abruptly from honest to dishonest. He calls this incremental development a "slippery slope to complicity" (see also Bazerman & Tenbrunsel, 2011).

Another related social phenomenon called "normalization of deviance" was identified as a cultural contributing factor to the Challenger Space Shuttle accident (Vaughan, 1996), see Chapter 2 and below in this Chapter.

When considering a culture for safety, more often than individual or collective wrongdoing we will find unintentional actions and **well-intentioned shortcuts and trade-offs**. However, the same mechanisms that are used to justify wrongful acts can be used to justify risky actions – and often these are the same actions. For example, the daily trade-offs where employees think "this small shortcut will harm no one" are at the same time deviations from rule and risky behaviors, but they are done to get the job done (or at least they are rationalized as such). Another example would be management who decides to cut corners on safety due to economic pressures "out of their control".

Poole and Dekker (2024, pp. 98–100) talk about "safety theatre" and "optic compliance" where people rig up a **façade of conformity** without genuinely believing in the procedures or sharing the values. This leads to cynicism and ultimately mental resignation. Poole and Dekker give the "numbers games" as a typical example of optic compliance, referring to the various indicators that organizations utilize (and optimize) in their attempt to somehow "manage" safety. Ironically, it can be the most loyal employees who are also willing to play the numbers game, if they believe that is what the management genuinely wants. Loyalty contributes to this kind of unethical behavior especially if the organization faces external pressure or competition (Hildreth et al., 2016; Bazerman, 2022). We will return to indicators later in this Chapter and Chapter 6.

> *There are several examples of failed internal oversight in cases of cost or schedule pressures. Palmer* (2012, p. 184) *describes how a Safety Surveillance Officer working in a pharmaceutical company was subtly ordered to remove reports of adverse effects concerning their new diet drug from their internal database. The coercion was executed both by the Officer's immediate supervisor as well as the company's Chief Safety Executive. Palmer refers to a book by Alicia Mundy (2001), where she describes also the challenges faced by the regulator in responding to the adverse side-effects. Another example concerns Enron[4] and its risk assessment group that was responsible for conducting due diligence. An analyst who opposed one of the earliest illegal Special Purpose Entities was transferred by the CEO from risk analysis to Enron's research unit, an Enron backwater. The CEO noted that "there have been some complaints ... that you are not helping people to do transactions. Instead, you are spending all your time acting like cops. We don't need cops." (Eichenwald, 2005, p. 250)*
> *Similar issue took place during the Challenger investigation (see the end of Chapter 2), when Alan McDonald, director of Solid Rocket Booster project at Morton Thiokol was accused by his own company of having no "corporate loyalty" when he told the truth of what happened to the Rogers Commission (McDonald & Hansen, 2009, p. 338). So-called corporate loyalty is thus dependent on culture, and being loyal to an unhealthy culture (or an unhealthy senior management subculture) may be detrimental to safety.*

Diane Vaughan (1999) analyzed mistakes, misconduct and disaster in organizations. She defines **organizational deviance** as "an event, activity, or circumstance, occurring in and/or produced by a formal organization, that deviates from both formal organizational design goals and normative standards or expectations, either in the fact of its occurrence or in its consequences, and that produces an unanticipated suboptimal outcome". She (1999) notes that much of organizational deviance is a routine by-product of the characteristics of the system itself and can thus be understood as **routine nonconformity**. She writes that "rule violations occur because of mistake, misconduct, institutional arrangements, informal organization, and cultural understandings" (Vaughan, 1999, p. 294) and further specifies that these are not anomalous events, but systematic products of complex structures and processes (Vaughan, 1999, p. 298). She concludes that in policy for preventive strategies against routine nonconformity and its manifestations as mistakes, misconduct, and disasters need to go beyond individuals to "institutional and organizational factors that shape individual cognition and action" (Vaughan, 1999, p. 298). Climate and leadership are among these factors.

An unfortunate feature of modern work life is that at a time when leadership is most needed, organizations are riddled with bad leadership. Often leadership at workplaces is just plain bad, but sometimes it is even worse, it is *destructive*, or toxic. **Destructive (toxic) leadership** has been defined as

> "The systematic and repeated behavior by a leader, supervisor or manager that violates the legitimate interest of the organization by undermining and/or sabotaging the organization's goals, tasks, resources, and effectiveness and/or the motivation, wellbeing or job satisfaction of subordinates" (Einarsen et al., 2007, p. 208).

In their elaboration on the definition the authors state that destructive leadership can be exhibited through both active and passive behaviors, by doing something bad or failing to do something necessary or good.

Several scholars emphasize that **laissez-faire leadership** essentially is a type of non-responsive leadership in situations where there may be a need for a leader's attention, and where the followers probably need a leader's help and support (Hinkin & Schriesheim, 2008; Skogstad et al., 2014). Laissez-faire leadership is the most common example of passive destructive leadership (Buch et al., 2014; Skogstad et al., 2007) and is defined as the avoidance or absence of leadership (Bass & Avolio, 1994). Laissez-faire leadership has been found to be negatively associated with various attitudinal, behavioral, and wellbeing outcomes (Hinkin & Schriesheim, 2008; Skogstad et al., 2007, 2014).

Bullying and harassment at work can take many forms (Cannon, 2023). Some forms of bullying are explicit (physical threats or attacks, verbal abuse, offensive jokes, stalking and coercion) but some are more subtle (such as setting impossible targets, criticizing "for the good of the subordinate", passive-aggressive behavior, isolation and neglect).

The so-called **dark triad**, Machiavellianism, psychopathy and narcissism have been shown to have a detrimental effect on organizational culture and employee wellbeing. Machiavellianism is a personality trait that is centered on manipulativeness, callousness and indifference to morality. Psychopathy typically manifests as superficial charm, manipulativeness, indifference, and lack of empathy. It is telling that psychopaths are overrepresented in prisons and in senior management (Almas & Lordos, 2025; Brooks et al., 2020). Senior leaders that act destructively promote toxic leadership at the lower hierarchical levels of the organization as well as toxic followership (see also Palmer, 2012).

Complex organizations experience tension because they depend heavily on leadership to foster a culture for safety while simultaneously needing to prepare for and mitigate the adverse effects of destructive leadership. A healthy culture for safety should survive an individual toxic leader. It can be argued that toxic leaders would not be promoted to leadership positions in a healthy culture, but that would be an oversimplification of how toxic leadership forms and how it can be anticipated or managed. There is truth to the saying "power corrupts", and often it is difficult to anticipate how individual copes with the increase in both power and responsibility.

Being a supervisor is demanding in many ways. They often face the various conflicting goals of the organization firsthand. Conchie et al. (2013, p. 112) write in their study of construction site supervisors that

> "multiple role responsibilities were perceived to hinder safety leadership because they dilute the perceived importance and salience of safety, increase cognitive load, and at a more basic level, reduce the amount of time that supervisors focus on safety".

This finding reflected the fact that supervisors saw safety as a distinct component rather than being embedded in daily work. Formal procedures around administration and discipline were regarded as strong hindrances on the engagement of the supervisors in safety leadership (Conchie et al., 2013).

This implies that poor leadership is not always, if ever, an individual choice, but rather the system influences what type of leadership capability is preferred, allowed, promoted and recruited.

Wrongdoing and violations are cultural. Groups form an opinion on what is right and what is wrong from their point of view, and, be it right or wrong, should they be doing it. Group climate has a significant influence on behavior. Negative safety climate leads to underreporting of near-misses and minor injuries (Probst & Estrada, 2010). Lack of psychological safety in groups (see Chapter 2) prevents people from speaking up and voicing their concerns before something happens (Edmondson, 2018). Based on evidence from healthcare and aviation incidents, it has been noted that hierarchical structures in teams may discourage subordinates from giving critical information to team members of higher authority and worsen communication between team member (Künzle et al., 2010a, p. 11). Toxic or destructive leadership may be harmful or even fatal for the followers (Saleem et al., 2021). Culture, climate and leadership clearly play a major role in both safety and accidents. However, to truly understand their contribution, we need to better understand the system and its complexity.

3.3 ORGANIZING COMPLEXITY

Organizing builds on condensing of information and specialization, and on the definition of reporting lines, authorities and "need to know". Often, these "need to knows" are apparent only in retrospect.

Studies of major accidents show that they are rarely caused by a single exceptional event. Rather they are a consequence of an **unexpected combination** of several quite ordinary events (Hollnagel, 2004). Complexity creates uncertainty and surprises (see Chapter 1). Problems that we face in complex organizations are seldom clearly defined and specified. Technologies are becoming more and more sophisticated, yet we rely on them more and more. Wherever there are groups of people working together, there is ambiguity due to communication issues but also different ways to make sense of reality, different values and goals, and different ways to react to each other. Sociologist Charles Perrow proposed that systems are prone to accidents when they are inherently complex and tightly coupled (see end of this Chapter). Managing risks that span organizational functions and professional boundaries is demanding for many reasons. Many accidents have been the result of the inability or unwillingness of different areas, functions or even organizations to adequately communicate information about their own work, its requirements and practical implementation.

Specialization to narrow fields of expertise leads to each expert understanding a limited part of an ever-expanding whole. Experts specialize in smaller and smaller entities and go deeper and deeper into their fields of expertise without having the opportunity to familiarize themselves with all the information outside their specific fields of expertise. In fact, even experts in the same field do not always understand each other if their specialization differs. And when fields of specialization further specialize, there are few in any organization who can challenge each other's decisions or expertise. Weick (1988, p. 311) writes that specialists "can do a few things well, which means that they search the world to see if it needs what they can do. If it doesn't, they do nothing else because they see nothing else".

The challenge for safety is that especially for controlling major hazards, a degree of centralization and specialization is needed (Hopkins, 2019; Monteiro et al., 2020a, 2020b). This challenge of organizing is further inspected in Chapters 6 and 8.

Journalist Claus Jensen (1996) points out that before the Challenger accident, narrow specialization hindered NASA's ability to see the whole. Further, he argued almost 30 years ago that it is becoming ever more typical of

> "engineers and scientists working within systems to be so specialized that they have long given up trying to understand the system as a whole, with all its technical, political, financial and social aspects" (Jensen 1996, p. 368).

There is no indication that the situation would have been improved. **Increased use of contracting** makes understanding of the whole system even more difficult.

Quinlan et al. (2013) noted that the outsourcing of aircraft maintenance was one of the causes of serious accidents and crashes in the United States during the 1990s and 2000s. In many other industries the increased use of contractors has raised concerns. There are also some indications that contractors have a higher accident rate than the regular employees. Outsourcing is further inspected in Chapter 6.

Alvesson and Spicer (2012) use the label "functional stupidity" to refer to seemingly sensible thinking within narrow, taken for granted assumptions combined with limited ability or interest in considering what makes sense from a broader perspective (Alvesson & Sveningsson, 2024, p. 133). Simply following orders and doing as one is told is often functional from the perspective of both workers and the management, at least on a short time frame. Thus, division of labor and other structural arrangements in the organizations may in fact support "functional stupidity" and hinder independent thinking, autonomy and a wider view of the topic of interest. This can lead to deficient understanding of the effects of one's own work on safety and risks.

Specialization also happens within the safety discipline. Safety culture professionals look at safety from a safety culture perspective, behavioral safety professionals from behavioral perspective. Process safety experts understand the hazards created by process disturbances but not necessarily occupational or environmental hazards. Often these representatives of different safeties do not even appreciate each other's work. Rollenhagen (2021) notes that safety related functions such as experience feedback, risk analysis, and quality assessments may develop into their own specific disciplines without contact with each other. This creates silos inside safety management. This was one of our main criticisms of the safety culture concept in our 2014 article "Does the concept of safety culture help or hinder systems thinking in safety?", that it easily contributes to the creation of silos instead of integration (Reiman & Rollenhagen, 2014). Specialization hinders system thinking when the system extends beyond the boundaries of the specialists' field[5].

Organizational structures may prevent risk related information from being adequately utilized. Organizations tend to **filter and distort information** (Bella, 1987; Smallman & Weir, 1999). Information is systematically and routinely filtered out according to the norms and structure of the organizational culture in which the communication process is set. In the routines of organizational life, information is filtered as it travels through formal reporting lines and informal networks. That flow is shaped by cultural norms, including expectations for how it will be handled, and habits of good-news filtering. Generally, more good news than bad news is reported upwards in any organization (Lee, 1993). Not everything can ever be reported, thus choices have to be made based on official reporting requirements, personal interests, social desirability, and whatever the employees have noticed or considered important in the first place. People usually do not know that they lack information, or if they do know, the nature and source of lacking information often remains obscure. The question is not only about information flow but also, and more importantly, about what is considered relevant information and how people interpret and look for information in their work environment.

Palmer (2012, p. 132) noted that **fragmentation of information** can facilitate organizational wrongdoing as employees may lack information available in other parts of the organization that would cause them to eschew inadvertent wrongdoing, or they may not feel obliged to stop or expose the wrongful behavior in other parts of the organization due to diffusion of responsibility. These mechanisms also apply to various safety related deviations and their normalization in different functions of the organization.

Due to complexity, views on what safety is, how important it is to the organization, how to improve it, and how good the level of safety is, typically differ inside the organization. Furthermore, also due to the complexity of the organization and its environment, no single group has a complete view of all the risks and vulnerabilities.

Safety professionals are not necessarily the ones that have the best view on safety. In fact, in complex systems no one individual alone has the perfect view, but rather everyone has a **limited and biased view** based on their competence, tasks, and typical interactions, including the way the function is organized. In the best scenario, the safety function can act as a facilitator, bringing the different views on safety inside the organization together and integrating them. The role of safety professionals and the safety department is inspected in more detail in Chapter 8.

Columbia *Space Shuttle accident demonstrates the dangers of condensing information (CAIB, 2003, p. 191). Edmondson et al. (2005, p. 232) stated in their analysis of the Columbia accident that the rigidity of communication protocols inhibited exchange of ideas, questions, and concerns, and encouraged the reporting of packaged, or summarized, results up the hierarchy. Uncertainties related to the information were lost when the information moved through the organizational boundaries. Especially relevant for the loss of information was the tendency to condense "key points" as bullets in presentations. The uncertainties and assumptions behind the bullets were not properly communicated as information moved upward in the hierarchy (CAIB, 2003, p. 201).*

If the top management shows that they only want to hear good news, that's what they will eventually hear – until the bad news are severe enough and cannot be dismissed anymore. Leaders are in a key position in creating psychological safety for personnel to speak up on issues they worry about, an issue that we will revisit multiple times in this book. **Open communication** and sharing of risk related information is especially important since people typically believe that a consensus exists among other people of their group on issues that matter to them (see consensus bias in Chapter 2). Especially for senior managers, it is sometimes hard to get information on possible dissenting opinions in the organization.

Vaughan (1996, p. 323) notes that at the end of the teleconference where the fate of the Challenger *was decided and where the contractor for the Solid Rocket Booster, Morton Thiokol first expressed concern, but then recommended launch, the personnel at NASA's end of the line were not aware that anyone at Thiokol still objected to the launch. They were also unaware of Morton Thiokol's decision-making process that led to the change of recommendation. At the beginning of the teleconference, engineers from Morton Thiokol presented their concern over the effect of temperature on the ability of the O-rings to seal and concluded that they could not recommend launch. NASA disagreed with the data and the launch delay recommendation. Managers at NASA used strong words in expressing their different opinion and in pointing out the flaws in Morton Thiokol's reasoning. However, they said they would not launch against a recommendation from a contractor. Morton Thiokol requested a brief break to review their data. During the break, a management decision was made at Morton Thiokol, meaning that only the four managers (out of 14 participants from Morton's team) voiced their final opinions. The teleconference resumed after a 30-minute break, during which the NASA participants were already preparing to call the launch off. However, to NASA's surprise, Morton Thiokol, represented by their vice president of the booster program, was now recommending a launch[6]. A critical piece of information - that the decision was a management decision made against the somewhat inconsistent and ambiguous analyses and worries of engineers - was lost.*

Organizational secrecy and hiding of information are built into the very structure of organizations (Vaughan, 1996, p. 250). The division of labor between subunits, the hierarchy and geographic dispersion segregate knowledge about tasks and goals. Physical as well as social distance interferes with the efforts to know the behavior of others in the organization. Specialization further aggravates the problem, as people in one department do not necessarily understand the work of the people in the other department. Structural secrecy may lead to information not being delivered to decision-makers. So called "**good news culture**" means that employees select what they report based on what they think the management wants to hear.

Organizations take various measures to increase the flow of information, making the process of information sharing more formal and impersonal. These formal organizational efforts to communicate can lead to information being either not read due to an overwhelming amount of it or to loss of details and impoverishing of information due to ready-made categories of computer systems and other communication forms (Vaughan, 1996).

Although employees may understand on a general level that operations include risks, it may be difficult to see how one's own work specifically affects those risks. Employees may feel emotionally committed to safety without fully understanding what safe practices entail. This knowledge and competence component of safety is often neglected in safety approaches that focus on individual attitudes and behaviors – or collective values as in the safety culture approach (see Chapter 5). Risky solutions are not always caused by people being negligent of safety. In addition to good intentions, an understanding of hazards and ways of preventing them is needed to act safely (Choudhry & Fang, 2008; Kletz & Amyotte, 2019; Pidgeon, 1991). Accident investigations have observed that the defective understanding of the connections between the specific task and safety of the entire system has been a critical contributing factor to different accidents (Hopkins, 2012; Kletz & Amyotte, 2019).

Understanding the safety significance of one's work is also important because understanding the interconnection between one's work and its effects, including its effects on safety, is a motivating factor at work, and it helps to commit to the quality and safety measures required by the organization. Therefore, understanding the connection between one's own work and its safety effects has an influence on the attitude towards the work. This understanding is not self-evident in complex work environments (Nævestad, 2008; Norros, 2004; Oedewald & Reiman, 2007). Complex organizational structures and work processes may lead employees to feel that their work has a smaller impact on safety than really is the case. They may also affect the sense of personal responsibility. In many organizations, safety is secured with multiple, partly overlapping, partly independent measures, the result being that no one feels the need to take personal responsibility for it[7].

One paradoxical consequence of increasing complexity of our systems is that the number of rules has also increased as an attempt to control these systems, including safety. There has been concern about the bureaucratization of safety (Dekker, 2021). We will return to the challenges of standardization, rules and bureaucracy in Chapters 6 and 10, and touch upon them when discussing leadership in Chapter 7 and safety professionals in Chapter 8. Finally, formal guidance, including safety rules, is discussed as one tool of safety management in Chapter 9.

3.4 UNDERSTANDING WORKPLACE REALITY

One of the challenges of complexity is that people are **unaware of all the effects** (and side-effects) of their behavior and decisions. Many organizations employ decision-making practices and other tools to make sure that safety-critical decisions are made, documented, communicated, and later assessed prudently (see Chapter 9). The challenge is that sometimes only hindsight reveals the safety consequences of all the mundane daily decisions. It is not feasible to utilize the formal decision-making tools to all organizational decisions; thus, a choice needs to be made; a choice to simplify reality into procedures.

There is always a gap between the world as it is described and the world as it unfolds in real time. Safety science calls this the gap between **Work-as-Imagined** (WAI) and **Work-as-Done** (WAD) (Hollnagel, 2015). This gap does not mean that people are breaking the rules, at least not intentionally, and not all the rules. It is also often the case that the actual way of performing the work is at least as safe as the described (imagined) way. This discrepancy and its reasons must therefore be analyzed without assuming a priori that any interference with the guidelines is a deliberate violation and that the guidelines are always correct.

Hollnagel (2017) describes work-as-imagined as what designers, managers, regulators, and authorities believe happens or should happen, which becomes the basis for design, training, and control. de Boer (2021, p. 31) describes Work-as-Imagined as dislocated in time and place from execution and therefore lacking immediate feedback. The circumstances surrounding Work-as-Imagined are assumed and static and will not always reflect the everyday variability and sometimes extreme circumstances under which work is done. The resources for Work-as-Imagined are predetermined and of assumed predictability. Training, knowledge, attention and motivation are all assumed to be appropriate, whereas during actual task execution, this might not be the case.

Work-as-Done is what people must do to get the job done in the actual situation; it is what actually happens (de Boer, 2021). A multitude of issues affect the conduct of work, from the availability of resources to the actions of other groups and the external circumstances (including weather, competitors, public, and a multitude of other factors). People adjust what they do to match the situation, but these adjustments will always be approximate. **Performance variability** is inevitable, omnipresent and necessary (Hollnagel, 2009a). Performance variability is the reason why things sometimes go wrong, but it is also the reason why everyday work is safe and effective. Deviations from rules are thus systematic, and systemic, rather than personal violations or errors (see above in this chapter).

The complex organizational structure also makes **managing change** challenging. Changing responsibilities of one function or department is reflected in many things and requires changes in other functions and updates to many guidelines and documents. In addition, it is difficult to predict the impact of the change on practical operations, as practical activities are often something other than strict adherence to instructions and official procedures. Furthermore, changes typically have unintended consequences which need to be managed during and after the change process. Complexity makes the identification and mitigation of the changes more challenging.

Paradoxically, since complexity makes it more difficult to find the most effective organizational structure, organizations create more complexity by regularly changing their structure. Managers, with the help of an army of consultants, desperately try to find the perfect organizational structure to support their business in today's volatile environment. Some organizations have become what Helms Mills (2003) calls "serial change companies", companies that follow the management fashions and seek to implement the "fad of the month". This may in part explain why organizational changes fail so often.

Changes, even those that ultimately meet their goals, may reflect negatively on culture and at least temporarily deteriorate climate. Especially the period of uncertainty that follows when changes in staffing are announced (typically reductions through co-operation negotiations) can be experienced as stressful by the employees. Communication is important, but at the same time there may be unclarities and confidentialities that hamper the communication attempts.

3.5 ASSUMPTIONS AND FAILURES OF IMAGINATION

Culture is learned by the group as it solves its internal and external challenges. This means that culture incorporates responses to the problems of the past. Often these responses are in the form of norms and narratives (see Chapter 2). They influence what we pay attention to, and how we subsequently respond to different contingencies. It is difficult for the personnel to imagine scenarios that the narratives do not account for. Thus, we simplify complex reality by creating stories, which in turn influence how we act in this still complex but in our minds much simplified reality. Sometimes these stories make causal assumptions between independent phenomena based on correlations or co-occurrence. As we discussed in Chapter 2, our everyday thinking is driven by patterns and plausibility, not objective facts.

This means that **culture blinds us to certain risks**. Due to simplified narratives (stories of success and failure), limiting assumptions, lack of safety imagination, or routinization, organizations do not realize all the risks in their activities. The stronger the culture, the stronger the potential ignoring (cf. Weick, 1998). Cultures can be too internally consistent, too strong in their views of the one best practice. As discussed in Chapter 2, people tend to hang around with people who share their views – this shows in how organizations recruit and promote people. Culture thus often reinforces itself, and stronger cultures are also stronger in reinforcing, creating a vicious circle.

As we more and more specialize in topics that few others in the same workplace understand, we tend to find our peer groups from outside the organization where we work (e.g., the internet). Whereas this can in principle act as a counterforce to a shared organizational culture, it often in practice leads to isolation and misunderstanding. The journalist / anthropologist Gillian Tett (2015) has studied the "**silo effect**" that is caused by the way we organize and specialize in organizations and society in general. She explains that "silos are cultural phenomena, which arise out of the systems we use to classify and organize the world" (Tett, 2015, p. xiii).

Our accident models and ways of thinking divert attention to out-of-the ordinary events. Failures stem from the same factors as successes – getting rid of these factors compromises safety. We often give inadequate attention to ordinary problems, and to the interaction of several minor "ordinary" issues. However, out of ordinary events, even though noticed, are often explained away.

> *At NASA, when the space shuttle finally became operational, Jensen* (1996, p. 202) *argues that "every single breakdown was regarded as an embarrassing exception, to be explained away and then be corrected, under wraps, as quickly as possible - so as not to damage the space shuttle's image as a standard piece of technological equipment". A somewhat similar phenomenon took place at Boeing, when they explained away the need for training for the 737 pilots transitioning to the new 737 MAX (see Chapter 8).*

Complex organizational structure and redundant safety systems may, in some cases, increase safety risks by making it more difficult to perceive the big picture and the safety consequences of various actions. This is one of the basic ideas of proponents of the Theory of Normal Accidents (NAT) (see Chapter 3), who argue that redundancy and increased complexity make it harder to detect errors and make the system more prone for surprises (Perrow, 1999; Rijpma, 1997).

Karl Weick (Weick, 2005, 2006) speaks about failure of imagination as one of the causes of accidents in complex systems. Weick (Weick, 1993, 1996) uses the phrase "**drop your tools**" to highlight the need to abandon our old concepts and ways of thinking. He analyzed the behavior of a group of firefighters in the 1949 Mann Gulch fire disaster. A group of 15 smokejumpers (paratroopers trained to fight wildfires) was dropped into the area to fight a rapidly spreading fire. The fire was more intense than they had anticipated, fueled by dry conditions, steep terrain, and strong winds. It quickly engulfed the men and in their desperate attempt to flee the fire, 13 of them died. Only those firefighters survived, who, against their training, dropped their firefighting tools and ran. Their training emphasized the importance of sticking to your tools, as these will save you. The lesson is that people sometimes become too attached to their existing knowledge, tools, or methods when facing new challenges, which can limit their ability to think creatively or respond effectively.

Weick emphasizes that in times of uncertainty or crisis, people often rely on familiar, habitual tools or strategies they've always used. However, these tools may not be effective in new or changed circumstances. Dropping one's tools is particularly difficult for experienced personnel. These tools do not need to be physical - and they seldom are in today's world - also our beliefs and assumptions are tools. In certain circumstances, it is necessary to let go of (drop) these mental tools. To facilitate this kind of mindful behavior, organizational culture should encourage reflection and dialogue, enabling personnel to move beyond traditional practices and routines when adaptability and novel solutions are required.

Long experience does not necessarily and automatically lead to more advanced models of thinking and action but may rather result in restricted routines that are difficult to change. Experience often narrows one's point of view to focus on some 'pet' theories or solutions that have worked well in the past. As experience accumulates, people learn what works well and what does not. These 'viable' solutions became personal preferences that are then applied to a wide range of situations. Relying on experience makes sense in many cases, but it has its drawbacks. Experience and the implicit beliefs formed from it have a strong influence on what the individual subsequently pays attention to, what he considers important, and what he learns from his experiences.

3.6 FAILURES TO LEARN AND ADAPT

Learning is more difficult in complex organizations than in simple systems or small groups. Feedback is delayed and the relationship between cause and effect difficult, if not impossible, to decipher. Furthermore, we are only able to get feedback on the actions that we take, not on the myriad possibilities we omit. We learn by simplifying and assuming in an environment that never stays the same but rather constantly evolves. Thus, even if we manage to learn something about the system in question, we need to constantly update that view since the system does not remain the same. The challenge for safety is that the safer the system, the less opportunities there are to learn from failures and develop the system (Amalberti, 2001).

Accidents happen in part because we fail to learn from the small, or even large, signals that precede them (Turner, 1978). These signals can be mistakes and so called near-misses in one's own organization or in the industry in general, or they can be actual accidents – again, in one's own organization or elsewhere. These signals can also be successes and triumphs of the organization or the industry – e.g. in the form of best practices disseminated by industry working groups that turn out to be fads or wishful thinking. Learning happens all the time, but such learning that would improve safety is surprisingly difficult. Organizational learning is always shaped by culture, power plays, and organizational politics. The daily learning that happens in organizations is not primarily motivated by safety considerations, but by numerous other factors, including the human tendency to avoid anxiety and uncertainty even by making up stories, if nothing else works. One particularly strong story is the story of human error. Looking for human errors after an event is a 'safe' choice, since one always finds them in hindsight, in almost any activity, regardless of its outcome. Of course, organizations typically only seek and find errors after failures, but not after successes.

Learning is also hampered by the misuse of the so-called **Heinrich pyramid** (see Chapter 4), which states that for every major event there are 100 (the numbers differ between sources but are not essential here) near-miss events. Studies show that statistically this is somewhat correct, but this does not mean that the causes would be similar[8]. Focusing on the immediate causes of near-misses does not help in preventing failures due to the systemic factors. And focusing on learning from near-misses that involve occupational hazards does not contribute to improved process safety (see Chapter 4). Finally, the number of near-misses perceived and reported depends on many factors, including the current safety level, vigilance of employees and the leadership response to reporting (see Chapter 6).

There is a tragic mismatch between typical human tendencies for learning and the requirements of organizational learning for safety. As discussed in Chapter 2, we tend to attribute successes to our own merits, but failures to the presence of external forces. This is not often good for organizational safety. Taking credit for success at an organizational level reinforces the current culture and the prevalent ways of seeing and acting in the world. After a period of success, the culture may turn so rigid that even an accident is not enough to convince its members to revisit and revise their beliefs and assumptions. If the accident was minor in its consequences, it is typically easier to explain away by referring to concepts that comfort but provide no real explanation of the event, such as human error, bad attitudes, or lack of commitment – even "safety culture" fits into this category. What is then perceived as needed is more of what the culture already prescribes, but just stronger than before the accident. This seldom leads to any real improvement in performance. (Baumard & Starbuck, 2005)

All too often, organizations **learn to repeat their mistakes** and to better justify why change is unnecessary. Small failures or errors reinforce the basic assumptions and conceptions of the personnel if the failure is attributed to a lack of commitment to official organizational practices and values. On the other hand, serious failures can be attributed to being due to the influence of external circumstances, thus requiring no learning or change of practices or thinking. Defining failure and success in organizational settings are always social and political processes, and by reinterpreting history, each can be turned into another (Baumard & Starbuck, 2005; cf. Weick, 1995).

Learning from experience in complex environments is challenging, and very different from learning from experience in simpler, more forgiving environments (think chess or golf). Complexity influences our ability to learn as well as the possibilities for learning in general. As we have noted, people think in narrative terms and are quick to make up stories that combine independent facts – and not always even facts. This happens when we try to learn from incidents. We typically underestimate the role that chance has played in the incident and seek to rationalize the decisions and choices that had been made before the incident. Likewise in the case of successes, the rationalization of the event typically disregards chance and disproportionately highlights the internal processes of the organization as responsible for the success.

Similar to how we view our own success, we tend to blame someone else's mistake on the characteristics of that person (laziness, indifference, lack of ability), instead of the situation or work conditions (see fundamental attribution error in Chapter 2). We overemphasize dispositional, or personality-based, explanations for behaviors observed in others while underemphasizing situational explanations (Fiske & Taylor, 2017). Thus, the errors and mistakes of others are perceived as being due to stable traits ('bad apples') rather than contextual reasons. Furthermore, these errors are typically attributed to the actions of people closest to the hazard source, such as operators or maintenance technicians. The fundamental attribution error makes it more difficult for us to learn from other people's (or organizations) mistakes as well as successes.

Social identity (Chapter 2) influences learning at both group and individual levels. Too strong a social identity can hinder learning from incidents that have happened elsewhere (at another organization or another work group in the same organization) in the form of "it cannot happen here" -thinking (cf. Cook & Woods, 2006). Thus, we learn little about the other groups' successes and failures because we explain their success / failure in a stereotypical manner.

As an example, Myers et al. (2019) studied the application of High Reliability Organizing (see Chapter 5) in healthcare settings and noted the following:

> "Professional fields—medicine, engineering, law, etc.—often suffer from … siloed approach to problem-solving and a lack of awareness of the lessons available from other disciplines or perspectives. This inability or reluctance to learn from the lessons of outside disciplines has been called the "not-invented-here syndrome" (NIHS), a term used in studies of organizational innovation to describe when useful ideas are devalued or dismissed simply because they come from external sources, either as an intentional or unintentional defensive strategy, or because a track record of success creates a belief that superior ideas can only be created internally." (Myers et al., 2019, p. 1416)

In the nuclear domain, one of the espoused values of nuclear safety culture is "nuclear is special / unique". While this may, in a good "safety culture" emphasize the utmost importance of nuclear safety and the need to prioritize it above all else, it may also be interpreted in a more unflattering light: Nuclear as an industry is so special there is nothing to be learned from the other industries – which is of course not true especially when it comes to organizational and cultural issues.

We often learn the wrong lessons from failures even in formal investigations. Our tendency to attribute causality to individuals (instead of e.g. systems) prevents us from seeing the systemic features. **Hindsight bias** (see Chapter 2) leads us to underappreciate the complexity of the accident. For this reason, we easily blame the front-line workers for missing the obvious (in hindsight) danger signs or for making the obviously (again, with the benefit of hindsight) wrong decisions, or just being plain incompetent. The hindsight bias has profound implications for our ability to make sense of accidents and learn the correct lessons from them. It can lead investigators to identify discrete causes or even the root cause, all of which can be controlled now that they have been identified. Locating the responsibility for incidents in individual decision makers allows quick remedies such as firing, transferring or retraining the individuals (Reiman & Rollenhagen, 2011). Combined with fundamental attribution error (Chapter 2), investigators are tempted to make dispositional attributions and view the event as evidence of individual defects or character flaws rather than considering systemic and situational influences.

If we consider ourselves, or our organization, as being excellent (as we often do, see Chapter 2), failures can cause **cognitive dissonance** as failure is not in line with our and our organization's self-image. When experiencing cognitive dissonance, we seek to alter that which is easier to alter. In this case it is easier (and better for our self-image) to alter the interpretation of failure. By focusing on the positive side of it, and even reinterpreting it as a success, we avoid dissonance. Individuals and organizations are quite innovative and resourceful when it comes to turning the experience of failure into a memory of success (Sagan, 1993).

In cases where the organization admits failure, a different mechanism kicks in, called *over-design* (Snook, 2000) or over-correction. The organization wants to make sure that the event will never happen again (what it would not do in any case, as all events in complex systems are unique) by (publicly) announcing a series of corrective actions to make a similar event impossible. Typically, these actions comprise strict procedures, employee training and clarification of the hierarchical chain of command. By over-designing the system (Work-as-Imagined) they make sure that the Work-as-Done will eventually differ from the plans.

Nippin Anand argues that our "quest for meaning in suffering is still the same as our hunter-gatherer ancestors – non-rational, inward, subjective and existential" and that we "reject failure because we do not want to be reminded of our mortality and fallibility" (Anand, 2024, pp. 319–320). In his view, accident investigations are often not so much about learning and change as they are about cultural retelling of events to maintain societal order. One of the main mechanisms of maintaining societal order is scapegoating, where the sins of the given culture are transferred to an individual. Anand goes as far as to argue that a no-blame culture is an impossibility, given our human tendencies and instinct to blame (Anand, 2024, p. 327). He describes the ritual of accident investigations as reflecting "rejection of fallibility and the delusion of certainty, comfort and control in making the world a safer place". According to him, this ritual is not so much about learning than comfort (Anand, 2024, p. 46).

Dekker (2015) has also criticized investigations that seek to find the "bad apples" hoping that by removing them we can redeem our sense of control over the technology and maintain the belief in the inherently safe system. Focusing on individuals and human error gives us easy (but wrong) answers to two difficult questions: How did that happen? Why are those individuals the ones who suffer? (Dekker, 2017b).

I argue that in addition to people we can use concepts as scapegoats. Safety culture has become one of the scapegoats. In fact, it started as one – being a convenient and comforting label to something the Western nuclear industry possessed but the Soviet Union lacked it offered an existential relief to Western societies. One more reason to talk about culture instead of safety culture (cf. Chapter 5).

We learn the wrong lessons from successes. Success trap refers to a process of specialization that fails when the specialty no longer applies as circumstances change. Through long periods of continuous success, organizations learn to eliminate activities that appear extraneous to optimize themselves, and e.g. focus on their core products, processes or customers. This often makes them rigid and susceptible to disturbances. As a consequence, these organizations become simpler, less aware of events outside their immediate domains and less capable of diverse actions. (Baumard & Starbuck, 2005).

Having competence in some action leads to this action being successful and improves competence through experience. Organizations are more likely to use the same action strategy in future. This can also lead to the "solutions looking for problems" strategy, or extreme specialization. When an organization overlearns one solution, it applies it in another domain, where it may not be as effective. Steele (1972) demonstrated how organizations tend to apply the same strategies that they use for dealing with their production task (external adaptation is Schein's terminology, see Chapter 1) to their internal adaptation. This tendency was termed *overlearning* by Steele (1972).

Starbuck and Milliken, in their examination of the Challenger disaster, discuss the concept of "fine-tuning until something breaks," cautioning that previous success fosters confidence but also illusion.

> "When an organization succeeds, its managers usually attribute this success to themselves, or at least to their organization, rather than to luck. The organization's members grow more confident of their own abilities, of their managers' skill, and of their organization's existing programs and procedures. They trust the procedures to keep them appraised of developing problems, in the belief that these procedures

> focus on the most important events and ignore the least significant ones." (Starbuck & Milliken, 1988, pp. 329–330)

This increased self-confidence often becomes generalized to other areas as well ("we were able to do this so we can do that as well"). The more successful the organization, the narrower its attention typically becomes. Personnel make more assumptions, and more things are ignored. Work becomes more efficient, but less adaptive. (Baumard & Starbuck, 2005; Starbuck & Milliken, 1988)

Starbuck and Milliken (1988) highlight how organizations often make small, incremental adjustments to their existing systems, processes, or strategies, without addressing deeper, more systemic issues or questioning the underlying assumptions. These fine-tuning efforts may seem reasonable or effective in the short term, but they can accumulate over time and create hidden vulnerabilities. This single-loop learning (Argyris, 1999) points out the risks of focusing solely on minor improvements without stepping back and evaluating whether the broader approach is still valid, and where in the chosen path the organization currently resides.

This all highlights how **we are tempted to make quick fixes and regain our sense (illusion) of control**. Looking at and finding human errors makes it seem easier to find out who should be held accountable, and where preventative measures should be targeted. Unfortunately, the preventative measures are usually off target if 'the cause' has been attributed to individual error. These measures can lead to a loss of adaptive capacity in the organization, including its ability to react to novel events.

Another typical quick fix is trying to close various open items on corrective action programs, deviation lists, or non-conformity programs. Organizations have campaigns to close items as they accumulate "backlog". There is nothing inherently bad in organizational safety or quality campaigns, but they easily fall into the same trap as rewarding and indicators – they end up focusing on **optimizing numbers instead of solving issues**.

Finally, even if organizations learn something, they tend to **forget the lessons**. Organizational memory tends to be short, and it depends on the key people remaining in the organization. Even embedding the learnings into systems, processes and practices does not shield them from being modified, changed, or outright removed as the memory of the accident and its reasons, or the rational of the corrective actions, fades from the organization. Actions already taken may be reversed and changes that are still in progress abandoned. For example, investigation of the Columbia Space Shuttle accident showed that although many important changes to NASA's organizational structure were made after the Challenger accident in 1986, many of these changes were undone over time by management actions (CAIB, 2003, p. 198), see Chapter 2.

3.7 DRIFT AND GRADUAL CHANGE

Safety-critical organizations are uniquely dynamic, and people are constantly changing and adapting to perceived challenges and opportunities. Perceived changes in the environment reflect on the organization. At the same time, technology and people age. This creates new organizational demands as well as potentially new technical phenomena. Finally, routines and practices develop over time even without any noticeable outside pressure or demand for change. People optimize their work and work practices, they come up with shortcuts to make their work easier and more interesting, they lose interest in the commonplace and reoccurring phenomena, and they have to make trade-offs between efficiency and thoroughness in daily tasks (Hollnagel, 2004).

Diane Vaughan (1996) presented the phenomenon of **normalization of deviance** as one of the main contributing factors to the Challenger explosion (see Chapter 2). Normalization of deviance means a process where small changes – new behaviors, technical anomalies, or variations that are slight deviations from the normal course of events – gradually become the norm, providing a basis for accepting additional deviance. Normalization of deviance produces disregard and misinterpretation of potential danger signals, effectively neutralizing them. This is what happened with the accumulating evidence indicating problems with the O-rings at NASA (Vaughan, 1996).

As these deviations become embedded in culture, they stop being viewed as deviant or dangerous and are seen as acceptable practices. The definition of what is considered "normal" changes. Normalization of deviance is reinforced by cultural beliefs and expectations, routines of daily work and commitment to past decisions and past lines of action.

Conflicting goals are one of the sources of risks in organizations. Safety and financial profitability, efficiency and thoroughness or occupational safety and process safety occasionally conflict in real life. On the other hand, the goals do not necessarily conflict, for example, in the sense that improvements to safety would always threaten financial profitability but the goals have different time perspectives and uncertainty in terms of return on investments. Investing in safety has a long-term return on investment, whereas economic goals are often actualized within a much shorter time span, or at least they are easier to divide into measurable subgoals. The challenge is not in determining whether safety and economy are in conflict, but in the ability to balance resources and focus attention on issues that most deserve it - be they safety, production or personnel-related.

People constantly adjust their practices depending on perceived demands and resources. They optimize by doing what they consider important and by devising new ways of achieving the same, or better, results. Every day, compromises must be made, goals have to be weighed up and prioritized, instructions interpreted, and improvisations made. These small adjustments are not malignant or negligent, but rather genuine attempts to get the work done and satisfy all goals, even if the goals are sometimes in contradiction. If these situational adaptations work and have no visible immediate side-effects, they will gradually become new informal norms or practice. From the perspective of the sociotechnical system, this can mean **drift in practices**. Snook (2000, p. 194) writes:

> "Practical drift is the slow steady uncoupling of practice from written procedure … After extended periods of time, locally practical actions within subgroups gradually drift away from originally established procedures … Constant demands for local efficiency dictate the path of the drift".

As has been highlighted in this book, effects in complex systems are often non-linear and delayed, making them especially vulnerable to drift. In fact, some form of drift is unavoidable and not necessarily even bad for safety.

The "Drift into failures" metaphor (Dekker, 2011; Rasmussen, 1997) explains how organizations can gradually move towards unsafe conditions due to various pressures and constraints. The idea underlying this metaphor is that humans working in an organization continually struggle to find a balance between workload, intellectual satisfaction in the job and avoiding failure. The struggle is due to the conflicting pressures, needs and constraints people face in daily work. Rasmussen (1997) identified three boundaries delimiting the acceptable performance space:

1. The financial boundary, i.e. what is acceptable from an economical point of view;
2. The workload boundary, i.e. the workload level humans are willing and able to manage;
3. The acceptable risk boundary, i.e. the amount of risks that is functionally acceptable.

Employees need to make situational decisions that balance each of the boundaries. Dekker (2006) calls them "sacrificing decisions". The "sacrificing" aspect is related to the impossibility of satisfying every decision criterion. The choice of taking care of financial pressure, as example, impacts the possibility to reduce workload or to be functionally safe or both. And the same happens when the decision is driven by workload and/or safety. The problem with these decisions is that they appear completely reasonable to the actors in the organization until the drift has gone too far and something goes wrong, either in safety, on the financial side, or on the workload front. Carrillo (2020, p. 52) emphasizes that drift is normal adaptive human behavior, and as such it is a natural phenomenon in complex organizations (cf. Dekker, 2011).

One of the main reasons for the danger of locally optimizing working practices are, first, the loose couplings prevalent in complex sociotechnical systems making it possible to change one part of the social system without immediate effect on the others, and secondly, the tendency of complex systems in some conditions to become tightly coupled. Loose couplings mean that, for example, some organizational units can change and optimize their work practices for a while without any effect on operations or other organizational units. This is as long as the quality of their work (their output) remains approximately the same (or is enhanced).

One often downplayed risk is the risk of **non-permanent changes**. Due to the complexity of the work environment, several adaptations and fixes need to be made. Often a non-permanent solution is implemented with the intention to replace it with a more permanent solution in due time (after the permanent solution is carefully planned, its risks are analyzed and resources reserved for its implementation). However, sometimes these non-permanent changes become semi-permanent or de facto permanent as time goes by and the permanent solution gets lost in the organizational (or regulatory) bureaucracy – or is abandoned altogether as the current solution seems to be working (and the original reason for the change is forgotten). Typically, non-permanent changes are not analyzed as comprehensively as changes that are intended to be permanent. And if they are analyzed, the premise is that they will be used less, or only for a limited amount of time, reducing e.g. their quality requirements.

The tragedy of safety is that increases in efficiency are easier to measure than decreases (or increases) in safety. If we remove slack resources from the organization, we can calculate how many person-hours of salaries we have saved. If we accomplish a task faster, we can calculate how many minutes, hours, or days we saved. It is, however, much more difficult to calculate how much, if any, we had to borrow from safety to do it. And to make matters worse, as safety manifests as a dynamic non-event (Weick & Sutcliffe, 2001), investing in safety at the expense of efficiency (e.g. cancelling a hazardous task, taking extra time for planning, using more conservative margins, or utilizing extra resources) shows as an expense when nothing happens – when the task was accomplished and nothing out of the ordinary happened. This is one more reason why rewarding and other leadership practices are so important for safety, and why they so often direct organizational performance in the wrong direction: without conscious planning otherwise, rewarding in organizations tends to emphasize efficiency and short-term goals instead of safety and long-term goals (See Chapter 6 and 9).

Setting specific numeric goals and then not caring how those numbers are reached is a sure way for senior leaders to contribute to drift. The BP Texas City Refinery explosion described in the next Chapter illustrates how specific cost-cutting goals set by the headquarters without specifying the means to accomplish them led to local savings that proved detrimental to safety.

Normalization of deviance can become institutionalized as formal procedures are changed based on the new accepted practice. In the case of drift, practices typically remain informal, and the gap between formal practices and actual work widens (Dekker, 2019).

3.8 STRATEGIC CHOICES AND POWERFUL ELITES

This chapter began with an examination of errors and violations committed by employees within safety-critical organizations. We now shift our focus back to individuals, but this time to individuals in senior positions, such as executives, owners, and shareholders of these organizations.

The strategic decisions, investments and priorities of the senior management and the owners influence the possibilities of the organization to perform safely. Notwithstanding what the official policies say, senior management's safety commitment (or lack of it) shows in their decisions. At the same time, senior management and owners have a highly simplified and condensed view of the system they are managing. They suffer from the same challenges of complexity, imagination, learning, and gradual change as others in the system, but they have far more power than the other actors.

Their decisions have wide and systemic influences, most of which can never be fully anticipated by them or their subordinates.

Many accidents illustrate the significance of perceived production pressures. The Three Mile Island (TMI) and Chernobyl accidents (see Chapter 2 and 5 for more details) both happened at new power plants that had just been commissioned. At TMI, the pressure relief valve that jammed during the accident had been leaking during the commissioning period, but since the company was in a hurry to connect the plant to the power grid before the end of the year 1978, troubleshooting the problem was postponed. The accident took place three months later, in March 1979. Similarly, in Chernobyl, the test that was conducted with disastrous results was supposed to be completed before the plant went operational in 1983, but again production came first. Pressure to launch was one factor in the Challenger *accident (see Chapter 2), and financial pressures greatly affected decision making at Boeing during the 737 MAX certification process (see Chapter 8).*

Despite the high significance of strategic choices to safety, strategy has received surprisingly little attention in safety research (Le Coze, 2021a). Le Coze (2021c, p. 67) highlights that "disasters, and conversely, safety, cannot be studied without a dedicated attention to strategy". He lists several strategic tasks and activities with potential impact on safety, including mergers, acquisitions and ventures, profit targets, cost-cutting, downsizing, budget allocations, choice of organizational structure, choices of geographic or product (services) diversification, choice of top personnel, design of work methods, motivation and rewards systems, management, and conflict resolution style (Le Coze, 2021c, p. 83). We will return to the issue of safety and strategy in Chapters 6 and 10.

The excessive short-term profit orientation of senior managers as causes of accidents was already noted by Perrow in his Normal Accident Theory as well as in Turner's Man-Made Disasters theory (see the end of this Chapter). Hopkins and Maslen (2015, pp. 1–2) explain why short-term orientation and risk taking may seem rational from a senior manager's perspective:

> Major accidents are rare, and underinvestment can continue for years without giving rise to disaster. On the other hand, managers are judged on their annual performance, especially with respect to profit and loss. Consequently, spending money on the prevention of major accident events is not necessarily in their short-term interest. On the contrary, cutting expenditure on maintenance, supervision and training may enhance short-term profits, while inexorably increasing the risk of disaster in the longer term. Moreover, business unit leaders tend to think in the short term because they may only be in a particular management position for a couple of years before moving on. They may thus be long gone before the results of their cost-cutting decisions become apparent.

Short-term considerations may lead managers to downplay safety and emphasize economic gains. Organizing and leadership is key also when managing conflicting goals (see Chapter 6 and 7).

McDonald and Hansen (2009, p. 12) note that in Morton Thiokol, all managers participating in the Solid Rocket Motor program had to wear two hats, a program manager's hat and a project engineer's hat. McDonald, who acted as the director of the Solid Rocket Motor Project at Morton Thiokol at the time of the Challenger *accident, further notes that the project structure was such because NASA wanted such a cost-efficient organization (McDonald & Hansen, 2009, p. 12). Principal companies have the power to define the level of safety they require from their network. They also use power in distributing and redistributing tasks and responsibilities based on competitive bidding. It is noteworthy that during the time preceding the* Challenger *accident, NASA had decided not to sign a major contract with Morton Thiokol but rather look for opportunities to second-source parts, or as Morton Thiokol feared, all of the SRB activities. McDonald (2009, p. 83) writes:*

"the sad truth was, if Morton Thiokol management had not feared that SRB activities might be second-sourced to our competitors, a key domino leading to the Challenger *disaster of January 28, 1986, might never have fallen."*

He argued that when Morton Thiokol reversed their launch recommendation after pressure from NASA, they were succumbing to pressure from their most important customer with whom they had an unsigned contract and leaned on the "customer is always right" decision-premise (McDonald & Hansen, 2009, p. 111).

Senior management also defines the priorities and practices of approving investments and development projects. Return-on-investment (ROI) is a typical indicator utilized. Focus on ROI proved challenging for BP Texas City refinery (the explosion of which is described in Chapter 4), since the refinery was very old. Further, the investment approval structure of BP made it very difficult to carry out purely safety-related investments that did not have a clear regulatory requirement.

McDonald (2009, p. 70) notes that prior to the Challenger *accident, Morton Thiokol, the designer of the solid rocket boosters (SRB) that failed, proposed a design change that would have reduced the O-ring problem considerably. NASA required Morton Thiokol to conduct a cost-benefit analysis of the modification and ultimately rejected the change proposal as it did not increase the service-life of the SRBs but increased the costs.*

Corporate power also affects learning as it is often in the short-term interest of corporations to downplay their own involvement in accidents. This is yet another reason for the popularity of human error, bringing us back full circle to where we started this chapter. Attributing accidents to individual human errors conducted on the front line serves the interests of powerful elites but seldom contributes to real safety improvements. Alternatively, there is some evidence that corporations are positive towards the more post-modern investigations that shift the blame to the system instead of any individual, including senior managers. The journalist James Hansen notes that Diane Vaughan's conclusions were readily accepted by both NASA and Morton Thiokol as they seemed to absolve both companies and treat the wrongdoing as normal. Powerful elites can influence safety not only prior to accidents but also afterwards, by shaping how they are perceived and who is deemed responsible for them (cf. Turner, 1978). Organizational learning, as all learning, always serves a purpose. And this purpose is heavily influenced by whoever is in charge.

3.9 SUMMARY

The chapter explored the various reasons behind accidents from a cultural perspective, focusing on human errors, organizational shortcomings, and systemic complexities. Accidents in complex systems are rarely caused by a single event but by an unexpected combination of ordinary events. A complex environment makes learning and predicting failures challenging. The unacknowledged gap between Work-as-Imagined and Work-as-Done affects safety. Effective information flow is crucial, but structural secrecy and good news culture can hinder this process. Complex organizational structures and redundant safety mechanisms can increase risks by making it difficult to perceive the big picture and manage changes effectively.

Conflicting goals, such as safety versus financial profitability, create risks. Balancing these goals is challenging and requires careful attention to resource allocation. Normalization of deviance and drift into failure occur as small deviations become the norm. These gradual changes can lead to unsafe conditions if not managed properly. Organizations often fail to learn from small signals preceding accidents. Cognitive biases and the complexity of systems make it difficult to derive the appropriate lessons from incidents.

Powerful elites define the organizational strategy and create the conditions for safety. They also affect learning from both successes and failures. Senior leaders set up the structures, including the safety management system, that affect the possibilities of all employees to ensure safety.

Reflection Questions

What are the critical events that have shaped the culture of your organization? Can you identify both positive and negative critical events? How have they impacted your organization and its culture?

How does the way you organize and lead your organization make it more susceptible to accidents? How can you improve the situation?

Think of a problem in your organization that has existed for a long time, but does not seem as critical anymore – which one has changed, the organization perception or the problem itself?

Think of an issue or area in your organization that was considered a strength but that started to show problems – why did it happen, and why it was a surprise?

Think of a problem in your organization that was discovered and solved by setting, monitoring, and optimizing an indicator. Was the original problem solved?

How does the strategy of your organization create constraints and possibilities for safety? How does the strategy reflect in safety related decisions?

MAN-MADE DISASTERS AND NORMAL ACCIDENTS

Man-Made Disasters

Barry Turner, in his ground-breaking book *Man-Made Disasters* (1978), investigated several major accidents in areas such as mining, railways, and maritime from a sociological and cultural perspective. He concluded that organizations have serious difficulties in recognizing deviant signals before an accident. Turner argued that disasters are not "bolts from the blue". They are preceded by sometimes lengthy periods of increasing risks. Yet this buildup of risk goes unnoticed, or the signals are misinterpreted. Turner referred to this as the "incubation period" of the accident. According to Turner, disaster equals energy plus misinformation.

Misinformation has many sources. Turner argues that the signals (later labeled "weak signals") that precede accidents often went unrecognized since they were at odds with taken-for-granted beliefs about hazards, and with norms for their control and avoidance. During the incubation period, there is a steadily growing discrepancy between what is happening in the world on the one hand and how it is thought to operate on the other.

This growing gap was due to problems in communication and interpretation of information. More specifically, he identified the following issues in the cases he studied:

- Erroneous assumptions and misunderstandings;
- Rigidities of human belief and perception;
- Disregard of complaints or warning signals from outsiders;
- A reluctance to imagine the worst outcomes;
- Decoy phenomena that suck up organizational resources but that do not represent the real growing threat out there.

In addition to signals that went undetected or were misinterpreted, Turner identified signals that were detected and assessed but never acted upon.

The challenges of communication and information use relate to the issues of organizing. Turner pinpointed the dilemma of organizations:

> Where the pattern of routine and regular communication sets up boundaries and barriers to the flow of information relevant to the understanding and prediction of the emergence of hazards, these barriers may make it difficult to stop a disaster occurring. There is a dilemma, of course, about this issue, for it is not possible to ensure that everyone communicates everything to

> everyone else; and the effectiveness of all organizations and institutions lies, in part, in the manner in which they restrict communications to those which they believe to be the most essential.
>
> **(Turner, 1978, p. 197)**

The incubation period ends with a "fundamental surprise". This is an event that unveils the latent structure of the prior events and shows people's prior beliefs and assumptions about the world to be wrong. The event is so evident and undeniable that it cannot go unnoticed – it is the man-made disaster.

Accidents are preceded by sometimes lengthy periods of gradually increasing risk, during which latent errors and events accumulate which are culturally taken for granted or go unnoticed because of a collective failure of organizational intelligence (Turner & Pidgeon, 1997). By heeding these signs, it is possible to prevent accidents, but this seldom takes place. The accumulation of more data per se does not prevent accidents; rather the processes through which information is disseminated, combined, and interpreted need to be developed (Turner, 1978, p. 198).

Turner notes that there is an overlay of differential power distributions which will affect knowledge, perceptions, and expectations of accidents. For example, there are typically one or two highly influential groups in an organization who are able to influence the way in which these dangers are seen and understood by others. Further, the information regarding risks and hazards is distributed differentially between different individuals and groups in organizations and society. This makes it more difficult to carry out "reasonable" or "well-informed" decisions.

Turner questioned the image of organizations as rational. He noted that decision-making in organizations is always driven by assumptions, incomplete information, and lack of knowledge. This also means that the actions of decision-makers should not be judged too harshly with the benefit of hindsight. One should rather aim to determine whether the decision was rational or reasonable, "given the information available to the decision-maker at the time" (Turner, 1978, p. 162). Turner notes that after an event, it is important to consider how the erroneous decisions were made and

> *if we have concluded that the decision-maker did act reasonably, so that other competent persons faced with the same choice in the same circumstances would have been likely to have made the same decision. ... it is clear that the source of error is not to be found in the behaviour of the individual, who has conformed to the norms of rational or intendedly rational behaviour. What then needs to be assessed is the question of how a situation could come about in which reasonable men, attempting to behave rationally, could still be in error. To answer this question, it is necessary to look at the manner in which rationality becomes established and embedded within organizational procedures and habits, and to gain an understanding of the hierarchy of decision-making within which the individual administrator finds that he has to operate.*
>
> **(Turner, 1978, p. 163)**

Turner raises several issues that make "rational action" challenging in organizations. One of these issues deals with the problem of integration, namely with the need to balance organizational constraints and individual capabilities and diversity. Turner notes the importance of shared decision-premises for maintaining this balance. These decision-premises contribute to organizational integration and coordination by "ensuring that all of its key decision-makers are operating with variants of the same organizationally approved bounded rationality" (Turner, 1978, p. 166). He also notes that, as a caricature, a way for organizations to achieve a minimal level of co-ordination is for decision makers to agree that they will all neglect the same kinds of considerations when they make decisions (Turner, 1978, p. 166). Turner's shared decision-premises resemble the basic assumptions and underlying logics depicted in Chapter 1.

This also means that rather than looking at individual rational action, it is more fruitful to look at groups of decision makers and their rationality. Furthermore, decisions made at the higher levels of the organization will most likely be more far reaching in their consequences and produce more complex forms of unintended consequences (Turner, 1978, p. 179). Simple errors are typically not enough to cause a large-scale accident, and as Turner notes, disasters require much more organizing ability (Turner, 1978, p. 180). Turner argues that

> *errors are only likely to develop into large-scale accidents or disasters if they occur in an organizational decision hierarchy or power hierarchy at a point at which they are likely to be magnified, and possibly to be compounded with other smaller errors, by the operation of the normal administrative processes.*
>
> **(Turner, 1978, p. 182).**

Disasters are thus undesired consequences of organizing, resulting from "anti-tasks", which are orderly errors made possible by the organizational hierarchy established to carry out those same tasks.

> Disasters, accidents and all forms of unintended consequences may be considered as the results of energy which at some point was combined with misleading or inadequate information from the point of view of the individual or the organization concerned.
>
> **(Turner, 1978, p. 183)**

To prevent accidents and disasters, Turner (1978, p. 194) notes that organizations need to clarify ambiguities, make sure that information is not overlooked, and provide both the information to control known complex situations and the information needed to foresee unknown, unprovided for, situations. He emphasizes that the countermeasures to incubation periods are not so much about supporting split-second action but rather providing more advance information. He argues that to prevent accidents,

> We would need continuously to adjust the incipient discrepancies between the picture of the world envisaged in someone's plan, and the way the world really is. We would have to provide a perfectly accurate, continuous feedback which would ensure that all informational adjustments made by those pursuing the plan would be infinitely small shifts of view, so that the information acquisition process proceeded perfectly smoothly.
>
> **(Turner, 1978, p. 194)**

Turner recognized that the incubation period is a consequence of normal organizational processes such as information gathering, communication, and decision making, not extraordinary wrongdoing, mistakes, or even negligence by the workers at the sharp end. He noted that accidents may happen despite the best intentions of everyone.

Turner's ideas were in many ways ahead of their time, and it took safety science quite a long time to really appreciate his cultural view. However, Turner has left a long-lasting legacy to the science of safety as well as organizational research, including the study of cultures (Le Coze, 2021b). The idea of latent organizational conditions or latent weaknesses was reintroduced by James Reason as part of his famous Swiss Cheese accident model. The cultural approach to accidents has been further developed especially in ethnographically oriented accident investigations (Hopkins, 2008; Vaughan, 1996). The information-processing challenges have been reintroduced in the 2000s by the Resilience Engineering paradigm (see Chapter 10).

Charles Perrow and the Normal Accident Theory

Sociologist Charles Perrow studied accidents in different industries such as process industries, the maritime and aviation industries, space exploration, mining, nuclear power plants, dams, and even in the use of DNA techniques. He illustrated the potential dangers of interactive complexity and tight couplings. His argument was that interactive, non-linear complexity produces bizarre and unanticipated failures whereas tight, time-dependent couplings cause the failures to escalate rapidly out of control. Systems with complex interactions *and* tight couplings will inevitably suffer from accidents. Perrow called accidents in such systems "normal accidents" and differentiated them from component-failure accidents (Perrow, 1999).

'Interactive complexity' refers to unfamiliar sequences, or unplanned and unexpected sequences, that are either not visible or not immediately comprehensible (Perrow, 1999). Complex interactions require decentralization, as specialized and localized responses are needed, because the system is too complex to be understood and controlled by management alone.

Tight coupling means that system components are interlinked, without any slack or buffer or give. When things go wrong in tightly coupled systems, there are limited options for corrective actions. Thus, tight coupling requires centralization and preplanned responses due to rapid escalation of failure situations.

Perrow's classification creates four types of systems: Linear and loosely coupled systems include most manufacturing, assembly-line production, and single-goal agencies such as post-office. Complex and loosely coupled systems include research and development (R&D) organizations, universities, and military and mining operations. Linear and tightly coupled systems in turn include power grids, dams, rail, marine, and airways. Finally, examples of complex and tightly coupled systems according to Perrow are nuclear power plants, nuclear weapons, aircraft, gene technology, chemical plants, and space missions.

According to Perrow, complex but loosely coupled systems should be managed in a decentralized manner, and linear and tightly coupled systems should be managed in a centralized manner. Linear and loosely coupled systems can be managed effectively in either way, but complex and tightly coupled systems cannot be managed in either centralized or decentralized manner. Complexity prevents centralization and tight coupling decentralization (Perrow, 1999).

In complex and tightly coupled systems, accidents arising out of complexity and tight couplings can be characterized as normal. Interactive, non-linear complexity produces bizarre and unanticipated failures whereas tight, time-dependent couplings cause the failures to escalate rapidly out of control. The Normal Accident Theory (NAT) also contends that standard fixes, such as the addition of redundant safety systems, will be ineffective, and might even increase the likelihood of accidents because they increase the system complexity. And in systems that are interactively complex and tightly coupled, unexpected interactions of several failures will defeat the training, the planning, and the designing of safety devices (Perrow, 1999). Adding safety systems does not help either. Rijpma (1997, p. 19) states, paraphrasing Perrow (1984) and Sagan (1993):

> Redundancy makes the system opaque and thus more complex ... Component failures and human errors can be harder to detect because they are compensated by backups ... Backup systems are not always as independent as imagined. At the same time, their inoperability is possible.

Rijpma also suggests that multiple data collection can lead to ambiguities and divergent views, which are considered counterproductive in NAT.

As accidents in highly complex and tightly coupled systems are inevitable, the only way forward is to dismantle them and replace them with technologies that are less complex and less tightly coupled. Attempts to prevent accidents should aim at reducing the complexity and couplings of our systems. Perrow (2007) has further specified that three concentrations in our society make accidents more likely to occur and more severe in their consequences:

- Concentration of energy;
- Concentration of power;
- Concentration of people.

By reducing these concentrations (e.g. smaller warehouses, more diversity in technology, smaller transports), we can reduce the severity of accidents, if not their probability.

NAT illustrates that the structural features of sociotechnical systems make some systems more susceptible to accidents than others. And that the typical accidents differ in these systems. This has interesting implications for how we should manage safety in different contexts and what can be learned from accidents based on the system characteristics.

NAT also emphasizes that politics and power plays in organizations have a significant influence on safety (see Chapter 1). These influences reach learning after accidents, as Perrow notes:

> Formal accident investigations usually start with an assumption that the operator must have failed, and if this attribution can be made, that is the end of serious inquiry. Finding that faulty designs were responsible would entail enormous shutdown and retrofitting costs; finding that management was responsible would threaten those in charge, but finding that operators were responsible preserves the system, with some soporific injunctions about better training.
>
> **(Perrow, 1984, p. 146)**

NAT has also been criticized for its determinism as well as downplaying the human and organizational issues in general (Evan & Manion, 2002; Hopkins, 1999; Le Coze, 2021c). Le Coze asks whether Perrow was right in that accidents are normal, but for the wrong reasons (Le Coze, 2021c). He notes that one of the key insights of NAT was that safety is also a question of power relationships between actors and institutions involved in the performance of these systems rather than only an intrinsic technological feature of these systems. These accidents were not normal accidents, according to Perrow, but rather component failure accidents. Le Coze has extended on the key role of power in his Post Normal Accident Theory (which we will revisit in Chapter 10) and concludes that "accidents are normal because they repeat despite knowing that they could be prevented in principle" (Le Coze, 2021c, p. xii).

NOTES

1. There were also many other contributors, including the inherently unstable design of the ship and a lack of warning signal on the bridge concerning the position of the bow doors, as well as a top management focus on financial issues. The doors were also visually invisible from the bridge. In many cases the crew had been adapting to and compensating for inadequate design, cf. Rollenhagen (2010).
2. Rollenhagen (2010) proposed that the concept of safety culture has been similarly misused.
3. For decades it was the most severe oil spill in U.S. waters, but BP took the lead after the 2010 Deepwater Horizon oil spill.
4. The Enron Corporation engaged in widespread accounting fraud that ultimately led to its bankruptcy in December 2001.
5. I am not arguing against specialization but against isolation. Specialization is important for developing expertise, and a source of professional identity. However, we need to always be aware of the potential side-effects of individual specialization on an organizational level.

6. NASA wanted the launch recommendation in writing, an untypical measure (McDonald & Hansen, 2009, p. 2).
7. See the description of NAT at the end of this chapter.
8. The relative frequencies differ between different types of hazards. For example, in a study by Bellamy (2015), the relation between explosion accidents was 1 fatality per 5 serious recoverable injuries, whereas contact with handheld self-operated tool had 1 fatality per 280 recoverable events.

4 What is Safety – A Cultural View

In this chapter I discuss the concept of safety from a cultural perspective. I will first provide an overview of the evolution of safety science, describing how the view on safety has developed. The history of safety science is traced from the Technological Age, focusing on technical reliability; through the Human Factors Age, which highlighted human errors and ergonomics; to the Management Systems Age, which introduced safety culture and leadership. The Adaptive Systems Age underscores the importance of resilience and adaptability in complex systems. I will then discuss various types of hazards, including physical, chemical, biological, psychosocial, and ergonomic, and the need for a systems view of safety as an organizational capability. I highlight that safety is not a static state but a dynamic process that requires continuous adaptation and learning. Organizations must balance various tensions, such as stability and change, to maintain safety in a complex environment. I also underscore the need for a cultural shift in how safety is perceived and managed and present that there is a need for a more holistic approach that recognizes the intricate interplay of human behavior, technology, and organizational dynamics. The chapter concludes by looking at BP's two accidents in Texas City 2005 and Deepwater Horizon in 2010, which demonstrate the challenges of understanding what safety is and how it can be monitored.

4.1 HISTORY OF SAFETY

Throughout the history of safety science and safety practice, the role of humans in safety – and in accidents – has been debated, revised, and reconceptualized. The view has evolved from an almost complete neglect of the human role to seeing humans as a threat to safety in the form of errors and violations, to slowly widening the scope of human influence to positive aspects as well, and eventually accepting their significant role in creating safety.

Errors have been seen (and sometimes still are) as a sign of incompetence or inherent traits such as laziness or indifference. In the 1980s, the science and practice of safety gradually started to view errors as part of the normal variation of human performance at work instead of something abnormal (Rasmussen, 1982). Nowadays the concepts of Resilience Engineering, Safety Differently, and Safety II add an additional flavor by questioning the usefulness of the concept of error in understanding human behavior in complex systems.

Errors are just one example of how the science of safety has developed over the years. The development of the science and practice of safety has relied on an increase in understanding of not only human behavior, but also the behavior of complex social (and sociotechnical) systems. What the different ways of seeing the human contribution to safety had in common is that they were all cultural – cultural in a way of being embedded in methods and practices of accident investigations and safety management in general; cultural in a way of reflecting the values and salient social questions of the wider society; cultural in a way of reflecting our understanding concerning technology, human behavior, and social systems; cultural in terms of several contradictory approaches and opinions co-existing simultaneously and ideas being selected for subsequent utilization based on several factors, with scientific validity being only one factor among them (e.g. the Swiss Cheese model was widely adopted despite its empirically non-validated nature); and finally, like culture, they were not permanent but slowly changing in response to several external and internal forces.

Our ideas on accidents and safety have developed over the years. Different characterizations of this development exist (Borys et al., 2009; Hale & Hovden, 1998). Below I draw on these

DOI: 10.1201/9781003491156-4

characterizations and present a division into four ages of safety (see Table 4.1) including their key focus, methods, underlying theory, assumptions, and legacy. Each subsequent age does not completely reject, but rather builds on, the earlier accomplishments (successes and failures).

The first age of a systematic development of safety can be called the *Technological Age*. During that period of rapid industrialization, safety concerns were mostly related to technical reliability. The objectives were, for example, to prevent explosions or the collapse of structures, and to guard machinery (Hale & Hovden, 1998). Between the world wars, attention was increasingly devoted to issues such as personnel selection, training, and motivation as a means of preventing accidents. This ushered in the second age of safety, which led a rather separate life from the technical measures until the period of the 1960s and 1970s, when developments in probabilistic risk analysis and the rise and influence of ergonomics led to a merger of the two approaches in health and safety.

TABLE 4.1
Different Ages of Safety had Different Assumptions about the Causes of Accidents at Work

	Technological Age 1750–	Human Factors Age 1940–	Management Systems Age 1980–	Adaptive Systems Age 2000–
Focus	Focus on technology, hierarchy of controls	Focus on human behavior, errors, ergonomics	Focus on safety management systems and organizational factors	Focus on complexity, variability, and adaptations in daily work
Methods	Technical barriers, recruitment and training, design	Human performance on an individual and team level, decision making, human–machine interface issues	Safety assessments, human, organizational, and technical barriers	Modelling of system dynamics and interactions
Underlying theory	Engineering	Behavioral sciences	Open systems view, management science	Adaptive systems view
Assumptions	"Technology is the source of hazards to be controlled" "Safety is hazard control"	"Humans are the hazard to be controlled" "Safety is error-free performance" "Accidents have a root cause"	"There are human, technical, and organizational hazards" "Safety is management of human, organizational, and technical factors" "Accidents can have multiple contributing factors"	"Complexity creates hazards that humans need to control" "Safety is the ability to handle variability and succeed under varying conditions" "Accidents and successes stem from the same factors"
Legacy	Hierarchy of controls, accident proneness	Human errors and violations, decision heuristics and biases, situation awareness	Safety management systems, safety culture, weak signals, latent conditions, normalization, drift	Resilience, adaptation and adaptive capacity, complex adaptive systems, Safety I and Safety II, sensemaking

An early contribution to the safety field came from Heinrich, who in 1931 proposed that accidents are not random, but they are rather caused by unsafe acts of humans. He also proposed the idea that for every serious accident (or "major injury"), there are many more minor accidents or near-misses, and underlying these there are even more unsafe acts of personnel (see Chapter 3). This "Heinrich's pyramid", although it is highly problematic and mostly false as a causal model, still lives in safety practice.[1] The Swiss Cheese accident model is an extension of Heinrich's ideas, as is Root Cause Analysis (see below). Approaches based on Heinrich's ideas have targeted the unsafe acts of personnel with the aim of preventing fatalities.

The most substantial development in safety perspective and models during the so-called second age of safety consisted in merging technological safety assessment and ergonomics/human factors studies. The Three Mile Island (TMI) nuclear power plant accident in 1979 (see Chapter 2) was a watershed moment for the Human Factors Age, especially in the high-risk industries. The concept of *human error* became established among researchers and safety professionals after TMI. Prior to this, errors were studied especially during World War II in order to develop user interfaces for aircraft and radars.

The partial meltdown of the nuclear core in the TMI was due, according to the accident analysis results at that time, to equipment malfunctions, human errors, and design problems. It became clear that it's crucial to consider the human aspect along with the technology in complex organizations.

In aviation and later in many other safety-critical environments, Crew Resource Management was implemented first in aviation after the high-profile accidents such as the Tenerife runway collision in 1977 indicated communication problems within the crew and between the crew and the air traffic control. These soft, or non-technical, skill programs are widely utilized nowadays in several safety-critical fields ranging from healthcare to the nuclear industry (Flin et al., 2008).

The idea of *defence-in-depth* was developed during the 1970s and 1980s. It is a safety strategy that uses multiple layers of protection to prevent accidents. If one measure fails, others still provide protection, lowering the risk of catastrophic failure. These redundant systems may include physical barriers, administrative controls, and automated safety mechanisms. Critical safety functions are typically backed up by secondary systems to ensure safety, often based on diverse operating principles, materials, and physical location. The assumption embedded in defence-in-depth is that safety is best centrally controlled. This assumption has been challenged by, for example, the Normal Accident Theory (see Chapter 3), HRO group (see Chapter 5), and the Resilience Engineering school (see Chapter 11).

The age of Human Factors continued its development from the late 1970s till the 1990s, and many of the developed concepts and tools are still used today. It also produced general information about human decision-making (see Chapter 2) and errors in organizations (see Chapter 3) and ways of identifying them, guarding against them, and mitigating their consequences.

Similar to the "second age" of safety, when it was gradually discovered that technical risk assessment and prevention were insufficient for managing safety, in the 1980s, dissatisfaction arose with the idea that safety could be improved simply by reconciling the individual and the machine. This ushered in the third age of safety, the *Management Systems Age*.

The third age focused strongly on the operations and management of organizations. The development of safety management and safety management systems were key themes of this era. In addition to safety management, the Chernobyl nuclear power plant accident of 1986 led to the development and application of the concept of safety culture (see Chapter 5). Safety leadership was also introduced as an essential part of safety culture and safety management systems.

James Reason (1990, 2000) created the well-known Swiss Cheese model of accidents during the 1980s and the 1990s. According to the model, accidents are rarely caused by a single failure, but rather by a combination of multiple, often minor, mistakes or weaknesses that align in a way that allows a disaster to happen. In the model, different *layers of defense* are represented as slices of Swiss cheese. Each layer has holes in it, symbolizing potential weaknesses or flaws. Under normal circumstances, these holes don't align between the layers, and the system remains safe. However, when the holes in different layers line up, it creates a direct path for an accident or failure to occur.

This alignment of weaknesses is what leads to catastrophic outcomes. People are "inheritors of system defects created by poor design, incorrect installation, faulty maintenance and bad management decisions" (Reason, 1990). The role of human error is "that of adding the final garnish to a lethal brew whose ingredients have already been long in the cooking" (ibid.). An important characteristic of the model that is often lost due to its static visualization is the idea that the "holes" develop and change gradually.

The introduction of latent conditions was an important addition to safety management. Reason (2016, p. 3) noted that:

> Latent conditions possess two important properties: first, their effects are usually longer lasting than those created by active failures; and, second, they are present within the system prior to an adverse event and can – in theory at least – be detected and repaired before they cause harm. As such, they represent a suitable target for safety management.

What the Swiss Cheese model brought to safety is the idea that accidents require multiple things going wrong, and that these issues are layered, starting from management decisions or design of technology – and that these defects can remain latent in the system, undetected, for a long time. There are several weaknesses in the model, though. It gives a simplified picture of how accidents occur, and it does not define the dynamics of accident causation: how different layers of defense interact with each other; how systems can adapt in response to unexpected situations; how the holes are formed and reformed; how different failures, adaptations, and interactions can combine to cause an accident; and how technical, organizational, and human layers differ from each other in terms of functionality. The model also looks at only one hazard, whereas all organizations have to deal with multiple hazards simultaneously (see *hazard control* later in this chapter).

Linear and epidemiological models of the Human Factors and Management Systems Age are based on the idea that organizations are rational or at least boundedly rational systems, where humans are the weakest link. Accidents are caused by a broken component (linear models) or a combination of several broken components (epidemiological models). Both models subscribe to the idea of safety management systems, auditing, and supervision as key safety mechanisms. They also embed an assumption that the barriers fail independently (Leveson, 2023).

During the 1980s and the 1990s, two schools of thought had a strong influence on the science of safety. Normal Accident Theory (NAT) and High Reliability Organizations (HRO) theory acted as a bridge between management systems and adaptive age. Another theory, Barry Turner's Man-Made Disasters (1978), was another influential model, specifically by introducing the concept of culture into safety science. NAT and HRO were presented in detail at the end of Chapter 3, and HRO will be presented at the end of Chapter 5.

The fourth age, the *Adaptive Systems Age*, emphasizes the phenomena of organizational activity and the importance of understanding organizations' daily reality and normal adaptations. The variability and flexibility of human activity, perceived as negative by previous eras, is perceived as an important characteristic of the safe functioning of organizations and, as such, it should not be eliminated (Hollnagel et al., 2006). We need to look at the dynamics (tensions, pressures, conflicts) within the system to understand how they develop, not only at individual outcomes labeled human errors with the benefit of hindsight. Furthermore, we need to look at actual work and how it is done, not only at various defenses and mitigation methods built around the work.

The Adaptive Systems Age highlights the importance of adaptiveness and resilience as essential skills for maintaining safety, necessitating a shift in viewpoint from one that sees human variability as a risk that must be managed to one that recognizes the variability as an asset for safety. The complexities and variations require adaptive behaviors that cannot be entirely dictated by procedures.

The idea of human error was reinterpreted. Sources of error were understood to be structural (aka system-induced), not personal.[2] The local rationality principle was introduced to safety science with the idea that the point was not to look for where people went wrong, but why what they did made sense to them at the time (Dekker, 2005). Errors and accidents were also discovered to be only remotely related. Accidents are caused by multiple factors, and the occasional human contribution to failure occurs because complex systems need an overwhelming human contribution for their safety. It other words, accidents are not the result of a breakdown of otherwise well-functioning processes, but they are rather structural by-products of the system's normal functioning (Dekker, 2011). Sidney Dekker (ibid.) highlighted that the enemy of safety is not human error, but rather the system's complexity.

Normal variability in some parts or elements of the system resonate with variability with other parts causing a stochastic event. Unexpected combinations create hazards. Organizations gradually drift and develop routines and normalize and simplify their environment until some previously recognized or completely new hazard actualizes. Accidents are normal by-products of doing normal business under normal pressures of resource scarcity and competition (Dekker, 2014, p. 135). This view emphasizes the dynamic and complex nature of organizations and the impossibility of describing them in detail in management systems or standard operating procedures. There is always a gap between work as imagined and work as done (see Chapter 3). Resilience Engineering and Safety II (Hollnagel et al., 2006; Hollnagel, 2014), are among the main concepts that characterize the adaptive age. Resilience Engineering is discussed in detail at the end of Chapter 11 and Safety II in Section 4.3 of this chapter.

4.2 SAFETY AS HAZARD CONTROL

Safety is always related to hazards and the risk of harm caused by them. Thus, to understand safety, we need to start with understanding hazards. Hazards can be classified in many ways. OSHA has five categories of workplace hazards: physical, chemical, biological, psychosocial, and ergonomic. ILO separates between physical, chemical, biological, ergonomic, psychosocial, and organizational factors, as well as accident hazards. Environmental protection agencies would add environmental hazards to the mix. ILO's "accident hazards" are also sometimes called "safety hazards", adding to the confusion between different hazards and different types of safety – and even complicating the question of what an accident is, if accident hazards are separated from some other hazards.

As safety depends on the hazard in question, safety looks very different if we are dealing with physical hazards in comparison to, for example, psychosocial hazards. When speaking of safety, we thus need to remember to ask "safe from what?" and "safety for whom?" And as we will see later in this chapter, safety requires more than absence of hazards and their effects. Thus, we also need to ask "safe to do what?"

Some classifications also include *behavioral hazards*. I do not think it is beneficial to treat normal variability in human performance as a hazard. The same reservation applies in principle to psychosocial issues such as bullying or poor working climate. For practical purposes and as a specific issue to be managed, it makes sense to treat psychosocial hazards as a separate (but interconnected) group of hazards. This group of hazards is something that raises a lot of discussion in the safety community, probably due to the close connection to traditional HR issues such as stress, and the fact that many safety professionals do not have education in psychosocial issues (see Chapter 8).

Hopkins (2000, p. 70) emphasizes the distinction between high probability and low severity scale hazards and low probability and high severity scale hazards. According to the traditional definition of risk, the two pose approximately equal risk to the company. Low severity scale and high probability hazards (minor hazards) include slips, falls, and trips that result in one person being injured. High severity and low probability hazards (major hazards) include explosions and major fires that can claim multiple lives. According to Hopkins, metrics that measure an organization's ability to manage minor hazards say nothing about an organization's ability to manage major hazards. As an example of a measure of minor hazards, Hopkins cites the Lost-Time Injury (LTI) measure. Different types of safety can be thought of as competing goals similar to the thoroughness and efficiency highlighted by Woods and Hollnagel (Hollnagel, 2014; Woods, 2009; see also Cameron & Quinn, 1999).

Safety is a concept that aims to describe the process of making sure that a certain type of hazard or certain types of hazards do not harm a certain type of entity. The fact that we have several hazards and various entities that can be harmed makes defining safety difficult. From a cultural perspective, the concept of safety turns out to be another narrative, a simplification that superficially eases communication and classification, but in fact hides potentially harmful aspects of reality. Cultural assumptions are another hazard to be managed!

Safety is often taken for granted, something that does not need to be defined in more detail. This has led to simplifying metrics and misdirected focus. A wrong focus can mean a focus on wrong safety, as happened at BP (see end of this chapter). At minimum, we can distinguish between occupational safety, process safety, fire safety, environmental safety, information security, and physical security. In each specific high hazard domain, there is also a type of safety that is of concern, for example, patient safety in healthcare, flight safety in aviation, rail safety in rail transport, or nuclear safety in the nuclear industry.

Consider the following example events and what type of safety or safeties they deal with:

- Severe injury from falling when working at heights;
- Process release of flammable gas to the environment in a processing plant;
- Multiple injuries resulting from chemical material during transportation;
- Major fire in a corporate headquarters;
- Process event forcing a power plant to an emergency shutdown.

The first example is clearly an occupational safety issue, where the harm is directed towards people. The second example is already a little bit more complicated. It can be an environmental safety issue (especially if in hindsight no people were harmed), but it is also a process safety issue that threatens production and infrastructure. In all likelihood this example would be classified as a process safety issue, but if the release would be less flammable but still hazardous to environment, the attribution would probably change. The third example concerning exposure during transportation can be inspected from multiple angles depending on the actors: if the exposed people were from another company handling hazardous material transported by a contractor, the organization in question would probably treat the matter as an occupational safety incident. The transportation company can look at the issue from the safety of logisctics perspective (equals process safety for transportation companies), but if the driver was among the exposed, the issue becomes an occupation safety issue too. The two final examples, the fire and process event, are at least as complex as the transportation example, and I leave them to the reader. The point should be clear by now: that our view on safety depends on both who we are protecting and from what.

Table 4.2 shows the five main issues we typically protect from (people, environment, production, infrastructure, information[3]) and four hazard types that threaten them.

Table 4.2 makes a somewhat simplified distinction between:

a) *Core task related hazards* – these are hazards that relate directly to the core processes of an organization and the technology it utilizes, the chemicals it uses, the products it designs or manufactures, and so on. These can also be called *system hazards*, but as a system can be defined on several levels of society, I prefer the term *core task related hazards*. Managing these hazards typically requires a centralized approach. Process hazards are a subset of core task related hazards.
b) *Local hazards* – these are hazards that relate to the interaction between the working conditions, work methods, and the people who conduct the work. These include falls, trips, electrocution, physical forces, chemicals, and so on. These hazards vary between workgroups and situations, and thus the hazards can be managed best in a decentralized manner. There are, however, potential links between the local and core task related hazards, and these need to be again coordinated in a centralized manner.

TABLE 4.2
Different types of hazards and what they threaten, primarily (X) and secondarily (x)

	Type of Hazard			
ASSET	**Local hazards**	**Core task hazards**	**Psychosocial hazards**	**External hazards**
People	X	X	X	X
Environment	X	X	(x)	X
Production	(x)	X	(x)	X
Infrastructure	(x)	X	(x)	X
Information	(x)	(x)	(x)	X

c) *Psychosocial hazards* – these are psychological, social, and leadership-originated hazards, and so on. They refer to issues that are difficult to observe as they take place inside the personnel's minds, such as stress, fatigue, or exhaustion. Their physical manifestations can be visible, though, including irritation or reduced cognitive skills. Even though the hazard is internal, the mechanisms by which it manifests are organizational and systemic. It is important to differentiate local hazards from psychosocial hazards due to the abstract nature of psychosocial hazards.
d) *External hazards* – these are hazards that originate outside the organization. They can be dangerous as such (e.g. floods, earthquakes) or they can facilitate the actualization of a core task related or local hazard (e.g. bad weather affecting landing conditions for an aircraft). The risks associated with these hazards are sometimes called "systemic risks".

In process safety,[4] the hazards and the risks to be managed are directly linked to the core task of the organization, for example, producing energy with nuclear power, producing chemicals, or operating on people with surgery.

This is by no means an exhaustive list of hazard types, and many hazard types fall between these three "idealized" types. For example, in many contexts, customer safety and process safety are intertwined. For example, in healthcare, customers are the process. In rail, aviation, and maritime, the combination of customer safety and process safety is called rail safety, maritime safety, and aviation safety, respectively. All these can be considered hazards that stem from the core task, and as such, most if not all employees influence how they are managed. Additionally, in each context, there are hazards that mainly threaten customer safety, but these typically at least potentially affect also production as well as occupational safety. Examples include railway or airplane passengers acting violently towards other passengers and/or the crew. Flying, in turn, would be a core task related hazard, even though an aviation accident threatens the pilots as well as passengers. Local hazards in aviation would include various occupational hazards for the crew as well as for the maintenance personnel and other service personnel.

Local hazards naturally vary in scope, and they have different potential for severe injuries and fatalities. This fact should be considered when utilizing the various Heinrich-inspired pyramid models depicting near-miss, incident, and accident frequencies (see the history of safety science above). This serves as a reminder that in addition to the type of hazard (local, core task related, psychosocial, external) it is always necessary to consider the exact hazard to understand the potential impact it may have on people and the sociotechnical system.

In *occupational safety*, the harm targets the workers. The hazards are not necessarily directly linked to the core task (although they might be). At a chemical plant, for example, the core task relates to the manufacturing of the chemical products, but many of the occupational hazards are not directly related to the chemicals, but to issues such as working at heights, lifting heavy loads, or operating heavy machinery. Common occupational injuries, such as cuts and bruises, may even result from activities not directly related to work at all, like walking up or down stairs. Insufficient management of those hazards may cause injuries to workers but not necessarily pose significant risk to operations or civilians (Grote, 2012; Reiman & Oedewald, 2007). There are typically several (relatively speaking) minor hazards to be managed in occupational safety, whereas in process safety, the challenge is to manage few major hazards. Hopkins (2019) has argued that occupational safety can be appropriately managed in a decentralized (local) way, but process safety requires a more centralized approach.

Another type of safety that is difficult to categorize according to the hazard type is "patient safety". Safety of patients is threatened by the inherent uncertainties associated with treatments and illnesses/ injuries (core task related hazards), but also to the local conditions, expertise, and cooperation of the medical personnel (local hazards). Finally, each individual patient is unique but comes from outside the system, bringing their history and specific vulnerabilities with them (external hazards).

Product safety is a type of safety that has recently been a source of increased attention (Maier et al., 2024). The hazards that threaten the safety of products can be considered primarily core task related hazards for those who design and manufacture them, but the extra challenge comes from the need to consider the future users and use contexts of the product.

The classification of hazards in this book is done from the safety perspective, but many of the hazards relate to security issues also. Security and safety are distinct topics with strong interconnections. While this book focuses on safety many of the topics and issues can be applied to the security domain as well. On the other hand, there are some distinct features of security domain that differentiate it from safety. One of the most important differences is that security deals mostly with intentional harm (cybersecurity is a borderline case here) and safety deals with non-intentional harm. However, as Chapter 3 demonstrated, violations and organizational wrongdoing are intentional and can be considered a threat from safety as well as security perspectives.

In many languages, such as in Finnish and Swedish, the word for safety and security is the same, which sometimes causes confusion. *Security* usually relates to intentional acts where the risks to manage are not those inherent in normal core task activities. When talking about security, the focus is on how intentional activities of either external or internal parties can threaten the assets of the organization. These threats include, for example, sabotage or a theft of hazardous materials or classified information.

Insider threats are a security issue that partly relates to psychosocial hazards but also forms its own distinct hazard type. It is also connected to core task related hazards, as these typically seek to either exploit the hazards inherent in the core task or otherwise harm the organization's ability to carry out its core task. Insider threats include violence, sabotage, theft (of information or material), and fraud.

The distinction between process safety, occupational safety, and security is also a contentious issue. This is because the measures taken to improve one of them do not necessarily support the development of the others. In fact, the different types of "safeties" may even conflict with each other. For example, the extensive security measures taken to protect chemical or nuclear plants from terrorist attacks complicate entering and moving at the plant area. In case of an occupational injury or a fire at the plant, the security procedures might slow down rescue operations. Furthermore, since money and human resources are not infinite, companies tend to emphasize one safety over the other – whichever seems more acute or otherwise salient at the moment. A balance of attention to all the safeties is not always achieved. For example, some process industries have focused extensively on process safety (where other organizations focused on occupational safety) and while doing so have implicitly downplayed the importance of environmental and occupational safety. This typically lasts until something negative happens, and the organization (over)reacts and directs attention and resources to the given safety issue – as they interpret it.

Many tasks include both local and core task related hazards. For example, maintenance of critical equipment that includes lifting heavy spare parts. A safety management challenge is to balance the risks and countermeasures and maintain an adequate awareness of the various types of hazards.

It is easy to see that the classification of different types of safety is like all human classifications: it is meant to simplify, but in so doing it creates arbitrary boundaries to phenomena that do not have natural borders. Fire is dangerous to people as well as infrastructure, including production technology. Additional sources of complexity are brought by the different regulatory regimes for the different types of safety. In some companies this has led to having multiple safety management systems, one for each type of safety – and another one for quality. One can imagine the difficulties of implementing several safety management systems in an organization, but at least for the various regulators it is probably easier to verify them one system at a time (see Chapter 6) – as long as one is satisfied with Work-as-Imagined.

Different types of safety interact in practice. In April 2023, one person was killed and 19 passengers taken to hospital when first a freight train and then an inter-city train collided with a road-rail excavator and derailed near Voorschoten railway station in the Netherlands. The operator of the excavator was killed in the accident. The accident in Voorschoten took place during scheduled maintenance work. The work was performed in a four-track section. Two tracks were temporarily taken possession of to reach and leave the worksite. On a four-track section, trains may continue to run during maintenance (on two of the four tracks). In its investigation, the Dutch Safety Board (2024) *notes that railway infrastructure managers, railway companies, and contractors do not jointly and comprehensively analyze the risks of trains running alongside track maintenance. This results in an incomplete view of these rail safety risks. They focus mainly on occupational safety for workers. The investigation uncovered several other contributing factors, including the management of contractors, communication issues, and the risks of nighttime work. The example illustrates the systemic nature of accidents as well as the differentiation between local and core task related hazards. In this example, rail companies actually shared a core task and should have collaborated in order to manage the core task related hazards.*

Another example comes from the petrochemical industry. The accident investigation of the Texas City oil refinery explosion in 2005 revealed that the management had emphasized occupational safety and rewarded supervisors for low lost-time injury rates. The perception of a good safety level in the organization was based only on quantitative, lagging indicators of occupational safety. Process safety, however, did not get the same attention from the management, and many challenges related to the operation of the plant were left unnoticed (Hopkins, 2008). The Deepwater Horizon accident in 2010 had similar contributing factors, indicating that BP failed to learn from the Texas City explosion (see the end of this chapter).

The final example concerns aviation and the interaction of flight safety and security. In Chapter 3 we discussed the Germanwings flight that the co-pilot deliberately crashed in 2015. One of the reasons he was able to close the captain out of the cockpit was the fact that after the September 11 terrorist attacks in the United States, cockpit doors had been reinforced to prevent intrusion. The most plausible scenario at that time was that the person trying to gain access to the cockpit had malicious intent, but in this case, it was the captain trying to save his aircraft.

Safety as hazard control is a necessary but limited view to safety. It easily leads to seeing safety as absence of hazards (even though that is not what hazard *control* means), and a focus on either technical or behavioral solutions. These challenges are apparent in the widely used concept of defense-in-depth and the Swiss Cheese model (see above in this chapter), which both can be considered to be based on the idea of safety as hazard control.

This view needs to be complemented with a systems view of safety as organizational capability and a positive presence in the organization.

4.3 SAFETY AS CAPABILITY

Many safety models implicitly include the idea of safety as an absence of something or the lack of deficiencies; for example, the fewer the number of lost-time incidents, human errors, or unplanned outages or other events, the higher the safety level at the organization. Safety is then seen as the opposite of incidents and accidents. This definition does not explain how safety is achieved, but it implies that the absence of accidents proves safety.

Defining safety as an absence can lead us to false conclusions about safety of our sociotechnical systems. For example, NASA's Space Transportation System was not safe before January 1986, even though no shuttle had yet been lost. Thus, the system can be unsafe even though it is still free from accidents and losses. This notion reveals the problem of defining safety based on the (absence of) negatives, which directs the focus of safety work towards past outcomes: losses, injuries, and known hazards.

When looking at *safety as the organizational capability to succeed under varying conditions*, the focus on activity widens. Management focuses on increasing the potential for the organization to cope with the work daily. Safety requires the presence of certain attributes in the organization. This presence cannot be achieved by a strategy of control and limitation alone; it also requires the proactive development of the capabilities of the system. When safety is understood as something other than the absence of incidents and adverse events, the methods and models of safety management should also be able to focus on the positive side of safety – on the presence of something (Hollnagel, 2008). Hollnagel (2014) calls this approach Safety II, and differentiates it from the more error and individual focused Safety I.

Weick and Sutcliffe (2001), building on Weick's (1987) original ideas concerning reliability, note that safety is elusive because it is a dynamic non-event – it is produced by constant change and mutual adjustments. The problem is that when a system is operating safely and reliably, this outcome (non-event) does not attract attention. This leads to the paradox that the safer the system, the less there is to learn. It also means that safety needs to be achieved every day; it is not a stable outcome that can be achieved once and then forgotten.

Safety is created through proactive resilient processes, not only through reactive barriers and defenses (Woods & Hollnagel, 2006). These processes must enable the organization to respond to various disturbances, to monitor its environment and its performance, to anticipate disruptions, and to learn from experience (Oedewald & Reiman, 2003; Hollnagel, 2008).

The same processes and behaviors that create safety sometimes create accidents. The key point is that things succeed much more often than they fail, and safety efforts should focus on facilitating those successes rather than merely trying to prevent failures. Human performance – and safety – relies heavily on people's ability to navigate complexity and adapt their behavior as circumstances dictate (de Boer, 2021). This adaptability and flexibility of human work is the reason for its efficiency and safety. However, it is also the reason why failures sometimes occur, without or only rarely being the cause of such failures (Hollnagel, 2006, p. 13). This is due to the fact that unsafe states or accidents may arise due to normal system adjustments rather than abnormal failures. Hollnagel (2006, p. 14) points out that failure is the flip side of success, and therefore a normal phenomenon (see Chapter 3 on accidents and failures).

A safe organization needs adaptive capacity. Woods (2019, p. 53) defines adaptive capacity as follows:

> Adaptive capacity is a system's readiness or potential to change how a system currently works – its models, plans, processes, behaviors, relationships – to continue to fit changing situations, anomalies and surprises.

In Chapter 3 it was noted that to understand accidents, we need to separate individual behavior from cultural patterns. Individual behavior brings an additional source of variance to systems, which can be positive or negative from the perspective of safety. Surprises can happen in the system due to individual performance variability or various external factors. For this reason, a safe system has adequate resources for reacting to various contingencies, both expected and unexpected.

As was illustrated in Chapter 3, one of the major challenges in complex organizations is to know what is significant at the given moment, and where to focus the effort to guarantee future success. As we will discuss in the next chapter, a cornerstone of a healthy "safety culture" is the ability to identify significant issues and give them the attention warranted by their significance (IAEA, 1991).

Allan McDonald, director of the Space Shuttle Solid Motor Project at Morton Thiokol at the time of the Challenger *accident (see Chapter 2) noted how the attention was not on the O-rings before the accident:*

"The O-ring erosion problem was not something that was occurring on every flight. Nor was it considered by either NASA or Morton Thiokol to be nearly as critical as the pocketing erosion of the nozzle, which was still the primary topic of discussion for every flight readiness review. Compared to the nozzle-erosion problem, and compared to the issues associated with the development of the FWC-SRM,[5] *the matter of the O-ring was still on the backburner" (McDonald & Hansen, 2009, p. 31).*

Safety is not an end-state or a stable outcome of certain predefined factors. Optimization of individual components or subsystems will generally not lead to a system safety optimum. Safety is a side-effect which emerges while different parties interact, and normal work activities are carried out. This definition of safety has its origins in systems theory. The important implication of this view is that safety itself is not the phenomenon to be controlled and managed, but rather the sociotechnical system is. Reiman and Oedewald (2007) have stated that safety emerges when an organization is willing and capable of working according to the demands of its task, and when personnel understand the changing vulnerabilities of their work.

This view pinpoints that to improve safety, managers and safety practitioners need to understand how the system works in practice. Safety is an emergent property of the sociotechnical system, and management efforts need to be directed at the entire system, not on the emergent property as such. As we will discuss in Chapters 5 and 6, many safety management and safety culture approaches tend to consider safety as a factor separate from, or only loosely connected to, the work, organizing, and technology. This misunderstanding of the nature of sociotechnical systems leads to measures that further disconnect safety from work, while creating a façade of various management systems and barriers that in fact increase the overall complexity of the sociotechnical system. The view of safety as a separate goal also fails to consider, or seeks to solve in simplified manner, the inherent tensions in complex organizations.

4.4 SAFETY AS BALANCE

At the end of Chapter 1, we presented the view of modern organizations as complex adaptive systems. We emphasized that the management of these complex systems is an inherently contradictory activity and requires balancing between various tensions, competing demands and irresolvable dichotomies (McMillan, 2008). The essence of managing safety in complex systems is balancing between various conflicting approaches and principles (Reiman et al., 2015; cf. Jeschke, 2022):

1. Promote a global identity and shared core task (*system goals*) – support the realization of local and acute goals (*local goals*)
2. Create standard operating procedures and define roles (*structure*) – learn new and create capability for situational self-organization (*exploration*)
3. Monitor system activities and reduce variance (*uniformity*) – facilitate novelty and increase variance (*diversity*)

4. Set objectives and prioritize (*directive approach*) – facilitate interaction and involve personnel (*participative approach*)

The first tension highlights the potential conflicts between *global versus local goals*. Due to the tendency of individuals in complex adaptive systems to self-organize, top-down control is not sufficient in all situations (e.g. Reiman et al., 2015). Organizational goals and objectives form a general guideline for personnel when they act or make decisions. However, local conditions and acute issues always need to be considered. Too heavy an emphasis on system goals leaves too little room for maneuver for units, teams, and individuals. This means their adaptive capacity (Woods, 2019) to respond to new or unexpected situations decreases. Too much local freedom and local practices drift from the system goals. This decreases adaptive capacity on the system level. To effectively manage safety, the system needs to have an ability to meet the local requirements as well as overall long-term goals.

The second dimension deals with the tension between structuring activities based on the existing understanding of risks and the system boundaries versus exploring the system, learning and crossing the boundaries. In a safety-critical context, a complex adaptive system requires normative boundaries to contain its self-organizing tendency (Reiman et al., 2015). In practice, this is achieved through the definition of procedures, instructions, rules, regulations, and other more or less rigid structures. This simplifies reality and helps employees to avoid ambiguity and uncertainty in their work by having ready-made solutions available to various organizational contingencies. On the other hand, safety also requires the ability to appropriately adapt to various situations and unanticipated disturbances. And to keep the system performance stable, change is needed within the system. Thus, maintaining safety requires change and exploration as well as stability and structure. Farjoun (2010) notes that "to survive and prosper, organizations must reconcile stability, reliability, and exploitation with change, innovation, and exploration".

The third tension focuses on *uniformity versus diversity* within the organization. Enough diversity and variation in an organization contributes to the development of a richer understanding of the state of the organization and creates innovative options for behavior and decisions. Thus, facilitating novelty and diversity in the organization is essential for its development (Reiman et al., 2015). On the other hand, not all behaviors are acceptable (e.g. risky practices, harassment, hiding of information), which means that the organization also needs to make sure there are adequate uniform requirements for behavior and shared values and attitudes concerning, for example, safety. This relates to Chapter 2 and the tension between agreement and conflict as well as to what Turner (1978, p. 163) called "the problem of integration" (see Chapter 3).

The fourth tension relates to the *directive versus participative approach* to managing the organization. On one hand, defining objectives and prioritizing them within a selected group people (e.g. top management or other leaders) is essential to effective safety management because it allows making decisions that might not be accepted democratically by the personnel, but which are important for the development of the organization or for achieving essential organizational goals (Reiman et al., 2015). Decentralization and participative management taken to extremes would also mean communicating everything to everyone else, which Turner (1978, p. 197) already pointed out would be impossible. Centralization builds on shared decision-making premises, limited delegation of authority, and formal information flows (top-down control) whereas decentralization relies on more emergent processes and bottom-up influences.

In essence, safety depends on balancing between these dimensions (see Figure 4.1).

It is imperative that culture does not constrain human behavior by enforcing concepts and norms of conduct that are too narrow. An organization's culture should provide a common set of decision principles and shared norms and guidelines yet leave room for situational adaptation and individual initiative.

The tensions in Figure 4.1 cannot ever be solved completely. Rather they are inherent features of complex organizations.

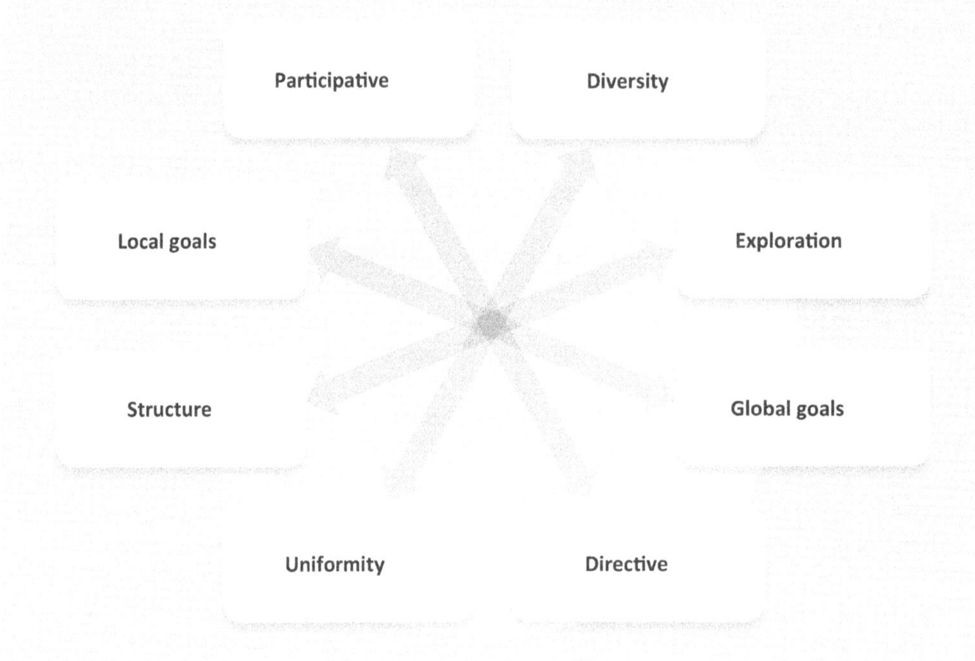

FIGURE 4.1 Conflicting requirements of managing safety in a complex system

Seeing *safety as a wicked problem* (Rittel & Webber, 1973) means that the problem of safety is unique for each and every organization, even in the same industrial domains. There is no one solution that fits everyone. Caution is needed when implementing various industry best practices. Learning and safety development should involve a qualitative increase in understanding rather than a quantitative increase in the number of tools at the organization's disposal. There is nothing wrong with having several tools, and those tools will also be presented throughout the book, but having the understanding to use them in a manner that contributes to safety is the real trick, as is realizing the limitations of the tools, including when and how to use them.

Wicked problems are a characteristic of complex systems (Rittel & Webber, 1973). They are typically difficult to formulate precisely, and correspondingly it is difficult to pinpoint when they have been solved. What is more, every attempt to solve them changes the nature of the problem, making learning by trial and error difficult. This also means that intuition (see Chapter 2) can lead us astray when applied to decision-making in complex organizations (Kahneman & Klein, 2009, p. 520).

In complex organizations, it is impossible to predict everything. Instructing too precisely may even lead to paralysis of the system in the face of unexpected situations. To compensate for this, some organizations also provide general instructions on how to act in situations for which there are no instructions. However, merely instructing something is never enough: the organization must also have sufficient competence, resources, cooperation, and coordination skills to cope with situations. The importance of employee engagement and creating a safe atmosphere is also emphasized in environments where everyone's contribution is critical to the overall operation (Edmondson, 2018).

In complex social systems, everything influences everything else. Nonetheless, certain influences hold more weight than others, and some become meaningful only when they occur alongside others and resonate together (cf. Hollnagel, 2014). Due to organizational complexity, many effects are understood only in hindsight. A complex systems approach is essential to balance the cultural approach, not for comfort or simplicity, but to reveal the pitfalls of oversimplification and linear thinking that culture formation naturally leans toward.

In complex systems there is a need for daily adjustments to match the work as imagined (by designers, planners, and writers of procedures) and the actual conditions under which the work is carried out. These conditions can never be known exactly in advance, and thus there is always a need to adjust activities based on the situation at hand (Hollnagel, 2015). The existing rules and practices may not apply in the new situation, thus requiring activities to be decided on the spot. Karl Weick (2009) has borrowed the French term *bricolage* (tinkering), first used by anthropologist Claude Levi-Strauss to describe how the natives (employees) utilize whatever tools and resources they have at hand and recombine them into new cultural products.

Management should aim at developing practitioners' skills in judging when and how to adapt guidelines to local circumstances (Dekker, 2015), how to make do with whatever resources and tools they have at hand (Weick, 1993), how to take action even when (and especially when) circumstances are unclear (Amalberti, 2013, p. 39), and how to not only passively adapt to the environment but also actively affect the environment. La Porte and Consolini (1991) noted that one of the characteristics of a High Reliability Organization (see Chapter 5) is the ability of the organization to reconfigure from centralized to decentralized structures in times of crisis.

Safety is not just about following instructions, and it is not achieved solely by restricting operations through various technical or administrative barriers. Safety is a broader entity that is connected to how the organization operates and fulfills its core task. Instead of the absence of accidents, safety is the organization's ability to function in varying situations. This ability refers to an organization's capabilities to anticipate, adapt to, and recover from various expected and unexpected challenges. Building adaptive capacity involves anticipating and preparing for potential safety challenges before they occur.

In complex systems, the same adjustments and trade-offs that make everyday work go right can, under different conditions, combine into failure, or an accident. Hollnagel's (2014) Safety-II is grounded in the observation that things succeed far more often than fail. Accordingly, safety management should consider how everyday work succeeds, with a focus on Work-as-Done rather than Work-as-Imagined and reinforce the conditions and adjustments that enable success.

Successful adjustments depend on effective sensemaking. It needs to build on a good understanding of the core task as well as the technology and the hazards that need to be taken care of (Reiman & Oedewald, 2007). Successful self-organizing requires adequate information, clear identity, and well-built relationships (Knowles, 2002). It builds on adaptive sensemaking (see Chapter 2) that allows both honoring and rejecting past beliefs and practices while interpreting an abundance of data into actionable knowledge (Weick et al., 2005, p. 415).

The key differences between earlier safety models (linear models) and the system models of the adaptive age can be summarized as follows (Table 4.3).

Definitions given in Table 4.3 can be explicit in the organization's formal statement or the management system, but typically they are implicit assumptions and embedded in the routines and other structures of the organizations. They affect how we make sense of work and its outcomes, and how we learn.

4.5 SUMMARY – AN ADAPTIVE VIEW ON SAFETY

The chapter explored the evolution of thinking on safety, emphasizing the role of humans in safety, and the development of various safety models and approaches over time. Thinking on safety has progressed from a technical perspective to an adaptive view. However, safety remains inadequately defined and often ambiguous concept. This highlights the futility of "safety first" slogans without clarification of what safety means in the particular context. Safety is related to various hazards, and its understanding requires asking "safe from what?" "safety for whom?" and "safe to do what?"

Safety was proposed to be an emergent property of the sociotechnical system. As an emergent property, its management requires understanding and managing the entire sociotechnical system,

TABLE 4.3
Two Ways of Looking at Safety (e.g. Dekker, 2011; Hollnagel, 2014; Leveson, 2023; Reiman & Rollenhagen, 2011) Make Different Assumptions about Key Concepts

Topic	Linear models	System models
Human errors	Human error is the cause of trouble.	"Human error" is an attribution given to unsuccessful behavior after-the-fact. It needs to be understood in context.
Accidents	Accidents are caused by chains of directly related events, which are often caused by human errors. A root cause can always be found.	Accidents are complex processes involving the entire sociotechnical system. There is seldom a single root cause.
Causes of accidents	Accidents indicate the system is ill. There is a major cause for major accidents.	Systems will tend to migrate toward states of higher risk – accidents are by-products of systems' normal functioning, with ordinary factors as causes.
The human factor	People cause accidents.	People create safety – by organizing, and in interaction with technology.
Learning	The best way to learn about safety is to check how many things are going wrong.	The best way to learn about safety is to learn how the sociotechnical system really behaves, how it succeeds, and how it fails.
Safety	Safety is achieved when everyone follows procedures, and no errors or violations are made.	Safety is the ability of the organization to cope with changing situations – its adaptability.
Leadership	Leadership means a position of authority and directing others. A leader knows what is best for the organization and for the employees; a leader decides.	Leadership means influence, regardless of formal position. A leader creates possibilities for the experts to work.
Change	Change is conceived at the top and is driven down the system. Change can be managed.	Change may emerge and/or be driven from anywhere within the system.
Culture	Culture is about attitudes; it is inside a person's head, directing behavior.	Culture is relational; it is about interactions and the shared beliefs and assumptions that both result from and steer interaction.

its work and inherent hazards. Safety cannot be understood in isolation of the system or by looking at its constituent components independently. Safety cannot ever be solved completely; it is a chronic issue.

A system with no incidents may not necessarily be safe since safety requires a presence of capabilities rather than an absence of incidents.

Safety can be viewed as a wicked problem, unique to each organization, requiring continuous learning and adaptation rather than relying on one-size-fits-all solutions. Like all wicked problems, the problem of safety cannot be solved; it can only be temporarily tamed.

Reflection Questions

How do the various ages of safety reflect in how your organization manages safety? What age is the most prominent in your organization?

Is safety defined in your organization?

How many different safeties do you manage in your organization? Are the main hazards local, core task related, psychosocial, or external?

Is it clear how the various safety management activities affect safety? Have you differentiated between what safety is affected by which method?

How do you know your organization is safe without relying on proof of absence (e.g. lack of incidents or accidents)?

How do you differentiate between safety and security in your organization?

Are there synergies or conflicts between safety and security in your organization?

BP TEXAS CITY REFINERY AND DEEPWATER HORIZON 2010 ACCIDENTS[6]

The accidents at BP Texas City refinery in 2005 and BP's Deepwater Horizon offshore platform in 2010 share some similarities when looked at from a cultural perspective. They both illustrate the problems of defining safety and the unintended consequences of rewarding. In addition, they both demonstrate failures in learning from the past (Hopkins, 2008, 2012).

BP Texas City Refinery Accident

The BP Texas City Refinery explosion on 23 March 2005 resulted in 15 deaths and more than 170 injuries and was among the most severe industrial accidents in the United States for decades. The refinery was the third largest oil refinery in the United States. It was originally built in 1934. BP acquired the refinery as part of its merger with Amoco in 1999.

The isomerization unit[7] (ISOM) was started after a maintenance outage and petrol was led into the raffinate splitter (distillation column/production tower). The ISOM startup procedure required that a level control valve on the raffinate splitter tower is used to send liquid from the tower to storage. However, this valve was closed by an operator and the tower was filled for over three hours without any liquid being removed. This led to flooding of the tower and high pressure, which activated relief valves that discharged flammable liquid to the blowdown system. The escaping petrol was released from the top of a blow-down stack. The stack did not have a flare for safely burning excess gases. Petrol fell to the ground where it accumulated as a large vapor cloud. The vapor cloud eventually encountered an ignition source (most likely an idle car) and exploded.

During the preparation for the start-up, many safety checks were omitted due to the pressure to startup the facility. For example, the BP Pre-Startup Safety Review (PSSR) procedure should have been used to ensure that complete and thorough technical checks were carried out and that all non-essential personnel were clear during the startup operation. However, the process safety coordinator was unfamiliar with the applicability of the PSSR, and therefore, no PSSR procedure was conducted. Many other checks were signed off as being completed even though they were omitted or ineffectively carried out.

The startup process lasted for several hours, and there was a shift change once the raffinate section equipment was filled. There was no proper shift handover, and the status of the startup was miscommunicated when the night shift left and the day shift started. The Day Board Operator did not realize how far the startup had already progressed. Further, the Day Supervisor arrived for work more than an hour late, at 7:15 am.

The splitter tower had two separate level alarms. However, when the raffinate splitter tower was filled beyond the set points of both alarms to a level reading of 99% on the transmitter in the early morning on March 23, 2005, only one alarm was activated. Filling the bottom of the tower until the level transmitter read 99% was not unusual, even though the startup procedure called for the level in the tower to be established at 50%. Operations personnel explained that additional liquid level was needed in the tower because in past startups the level would typically drop significantly later. To avoid losing the liquid contents of the tower and potentially damaging equipment, operators typically operated the tower level well above 50%. For this reason, they ignored the alarm.

The tower instrumentation continued to show a liquid level less than 100% of the range of the transmitter. The level sight glass, used to visually verify the tower level, had been reported by operators as unreadable because of a buildup of dark residue; the sight glass had been non-functional for several years. Knowing the condition of the sight glass, the Day Board Operator did not ask the outside crew to visually confirm the level.

As the unit was being heated, the Day Supervisor, an experienced ISOM operator, left the plant at 10:47 a.m. due to a sudden family emergency. The second Day Supervisor was devoting most of his attention to the final stages of a startup of another column. He had very little ISOM experience and, therefore, did not get involved in the ISOM startup. No experienced supervisor or ISOM technical expert was assigned to the raffinate section startup after the Day Supervisor left, although BP's safety procedures required such oversight.

The Day Board Operator continued the liquid flow to the splitter tower but was unaware that the actual tower level continued to rise. All three relief valves opened at 13:14 and stayed open for over six minutes. The blowdown drum and stack was overfilled in less than five minutes after relief valves opened.

Contributing Factors to the Texas City Accident

There were several contributors to the accident, including failure to learn from previous incidents as well as insufficient senior management commitment and support to managing process safety issues. Senior managers were subjected to individually constructed performance agreements that emphasized the financial performance of the business unit, with no process safety indicators. They did not understand the distinction between occupational safety and process safety and relied on personal injury rates as a safety indicator (CSB, 2007).

Bazerman (2022), who consulted BP right before the accident, noted that the senior management did focus on safety but they simultaneously set clear and measurable goals mostly on cost cutting issues only. As a result, the production sites prioritized cutting costs and overlooked the less quantifiable goals.

This was not the first time the operators had overfilled the tower. On six previous occasions in the preceding ten years there had been releases from the same tower, producing vapour clouds at or near ground level. The refinery failed to investigate and learn from these near-misses. The refinery did not have an open climate for reporting. The Baker Panel report notes that many employees feared repercussions if they raised issues, and others did not report because they believed that the refinery would not act on the reports. Further, potential safety effects of changes involving people, policies, or the organization were not assessed. Numerous surveys, studies, and audits identified deep-seated safety problems at Texas City, but no effective corrective actions were carried out. A "check the box" mentality was prevalent at Texas City, where personnel completed paperwork and checked off on safety policy and procedural requirements even when those requirements had not been met. Outdated and ineffective procedures did not address recurring operational problems during startup, leading operators to believe that procedures could be altered or did not have to be followed during the startup process.

The process safety management (PSM) program had serious deficiencies and a growing backlog of delayed or cancelled activities. Cost-cutting and failure to invest in the 1990s by Amoco and then BP left the Texas City refinery in poor technical condition. BP targeted budget cuts of 25% in 1999 and another 25% in 2005, even though much of the refinery's infrastructure and process equipment was in disrepair. Also, operator training and staffing were downsized. The central training department staff had been reduced from 28 to 8, and simulators were unavailable for operators to practice handling abnormal situations, including infrequent and high hazard operations such as startups and unit upsets. The technical condition of the plant was poor. Blowdown drums and atmospheric stacks were not replaced, even though a series of incidents warned that this equipment was unsafe.

The refinery was heavily focused on return on investment maximization. The site was generating in the order of 100 million dollars profit annually, but this delivered a rate of return on capital of less than 10%, which was generally seen as inadequate. The strategy adopted was cutting maintenance (leading to large maintenance back-log), cutting training, and cutting back on staff (leading to a lack of safety staff and a lack of additional competent staff during startup) in order to increase the return on capital. Furthermore, the budget priority categories in use at BP lacked a category for purely safety related or risk reducing expenditures. BP categorized investment into three categories: operation compliance, sustain production, or new commercial opportunities.

The process unit was started despite previously reported malfunctions of the tower level indicator, level sight glass, and a pressure control valve. The indicator showed that the tower level was declining when it was actually overfilling. The redundant high-level alarm did not activate, and the tower was not equipped with any other level indications or automatic safety devices. Further, the control board display did not provide adequate information on the imbalance of flows in and out of the tower to alert the operators. Supervisors and operators poorly communicated critical information regarding the startup during the shift turnover. There was a lack of adequate supervisory oversight and a lack of technically competent personnel during the startup, and the ISOM operators were likely fatigued from working 12-hour shifts for 29 or more consecutive days. Finally, the BP Board of Directors did not provide effective oversight of BP's safety culture and major accident prevention programs.

BP was unable to utilize its existing process safety competence. The safety organization was sidelined from the decision-making practices and information flow in general, and subsequently had a limited effect on the operational activities at the refinery. This was due to the highly decentralized structure of BP's operations.

Deepwater Horizon Accident in 2010

The April 2010 blowout in the Gulf of Mexico occurred during operations to "temporarily abandon" the Macondo oil well, located in approximately 5,000' of water some 50 miles off the coast of Louisiana. Mineral rights to the area were leased to BP, which contracted with Transocean and other companies to drill the exploratory Macondo well under BP's oversight, using Transocean's *Deepwater Horizon* drilling rig. This was a semi-submersible, mobile, floating, dynamically positioned oil rig. In addition to BP and Transocean, a third company, Halliburton, was involved in the event. They carried out essential cementing work.

The blowout was directly caused by the failure of the cement to seal the well and the failure of the safety mechanism, the Blowout Preventer (BOP), to activate. The crew at BP and Transocean also misinterpreted a critical pressure test, believing it to indicate that the well was sealed when it was in fact leaking.

The blowout caused an explosion killing 11 workers and igniting an inextinguishable fire that burned for two days before the entire rig collapsed. The well was left open, and the oil spill is regarded as one of the largest environmental disasters in American history. In September 2014, a U.S. District Court judge ruled that BP was primarily responsible for the oil spill because of its gross negligence and reckless conduct.

The blowout followed a failure of the cementing job to temporarily seal the well, while a series of pressure tests were misinterpreted to indicate that the well was in fact properly sealed. There were several safety systems that should have given advance warning of a blowout or should have promptly cut off the flow of oil. All these systems failed. The failures included a dead battery in the blowout preventer, suggestions of a breach in the well casing, and failure in the shear ram, a device of last resort that was supposed to cut through and seal the drill pipe in the event of a blowout.

Contributing Factors to the *Deepwater Horizon* Accident

BP was measuring its safety performance by occupation accidents (lost time incidents, LTI) and did not realize that LTIs revealed little about process safety – at Texas City a major program was ongoing at the time of the accident to improve (occupational) safety, and *Deepwater Horizon* had just won the company safety prize for seven years without reporting a lost-time incident.

Hopkins criticized the decision-making process of the Macondo engineering team for taking place extensively by email, and for their failure to adequately utilize the Management of Change process. For example, even though the process was formal and included reviewers for the decisions, these reviewers were the same people who had participated in the decision and thus did not provide any added value to the review process. Furthermore, the engineers were subordinated to the line managers in a decentralized arrangement. This meant the engineers were subject to the same financial incentives and pressures as the line organization (Hopkins, 2012).

Hopkins points out that the decentralized structure of BP's operations was already a contributing factor to the Texas City refinery explosions, since, citing the Baker Panel report, this structure left the refinery plant manager a tremendous amount of discretion in running the business. After Texas City, BP established a system of engineering authorities, who reported to a high-level engineering authority but were still accountable to the line organization.

The National Commission on the BP *Deepwater Horizon* Oil Spill and Offshore Drilling (United States Congress, 2011) noted strongly that the "immediate causes of the Macondo well blowout can be traced to a series of identifiable mistakes made by BP, Halliburton, and Transocean that reveal such systematic failures in risk management that they place in doubt the safety culture of the entire industry". The Chemical Safety and Hazard Investigation Board (CSB) found an "eerie resemblance" between the 2005 explosion at the BP Texas City refinery and the explosion aboard the *Deepwater Horizon* as well as some other accidents. CSB Chairman Dr. Rafael Moure-Eraso noted that "a number of past CSB investigations have found companies focusing on personal injury rates while virtually overlooking looming process safety issues – like the effectiveness of barriers against hazardous releases, automatic shutoff system failures, activation of pressure relief devices and loss of containment of liquids and gases. Furthermore, we have found failures by companies to implement their own recommendations from previous accidents involving, for example, leaks of flammable materials." CSB Investigator Cheryl MacKenzie concurred: "How safety is measured and managed is at the very core of accident prevention. If companies are not measuring safety performance effectively and using those data to continuously improve, they will likely be left in the dark about their safety risks."[8]

In addition to using wrong indicators to monitor their safety, there were also several other contributing factors to the accident. For example, information flow inside BP and especially between BP and its contractors (and between these contractors) was insufficient. The accident also questioned the adequacy of regulatory practices in the U.S. oil and gas industry, especially concerning the independence of the regulator from the industry it is regulating.

The Mineral Management Service (MMS) was established in 1982 under the U.S. Department of the Interior to regulate offshore drilling. In addition to regulation, it had an additional role in collecting revenue from oil and gas leases. Chernov and Sornette (2016) describe how the budget, competence, and authority of MMS dwindled under industry lobbying during the period leading up to the accident. Safety measures required by regulators in other major offshore drilling countries such as the United Kingdom and Norway were optional in the United States, including an acoustic emergency trigger for the blowout preventer, an environmental impact assessment, and an emergency plan in case things went wrong (as they did). MMS was dismantled after the accident and three new bureaus were formed, one for regulating, one for handling leasing, and one for collecting revenue.

What Did We Learn from the Two Accidents about Safety in Complex Organizations?

The challenges of measuring safety and dangers of oversimplified indicators were visible in both cases. By focusing on easily counted LTIs, BP missed monitoring the level of process safety.

BP failed to learn from previous incidents and even accidents. After Texas City, BP did not change its incentive scheme that focused on financial goals and occupational safety issues to the detriment of process safety issues. BP did reorganize its activities, but did not adequately centralize engineering and safety considerations.

Hopkins (2012) points out that even though BP did learn specific lessons from Texas City, this learning did not extend to other areas. For example, there were several cases of blow-related incidents before the Macondo accidents that could have taught BP how unreliable the blowout preventer was. However, these lessons were not heeded. BP thus did not learn to learn.

Organizational structures can also create barriers to communication, especially structures between organizations. For example, Halliburton had numerous challenges with the stability of the cement slurry in their tests before the accident, but they did not communicate these challenges to BP (Chernov & Sornette, 2016, p. 120).

Overall, the accidents show that organizing and organizational structures have an effect on safety. Management of major hazards (core task related hazards, see earlier in this chapter) requires a centralized or matrix structure to compensate for local pressures, goals, adaptations, and drift. Independent review and challenge are needed to ensure safety. These issues are elaborated in Chapters 6 (organizing) and 8 (independent challenge).

NOTES

1. A good critique of Heinrich is provided by Manuele (2013). An indepth analysis of Heinrich's legacy has been conducted by Busch (2020).
2. See also Reason (1997) for an early attempt to redefine errors as latent conditions
3. Reputation is sometimes mentioned as a sixth topic to be protected. This brings an additional angle to safety in terms of politics and public relations, but it is outside the scope of this book.
4. The term *process safety* has its roots in the chemical industry and is widely used in the oil, gas, and chemical industries. In the nuclear power industry, the term *nuclear safety* refers to the same phenomenon. In aviation the term is *flight safety* and in healthcare *patient safety*.
5. FWC refers to the new, lighter, SRB design (filament-wound-case), which was to replace the heavier steel case currently used, and increase the payload of the Shuttle. The project was cancelled after the *Challenger* accident.
6. Descriptions of the accidents are based on Hopkins (2008, 2012), BP Baker Report (2007), CSB (2007, 2014), US Congress (2011).
7. The isomerization unit is used to boost the octane level of gasoline – it rearranges the structure of molecules to turn naphtha into high-octane gasoline.
8. There was also an interesting interaction between the occupational safety measures and the evacuation measures. As described by Smith (2024, p. 153), the rig had banned the use of pocket knives due to hand injuries, yet during the evacuation, the crew faced challenges with tethered life raft as there was a shortage of cutting devices to sever the tether and get away from the burning platform. Banning the pocket knives surely made an impact on the minor injury statistics but threatened emergency operations in case of a major accident. Smith (2024) calls this the trade-off between trivial (but easy to identify and measure) and critical (requiring more work to understand and difficult to measure).

5 Safety Culture

This chapter examines the concept of safety culture, which is widely used in many safety-critical domains but also widely critiqued. The chapter suggests that the concept has some value, particularly as a normative model outlining issues that a healthy culture for safety should address. The chapter concludes by presenting an organizational safety model that emphasizes the significance of culture: High Reliability Organizations theory.

5.1 SAFETY CULTURE – A REMEDY OR AN EXPLANATION?

Safety culture as a concept originates from the nuclear industry, where it was first used to explain the Chernobyl nuclear accident in 1986. It was for the first time defined in IAEA (1986) "Summary Report on the Post-Accident Review Meeting on the Chernobyl Accident" as "that assembly of characteristics and attitudes in organizations and individuals which establishes that, as an overriding priority, nuclear plant safety issues receive the attention warranted by their significance". Several other definitions have followed. They all share similar components, but some disagreements and a lot of confusion still exist. Despite the ambiguities, the concept has become established in usage in many different industries. Some industries, such as the nuclear industry, nowadays require that companies regularly assess and develop their safety culture. Investigations of accidents in other industries have also identified cultural issues as contributing factors and have sometimes explicitly mentioned deficient safety culture as one of the root causes. Examples of these high-profile accidents include the King's Cross underground fire in London (Fennell, 1988), the explosion in 1988 on the offshore oil and gas production platform *Piper Alpha* in the North Sea (Cullen, 1990), and the *Columbia* Space Shuttle accident (CAIB, 2003). More recent examples include BP *Deepwater Horizon* (United States Congress, 2011) and Boeing 737 Max (DeFazio & Larsen, 2020).

The Chernobyl nuclear power plant accident was the worst nuclear accident in history in terms of costs and casualties.[1] *It is one of the two nuclear accidents classified as a level 7 event (the maximum classification) on the International Nuclear Event Scale (INES). The other is the 2011 Fukushima Daiichi nuclear accident in Japan (see Chapter 6). The Chernobyl accident occurred on April 26, 1986 in the Chernobyl Nuclear Power Plant situated in the village of Pripyat in the former Soviet Union (present-day Ukraine).*

Due to the rising electricity needs in the Soviet Union, the construction and operation of the RBMK reactors was started before the design was complete.[2] *One of the unsolved issues concerned potential power grid failures. Since cooling pumps require electricity to cool a reactor after a scram, in the event of a power grid failure, Chernobyl's reactors had three backup diesel generators. These could start up in 15 seconds, but they took 60–75 seconds to attain full power. To solve this one-minute safety gap, it had been theorized that rotational energy from the steam turbine (as it wound down under residual steam pressure) could be used to generate the required electrical power. Thus, the test was part of an engineering R&D project to develop the RBMK reactor type, even though it was already in operation in multiple sites around the Soviet Union.*

The test required shutting down some safety systems (e.g., the automatic reactor shutdown mechanisms) and operating the plant at a lower power than was normal. Since the city of Kiev was preparing for the May festivities, they required power for the night of 25 April. The test had to be moved to past midnight. Preparations for the test were continued, and power was lowered. During the preparation, the reactor power unexpectedly dropped to a near zero level. The operators were able to restore the power level by, for example, raising the control rods – unaware of the critical design error in the control rods and a classified document warning against raising all control rods from the reactor. The low power level led to a release of xenon-135 gas in the reactor, making the reactor unstable.

DOI: 10.1201/9781003491156-5

Due to the delay, the shift responsible for making the preparations and initially scheduled to conduct the test had to leave. A new shift then arrived to complete the test. Due to several factors, including the presence of xenon-135 gas, the shift had difficulties in maintaining stable reaction. Immediately after the test was (successfully) completed, the operators executed emergency shutdown of the plant by lowering all the control rods into the reactor. This action caused a dramatic power surge due a design error in the control rods combined with the inherently unstable nature of the reactor design. The destruction of the reactor happened 6–7 seconds after the operator pressed the scram button to insert all the control rods into the core. Fatality estimates of the accident vary considerably, from 4,000 to more than 10,000 fatalities.

IAEA's International Nuclear Safety Advisory Group (INSAG) reviewed the evidence concerning the accident and concluded that the "root cause of the Chernobyl accident ... is to be found in the so-called human element" (IAEA, 1986, p. 76), and that "formal procedures must be properly reviewed, approved and supplemented by the creation and maintenance of a 'nuclear safety culture'" (ibid., p. 77). The report offered little definition of what this "nuclear safety culture" could entail, but since many references were made to procedure violations as causal factors in the report, adherence to formal procedures seems to be an important aspect of the report's notion of safety culture.

INSAG updated its initial 1986 accident review in 1992 as a result of new information concerning the course of events. For example, it turned out that some of the so called "procedure violations" were approved by the Chief Engineer and were part of the test procedure. Also, some of the hazardous actions made during the accident sequence, and first presented to IAEA as procedure violations by Soviet experts, turned out to have been made without any existing procedural guidance. Thus, the "violations" were probably made without operators having any knowledge that they were actually violating something.

Information about the accident was uncovered gradually, and the emerging picture demonstrated lack of feedback of operating experience and problems in communication between designers, engineers, manufacturers, constructors, operators, and regulators. These deficiencies, coupled with a lack of clear lines of responsibility, deference to authority, insufficient procedures, and operators' insufficient understanding of the nuclear hazards and the inherent weaknesses in the RBMK reactors, were found to be critical contributing factors behind the accident. There was also a prevalent belief in the safety of RBMK reactors in the USSR, and consequently the emergency preparedness was insufficient.

Many of the contributing factors in the Chernobyl disaster echo those of the Three Mile Island accident, including the lack of competence of the operators, the lack of operating experience feedback concerning similar previous incidents in other companies/organizations, problems in human–machine interfaces, and inadequate procedures for dealing with emergency situations. The lessons from the Three Mile Island accident had not been acted upon in the USSR, as that accident was interpreted to demonstrate the failure of capitalism. Another explanation (rationalization) for not learning from TMI was that Soviet technology was unique, based on RBMK and VVER reactor types, and thus the lessons "from the West" would not apply.

Similarly, it can be argued that the world in many ways failed to learn from Chernobyl, as that accident was attributed by many Western actors to the futility of the communist regime, and even to Russian national cultural characteristics.

During the 1990s, *safety culture* became a popular term, even though a somewhat similar concept, "safety climate", had already been introduced several years before (Zohar, 1980). Safety culture development drew more on the general organizational culture literature than the safety climate literature. Nowadays the two concepts of safety culture and safety climate are associated with different definitions, research traditions, and methods (Reiman & Rollenhagen, 2018). *Safety climate* is often described as consisting of shared perceptions of safety related states of affairs (management commitment to safety, rule adherence, safety training, procedures, working conditions etc.) whereas *safety culture* is more associated with safety related values, assumptions, and norms (Reiman & Rollenhagen, 2018). In practice, the concepts of safety culture and safety climate are often intermingled.

Despite decades of research, no clear and widely accepted definition of safety culture and of the means to develop it exists. The definitions of safety culture emphasize to varying degrees the attitudes, behavior, or knowledge of personnel, with some definitions placing emphasis on the structural features of the organization and some on the social aspects such as practices and norms. Many definitions include dimensions from several levels of the sociotechnical system, from individual to group to organization level variables (Guldenmund, 2010; Reiman & Rollenhagen, 2014).

The various definitions of safety culture have also been very generic in nature. Thus, they have failed to adequately consider the varying demands of different industrial domains or different functions operating inside the given industrial domain (Guldenmund, 2000; Pidgeon, 1998), or the different hazards (see Chapter 4). The generic approach also shows in neglect of the "historical-political context, the structural relationships, and the interdependencies" (Silbey, 2009, p. 343) in the organization. Too generic definitions of safety culture concept can lead to overgeneralizations about what is the best for a given organization and thus to the limited usefulness of the approach in actual safety improvement (Reiman & Rollenhagen, 2014).

The exact dimensions comprising safety culture have been debated in safety science. Meta-analyses of questionnaire studies have identified management commitment to safety as the most common factor (Flin et al., 2000)[3]. Other widely utilized dimensions of safety culture include communication, competence, resources, quality of rules and instructions, personnel's risk perceptions, organizational learning practices, including reporting of near-misses, and the safety management system.

A further issue concerns the normative element of the identified dimensions: some studies or technical reports seek to identify the dimensions of good safety culture and report their dimensions in positive terms that apply to organizations having the good safety culture. As an example, management safety commitment and a questioning attitude among employees are clearly normative dimensions. Another way is to conceptualize safety culture in descriptive terms with dimensions that can vary from good to bad, for example, the level of management commitment or the safety attitudes of the employees. Thus, instead of key dimensions of good safety culture, there might exist key organizational elements or phenomena which every safety critical organization needs to solve. These elements might be such that it is impossible to define a priori optimal solutions for them. Instead, these would depend on other issues such as the domain of work, life-cycle of the organization, and societal constraints.[4]

Safety culture is often presented as both an explanation and the remedy in the prevention of accidents (Silbey, 2009). As a remedy, the concept loses the sight of those aspects of culture that are dysfunctional, fragmentary, or conflicted (cf. Martin, 2002). One of those neglected issues is the concept of power (Antonsen, 2009a; Silbey, 2009). Silbey notes that the neglect of power does not mean that safety culture concept would not recognize hierarchy: "The consistent valorization of clear lines of hierarchy accompanies a surprising failure to see how this same hierarchy undermines communication and self-reflection about hazards" (Silbey, 2009, p. 361). These underlying assumptions of unity and hierarchy make safety culture itself an instrument of power. They also imply that the adoption of the concept of safety culture in the industry (and research to some extent) has not led to an actual change in culture (cf. ages of safety in Chapter 4 and especially Table 4.3). In fact, many safety culture approaches do not essentially differ from the previous human error–focused traditions (Reiman & Oedewald, 2007). When safety culture is seen as a remedy, interventions and corrective actions often aim at this abstract whole and fail to achieve concrete and lasting change in the organizational practices, structures, and eventually, culture.

Despite the critique, safety culture as a concept can offer us a view of what issues are important to consider when managing safety in complex organizations, what the culture for safety could look like at the different levels of the organization, and what issues should be monitored (cf. Le Coze, 2019a). We turn to these characteristics of good safety culture next.

5.2 CHARACTERISTICS OF A GOOD SAFETY CULTURE

Reason (1997) proposed that an effective safety culture could be socially engineered by developing five components: a reporting culture, a just culture, a flexible culture, an informed culture, and a learning culture. In an informed culture, the organization collects and analyzes relevant data and actively disseminates safety information. A reporting culture means cultivating an atmosphere where people have confidence to report safety concerns without fear of blame. Employees must know that confidentiality will be maintained and that the information they submit will be acted upon. A learning culture means that an organization can learn from its mistakes, draw the right conclusions, and make changes. In a just culture, errors and unsafe acts will not be punished if the error was unintentional. However, those who act recklessly or take deliberate and unjustifiable risks will still be subject to disciplinary action. A flexible culture is one where the organization and the people in it can adapt effectively to changing demands.

Most industry models of safety culture focus heavily on artifact-level characteristics, seeking to describe what good safety culture looks like on the level of organizational processes and management. For example, the World Association of Nuclear Operators (WANO) and the Institute of Nuclear Power Operations (INPO) have published traits of a healthy nuclear safety culture and have identified the main organizational processes critical to such a culture (INPO, 2013; WANO, 2013)[5]. According to INPO (2013), the categories of the traits of a healthy nuclear safety culture include individual commitment to safety (personal accountability, questioning attitude, and effective safety communication); management commitment to safety (leadership safety values and actions, effective decision-making, and a respectful work environment); and management systems (continuous learning, problem identification and resolution, environment for raising concerns, and work processes). There is little information available on what evidence the dimensions and their specific contents are based on. Little empirical research has been conducted on the validity of the dimensions. An exception is the study by Morrow et al. (2014), who studied the relationship between safety culture and other indicators of safety in the U.S. nuclear power industry.

Their questionnaire study included 2,876 responses from 63 nuclear power plant sites. The relationships between safety culture and safety performance were more consistent when safety performance was measured concurrently, as opposed to being measured one year after the survey administration. However, some interesting correlations were found. Organizations where employees perceived less of a questioning attitude were more likely to receive higher numbers of allegations from the Nuclear Regulatory Commission in the same and the following year. Moreover, organizations with lower overall scores on the safety culture survey were more likely to have higher counts of unplanned scrams and have inspection findings related to inadequacies in problem identification and resolution (Morrow et al., 2014).

In many models of safety culture, there are dimensions from several levels of the organization, from individual to group to organization level. However, the qualitative differences in the dimensions are seldom made explicit. That is, it remains implicit what it means, for example, in terms of safety development that some dimensions deal with individual level issues, some with group level, and some with organizational level issues. How do the different levels interact? Where should interventions be targeted, and how does intervention on an individual level differ from that on a group or organizational level? Since safety culture models typically lack a theory of the organization and a model of safety, these questions remain unanswered – or worse, they remain unasked.

The following characteristics of good safety culture have been extracted from the VTT, IAEA, ERA, and WANO models of good safety culture. Characteristics are divided into four domains: individual, group, management, and organization.

Management and Leadership

Prioritizing safety means that managers, from front-line supervisors to senior management, must be visibly committed to safety, show interest in safety issues, and allocate time to discuss safety, risks, and working conditions in general. Leaders need to make their safety-related expectations and commitment visible, and act as personal examples of safety conscious behavior. Leaders adhere to company policies, and do not expect special treatment regarding, for example, rules or expected behavior. Leaders promote safety behavior by explaining the reasons for it, but also with incentives and rewards. They intervene in case of risky behavior or behavior that is not in line with organizational values. They lead by personal example.

Integrating safety into operations means that safety must not be separated from daily work but must be an integral part of operations. Commitment to safety is best promoted by explaining the broader purpose of the tasks (why), not just what and how. Further, it needs to be explained how safety is connected with the core task and strategy of the organization. The management also ensures that the organization's remuneration system, including key performance indicators, promotes a safe way of operating.

Participatory leadership means involving employees in the development of the organization and safety. Senior leaders delegate responsibility and authority and defer to experts when making technical decisions. Giving feedback is important for developing culture. Regular and meaningful feedback correlates with lower burnout, a better safety climate, and better teamwork (Tawfik et al., 2023a). Leaders must actively share knowledge with others and encourage communication between teams and departments. Leaders need to ensure their communication style conveys trust, not fear (Lloyd, 2020). They should encourage staff to report safety observations and share their ideas and suggestions for improvement.

Awareness of the workplace reality refers to an understanding of what is happening in work and how the work is carried out in practice (Dos Santos Grecco et al., 2014; ERA, 2020). This dimension is important for leadership to be grounded in operational realities and in understanding the gap between Work-as-Imagined and Work-as-Done as well as gradual change in practices in the field (see Chapter 3).

A respectful working environment means that different people and opinions are respected in the organization, conflicts are handled fairly, and relations between management and personnel are based on trust. Personnel are involved in the development of work and safety, and reporting systems are implemented and promoted that allow personnel to express their opinions or worries. Messengers are not shot, even if the message is negative or critical. Treatment of employees is equal, inclusive, and fair, even in error situations where managers and colleagues do not blame but try to learn according to a just culture (Dekker, 2017a). Work motivation and wellbeing issues are actively considered and developed. There is respect between different professional groups and organizational levels, and the importance of other groups in achieving a common goal is recognized. The lack of a respectful working environment can manifest itself as disruptive behavior, discrimination and harassment in the work community, which have a negative impact on the wellbeing of personnel as well as safety in general (Guo et al., 2022).

Organization

Decision-making practices are clear and documented. The safety impacts of decisions and changes are assessed. In decision making, the emphasis is on prudent choices and justifying safety with information instead of assumptions. In uncertain situations, the most conservative estimate is used. Decisions are traceable, which means that it is clear who decided what and how. Traceability extends also to products, requirements, and their fulfillment, and to other intellectual items such as competence. Decision-making practices in safety-critical issues have been described and communicated to the personnel. The organization has a method for transparent decision making in a situation where professional opinions differ, or concerns about safety are expressed.

Continuous improvement refers to an organization's ability to learn from successes and failures and to develop operations without visible signs of deterioration. The organization systematically and effectively collects, evaluates, and implements relevant internal and external lessons learned in a timely manner. An organization learns not only from its own operations but also from other organizations. Important lessons are shared within the organization. Safety is regularly monitored and evaluated using various techniques, including independent assessments and various self-assessments. The organization actively strives to develop and maintain the competence of its personnel, for example by providing training. A continuously learning organization employs a variety of approaches to stimulate learning and improve performance, including human, technical, and organizational aspects. Individuals and teams continuously seek opportunities for improvement.

Competence management means a systematic way to ensure that personnel have suitable knowledge, skills, and abilities to carry out their task safely. The organization provides effective training and ensures knowledge transfer to maintain a knowledgeable and competent workforce. Competent leaders are developed through leadership training and succession management processes. Competence management should ensure both depth and spread of competence. A diversity of ideas and thoughts is needed to ensure safety, and this requires diverse expertise.

Problem identification and resolution refers to a process of identifying, evaluating, and resolving issues potentially impacting safety and quality. The organization needs to have a system for recording and processing various deviations, near-misses, and other observations. The system should enable the classification of observations (e.g. causes, types of events, etc.) and the monitoring of the implementation of corrective actions and the provision of feedback. Issues are systematically identified, fully evaluated, and promptly resolved according to their significance. Observations are analyzed to identify possible patterns and trends. The effectiveness of the corrective actions is regularly assessed to ensure issues are adequately addressed. Self-reporting is expected and valued by the organization.

Work processes means that the organization needs clear working processes, including the definition of roles, responsibilities, and authority. The organization creates and maintains complete, accurate, and up-to-date documentation. Procedures are clear and their safety significance has been taken into account. There is a systematic approach to the selection, scheduling, coordination, and performance of work tasks, emphasizing safety and quality. In the planning of work processes, risks are identified and managed. This applies also for outsourced functions. Human Factors Engineering (HFE) is applied when writing procedures or designing new technology, tools, or interfaces. Human Factors knowledge is also applied to work planning by creating company-level guidance on issues such as work hours, stress and fatigue, and workload.

Work conditions refer to the preconditions that are needed for the work to be performed well and with high quality. These include personnel, equipment, procedures, and other resources. Lighting, temperature, and other physical (posture, etc.) factors related to work are also related to working conditions. The ability to work without interruption when the task requires concentration is also part of the working conditions, as is the possibility to take breaks between tasks. Work-related instructions and procedures are understandable, accurate, up-to-date, and easily accessible. The premises are clean and in good condition.

Resilience refers here to the ability of the organization to adapt to changes, either by reacting to disturbances or anticipating the disturbance and adapting proactively. Flexibility and the ability to change operating modes (and operating philosophy) are critical for resilience. Organizations develop the ability to adapt to unforeseen situations where no procedure or plan applies. Emergency preparedness is also included in resilience.

The new safety culture models, inspired by the Resilience Engineering ideas (see Chapter 11) have included resilience either as an entire dimension (ERA, 2020), or as a sub-dimension of an established dimension (e.g. the IAEA model). There is a risk that various workarounds and adaptations that bring resilience will become routine and lead to suboptimizing or drift. For this reason, resilience must always be examined at different levels of the organization (individual, group, organization). This is also the reason resilience is illustrated as an overarching dimension in Table 5.1.

Groups

Cooperation and the flow of information require well-defined roles and responsibilities of different actors. Groups must also understand the value of open communication and information flow.

TABLE 5.1
Dimensions of a Good Safety Culture Based on Research (Reiman et al., 2008; Reiman &; Rollenhagen, 2018) and Industry Experience (ERA, 2020; IAEA, 2020; WANO, 2013).

Management	Organizational level	Group level	Individual level
Prioritizing safety	Decision-making practices	Cooperation and flow of information	A sense of personal responsibility
Participatory leadership	Issue identification and resolution	Openness and reporting	Questioning attitude
Integrating safety	Continuous learning	Psychological safety	Risk awareness
Creating a respectful work environment	Work processes	Work planning	Sense of control
Awareness of the workplace reality	Work conditions		Systems thinking
	Competence management		
	Resilience		
	Shared meanings and beliefs		
	Shared assumptions about work, organizing, and safety		

Efforts are made to identify and eliminate structural or behavioral factors that weaken cooperation and openness. Efforts are also made to keep employees aware of safety issues. Employees and management believe that cooperation is necessary to achieve goals. Information needed for safety and work is shared openly, and those who hide information are not favored. Cooperation is promoted in various ways, including remuneration and feedback. If necessary, information is also shared across unit and organizational boundaries. Communication and cooperation with the authorities is open. The information to be shared and the channels for sharing information are well thought out and in active use. Information flows from bottom to top, top to bottom, and between organizational units. When planning technical or organizational changes, the aim is to involve the personnel in considering together the best working solutions from the perspective of safety and usability.

Openness and reporting. Openness of personnel in sharing information, talking about safety and hazards, and generally working with each other is important. All employees actively raise safety and quality issues and encourage others to do the same. People listen to each other and openly express their own considered opinions. Mistakes are also talked about without fear of consequences. Issues are reported actively and with a low threshold ("if in doubt, report").

Psychological safety means the belief that there will be no punishment or humiliation in that team or work community for raising ideas, questions, concerns, or mistakes (Edmondson, 2018). Psychological safety is important for team performance, including learning. Psychological safety varies within the same organization between different teams. Supervisors have a great impact on the psychological safety of their work community. If they present their own views too strongly and too early, it can limit the decision-making process by excluding possible dissenting opinions (Edmondson, 2003). Psychological safety was discussed in more detail in Chapter 2.

Work planning, including shift and resource management, is important to ensure both safety and smooth operations. Managers must clearly define roles and responsibilities, ensuring that staff understand these expectations and are able to meet them. It is also important that supervisors are aware of the resource situation, staff motivation, and work ability. Basic understanding of human performance is needed for planning (e.g. fatigue, vigilance, stress, motivation).

Individuals

Personal responsibility means the feeling that a person can influence the results of their work, and that they feel obliged to do their best to achieve goals. Responsible people know their roles and responsibilities, and if they are unsure about them, they will try to sort them out. They are committed to safety and demonstrate it in their actions. They adhere to rules and instructions and take responsibility for safety and quality of the entire system. They are willing to raise safety and quality issues even when they know those are unpopular or may slow down the schedule in the short term. Most models of safety culture include personal responsibility explicitly or implicitly.

A questioning attitude refers to the vigilance and watchfulness of employees. Vigilant employees are aware of the working conditions and warn of possible shortcomings. Employees with an appropriate questioning attitude stop when they are unsure and ask for advice. They also question assumptions and are willing to offer different perspectives when they believe that something is not right or something could be improved. They recognize the dangers of complacency and continually strive to avoid it in themselves and others.

A sense of control refers to the experience of being able to perform their work tasks and do them as well and as carefully as they judge necessary. It also means that the tasks are not too demanding for the employees' skills and that the work is not too stressful and that the employees can see a relation between the effort and the outcome (Bisbey et al., 2021). Often, problems with a sense of control are seen as problems at the individual level, and attempts are made to solve them as such (Kirk, 2024; Taylor et al., 2024; see also Väänänen, 2024). However, challenges to a sense of control should be seen as one of the features of the safety culture affected by decisions made at the organizational and group level, as well as by leadership. Interventions must also take place at different levels, not just at the individual level.

Systems thinking refers to a systemic view of safety that recognizes the impact of all activities on safety, including people and organizations. This view sees people as creators of safety and looks at human and organizational as well as technological factors. This view also recognizes that many types of safety are relevant (e.g. labor, environmental, industrial, process, safety, etc.). Mistakes are seen as learning opportunities for the entire organization, not as individual wrongdoings or personal traits. Systems thinking is seldom explicitly mentioned as a safety culture criteria, but, for example, IAEA requires in its leadership requirements a systems view on safety (IAEA, 2016a). Systems thinking is useful for all employees, but it is essential for leaders. This issue is elaborated in Chapter 7.

Risk awareness refers to the personnel's perceptions of hazards and the risk they pose. The personnel must be aware of the typical hazards related to their operations and the hazard mechanisms or general principles through which the hazards in their own organization can materialize. This information can help staff avoid the most typical, frequent problems and also pay proactive attention to various potential problems, that is, be actively aware of the potential for hazards and uncertainties in their operations. Similar to systems thinking, many safety culture models miss this dimension, which is associated with employee knowledge and skills (Bisbey et al., 2021). This probably has to do with the fact that knowledge is a difficult issue to measure with questionnaires, which are the typical method in safety culture studies.

5.3 MATURITY OF SAFETY CULTURE

Different types of maturity models, or safety culture ladders, are widely utilized in practical safety culture work and consultancy (Goncalves Filho & Waterson, 2018). They aim to visualize the development stage (a ladder) of a particular organization and provide guidance on the next steps in order to improve.

Vernon Bradley developed the first safety culture ladder in 1995 while working at the chemicals company DuPont. It is nowadays called the DuPont Bradley Curve. The ladder presents an organization's behavior moving from reactive to dependent to independent and, finally, to interdependent. When moving from one level to the next, employees' sense of responsibility changes. First, they take no responsibility (reactive stage), whereas in the dependent stage they see safety as a matter of following rules that someone else makes. Following rules is their main responsibility at this stage. In the independent stage, individuals start taking responsibility for themselves and believe that their own actions make a difference. An interdependent stage is characterized by teams of employees feeling ownership and collective responsibility for safety. The main cultural change thus concerns the feeling and locus of responsibility.

Ron Westrum distinguished in 1988 three types of organizations – pathological, bureaucratic, and generative – based on how they process information (Westrum, 2004). He identified three typical patterns of information processing. The first is a preoccupation with personal power, needs, and glory (pathological). The second is a preoccupation with rules, positions, and departmental turf (bureaucratic). The third is a concentration on the mission itself, as opposed to a concentration on persons or positions (generative). While in Bradley's model the change in maturity addresses feelings of responsibility, in Westrum's model the change concerns the flow of information and how information is viewed. Westrum's model is originally meant to be a typology rather than a maturity model.

Patrick Hudson (Parker et al., 2006), building on Westrum's typology, presented a maturity model with five development stages:

1. Pathological – safety is viewed as an expensive inconvenience;
2. Reactive – the organization responds to safety issues once an incident has occurred;
3. Calculative – the organization views safety as a set of standards to be complied with;
4. Proactive – the organization focuses on the inputs to creating safety;
5. Generative – the organization views safety as a value and people look out for each other.

The main issues that change as the organization progresses through the stages are the role of the safety department, the way leadership is practiced, and how trust and accountability are experienced.

Maturity models can be utilized in the assessment and development of a culture. However, assessment should be more than indicating the exact level or stage of the current safety culture. Clients sometimes ask me how they know whether they are on the fifth level or the fourth (or the second versus the third, depending on the model they use). My answer is that one of the best indicators that an organization is on the "highest level" is that they realize the futility of assessing safety culture levels with one number and focus instead on improving their qualitative understanding of the contents of their culture. These organizations are still a rare breed indeed.

Figure 5.1 illustrates a maturity model of how a culture for safety can develop from a pathological to an adaptive view on safety. It distinguishes seven main cultural features that gradually change as the organization develops a more mature culture for safety: assumptions concerning what safety is, ways of taking responsibility, trust, leadership, the nature of uncertainties, and how the organization deals with information and development. The main idea is that these cultural features change qualitatively as the organization's culture matures, not additively.

Reactive cultures typically value compliance, whereas in proactive cultures, efficiency and effectiveness are valued. Adaptive cultures recognize the potential conflicting priorities of safety and efficiency. In an adaptive safety culture, there is a shared sense of purpose and understanding of the organizational core task, including the importance of safety for fulfilling it.

Managers often ask what more they need to do to reach the next level, whatever the current level is considered to be. The question is not so much about doing more things than doing things differently. And maybe the organization does not even need to do more, but rather less. Maybe they need to stop doing something that they are doing currently. Our thinking of progress and development is too linear and additive. To learn, something always needs to be unlearned. Thus, when

FIGURE 5.1 Development stages of safety culture

organizations ask what more they should do to reach the next level, it is good to rephrase the question: what do we need to do differently to reach the next level – in what issues and questions do we need to think differently? And what do we need to stop doing to reach the next level – what kind of beliefs and assumptions do we need to get rid of? Only after answering these questions is it worth asking whether there is something more that needs to be done. There often is, but the main point is that additional measures follow and support the change in the mindset; they are not the change.

Lloyd (2020, p. 8) notes that in the more mature safety cultures, it is not *golden rules*, policies, procedures, safety signs, or slogans that increase in frequency or significance but rather it is trust, visible felt leadership, psychological safety, and authentic information sharing that make the difference. Again, in my view, *improved safety culture maturity is not about quantitative increase, but about qualitative change.* It is a change in how trust is experienced, how safety is conceptualized, how information is perceived, and how responsibility is taken. It is about cultural change. A mature safety culture can do more with less, rather than increase its arsenal by implementing one more new safety acronym.

Maturity models are also sometimes confused with typologies of organizational culture, such as the bureaucratic culture or innovative culture. These are, however, completely different things, as typologies do not represent cultures of higher maturity but cultures of different types. The ideal culture that best suits a particular company depends, for example, on its core task, current safety level, and the operating environment. Chapter 7 presents two typologies of organizational cultures, the Competing Values Framework and the Multi-Focus Model.

What the maturity models easily hide is the fact that each organization develops a unique culture which cannot and should not be copied by another organization. Furthermore, the ideal culture depends on many factors, including the nature of hazards (see Chapter 4) and the context. Thus, what is ideal for one organization may not be so for another. These contextual influences are discussed further in Chapter 10 and the assessment of culture in Chapter 11.

5.4 SAFETY CULTURE PROGRAMS

Some high reliability industries, such as the nuclear industry, have adopted the practice of implementing safety culture programs to make the development of safety culture systematic. There is much to say both for and against introducing a program to develop safety culture specifically. The drawbacks are clear: making safety culture its own target of development risks separating safety issues further from the daily business of the organization, and risks separating "company culture" from "safety culture", at least in the senior management's perceptions. The pros in turn are that by having such a program, cultural issues get attention from the senior management, and they can be systematically monitored, assessed, and developed. I would argue that such a program is needed when the organization strives to develop from a reactive culture into a proactive culture (see Figure 5.1). During the proactive stage, the program will most likely get integrated into a wider culture development process – with its accompanying risks of losing the safety focus due to drift and normalization (see Chapter 3). The need to keep discussion on safety alive never dies, whatever the "level" of the culture is. A program focusing on monitoring and developing culture, whether formulated as a safety culture program or culture development program, is one way of making safety discussions systematic.

The purpose of a formalized safety culture program is to ensure that organizational focus, time, and effort are directed at improving the awareness of existing cultural patterns and their inherent strengths and challenges, and to guide improvement actions that are aligned and integrated with the larger organizational improvement agenda. A safety culture program can help the organization to identify how the different activities of the organization contribute to the culture, and how these could be aligned to maximize their impact.

Safety culture programs should contain at least the following elements (cf. Viitanen et al., 2022):

- *Definition of key concepts*: The program should offer a definition of safety culture and an elaboration of the safety culture concept and its dimensions and how it may manifest in practice, as well as an elaboration of the needed competencies and an elaboration of how safety culture affects safety.
- *Purpose*: The program purpose can be defined, for example, as facilitating the priority of safety, or ensuring the maintenance and continuous improvement of safety culture.
- *Scope*: The program scope refers to the processes or activities that the safety culture program covers. For example, does the program apply to contractors? How about different types of safety – does the program cover only core task related hazards (see Chapter 4) or does it apply to local and external hazards as well?
- *Responsibilities*: The program should refer to the line organization and the senior management as having the overall responsibility for safety culture, and safety culture experts as supporting or coordinating the implementation of specific safety culture activities. Also, the role of supervisors and all employees in safety culture should be described.
- *Safety culture assessment and improvement activities*: The program describes what specific activities companies implement to improve, monitor, and periodically assess safety culture.
- *Interfaces*: Programs can refer to other management system processes or procedures for detailed descriptions of some activities (e.g., human performance programs, procedures for conducting assessments, corrective action processes, reporting system processes, etc.).

The program needs to be constructed and implemented in close collaboration with the senior management. Even though the practical implementation of the program typically falls on safety professionals, the accountability for safety and safety culture should always be clearly at the senior management level.

Similar to all plans, the definition of a safety culture program is at least as important a phase as the actual implementation. When the program and its contents are defined, it is important to make sure there is adequate involvement from all levels of the organization. The development of the program also offers a good chance for the safety professionals to have dialogue with senior management about the safety goals of the organization as well as about leadership aspects and other "soft" issues that may otherwise not easily find a place on the management agenda. Another important stakeholder to involve is human resources function (HR), as traditionally HR has a role in most culture-related issues in organizations, whether they have the competence or not.

5.5 THE USES AND MISUSES OF SAFETY CULTURE

The various normative safety culture models produced by the industry groups (WANO and IAEA in the nuclear, ERA in rail in the EU, ICAO in aviation, national authorities in healthcare, etc.) are sometimes misused in practice. These models typically list a series of safety culture dimensions, but in addition they typically include several traits, attributes, subdimensions, or behavioral expressions of the main dimensions. Many organizations use these as a checklist in audits and assessments, where they verify that all attributes of, for example, personal responsibility are implemented. These can include compliance with procedures (check the box), commitment to and ownership of safety (check), and collaboration with others (check). This type of check-the-box approach to safety culture may lead to small improvements in safety but seldom reveals anything surprising or unexpected. In fact, the better the safety culture is in the first place, the better the organization is at verifying it – and vice versa; organizations with problems in their (safety) culture may still check all the boxes without learning anything new. In any case, following the example above, the meaning that employees give to personal responsibility; or the associations they make between responsibility, leadership, and organizational arrangements; or even whether personal responsibility is seen by the personnel as a significant issue in terms of safety, may remain sidelined with this auditing-oriented approach.

One of the challenges with the concept of safety culture in both research and practice is that more attention has been paid to defining the factors or criteria of good safety culture than to defining the dynamics of the culture itself. To blame an organization for having a weak safety culture has become an easy response to problems (Reiman & Rollenhagen, 2018). As an inherently normative concept, the existing models of safety culture are typically surprisingly descriptive. They mostly list various dimensions and other manifestations of safety culture but do not explicate the mechanisms that shape their formation. Similarly, as these models seldom describe how accidents happen (other than not implementing all of the dimensions) and how safety is created (again, other than implementing all of the dimensions), one is left with a conceptually empty list of terms that all would require much deeper understanding. This is not a systemic view (see systems theories and culture in Chapter 1). It rather makes organizations look static. There is also an overemphasis on the possibilities of organizational design and the ability of a central controller (senior management) to analyze what the system needs and to implement the necessary actions. Finally, as the safety culture models and theories lack attention to mechanisms and processes that produce systemic meanings, including understandings of risk, safety, authority, and control (Silbey, 2009, p. 360), including how the different levels interact (cf. Table 5.1), they in the end leave the responsibility (and blame) on the individuals.

An argument has been made that safety culture is first and foremost a management tool, a means to improve safety. However, utilizing safety culture as a development tool in organizations can come with some risks. One such risk is the assumption that having a shared view on safety matters is beneficial while differing opinions pose a risk. To remain vigilant and evaluate one's own thinking, it is often useful to engage with views that challenge prevailing organizational beliefs and assumptions. An overly uniform culture may become oblivious to its own weaknesses and seek to validate old and familiar opinions (Sagan, 1993; Weick, 1998). For a questioning attitude to be effective, the organizational climate must support open discussions.

Another problem arises when safety becomes a topic of discussion, and the concept of safety culture is used as a tool for organizational discourse. These discussions might become detached from normal work practices. This poses a risk because many safety culture tools have clear-cut answers, attitudes, and practices that are easy to identify through questions and exercises. Organizations and employees might focus on safety culture aspects that are easier to discuss or address with corrective measures. Conversely, an organization might collectively blame its issues on poor safety management or inadequate safety values, or some other abstract entity that does not take offense. This diverts attention from the original purpose of introducing a safety culture concept in high-risk environments: enabling the discussion of subconscious, unspoken principles that influence practices and daily decision making and that are embedded in structures and technology.

The main challenge in the concept of safety culture is that its use very easily drifts into consideration of those "extra" human-related issues that safety is considered to require. This is typically done without considering the requirements of the work itself or the organizational core task.

Hopkins (2005, 2018) points out that the concept of safety culture has often been misused in the industry to denote individual attitudes instead of organization-specific phenomenon. He notes that the prevalent approach to safety culture in the industry often prioritizes changing individuals' attitudes and their "hearts and minds", which helps reduce personal injuries but overlooks major accidents caused by loss of control of process or containment.

Safety culture in practice and in research often neglects the relationships between culture, technology, and structure (Antonsen, 2009a; Reiman & Rollenhagen, 2014; Rollenhagen, 2010). The practical implementations of the concept also often neglect the separation of different hazards (see Chapter 4) and may thus contribute to a simplified view on safety (see the BP accident at the end of Chapter 4).

The existing safety culture models are quite generic, and certain industry-specific or life-cycle specific dimensions may be missing. For example, asset management, configuration management and other domain-specific processes that focus on the core technology are missing from the current safety culture models. Even in a certain industry, the models typically are done so that they are applicable to different functions (e.g. maintenance, operations, design), making them somewhat acontextual. The models of safety culture are also typically organization-centric. Supply chains and the networked nature of modern business are not very well covered by the current models.

Paradoxically, the existing models of safety culture also typically miss the cultural aspects of organizations such as stories, myths, rituals, symbols, and assumptions. The studies that characterize "safety cultures" in terms of selected dimensions thus end up producing correlations between various superficial features of the organization without a deeper understanding of what is actually going on in the organization. Some of these studies become so frustrated with the entire concept that they propose to abandon it. I agree, but only to the degree of abandoning the concept as a superficial reflection of the culture for safety. A cultural approach to safety is valuable, and the concept of safety culture as it is currently used has a value too, but it is much limited by the fact that it paradoxically misses the cultural dimension. This problem haunts the organizational culture concept also, as described by Alvesson and Sveningsson (2024, p. 42):

> A problem in much of the literature on organizational culture is that the potential value of the culture concept easily disappears behind rather thin and superficial descriptions. Organizational cultural characterizations are often used as slogans, wishful thinking and fantasies rather than as a way of gaining a deeper understanding of organizational life.

All categorizations and typologies suffer from this reductionism in some degree, including the ones proposed in this book.

5.6 THE WAY FORWARD

Safety culture has been considered a specific application of the concept of organizational culture. The concept of safety culture is currently used in many fields of safety, including aviation safety, occupational safety, patient safety, nuclear safety, rail safety, and process safety. Many industries, such as nuclear (IAEA), railways (ERA), and aviation (ICAO), have international requirements for organizations to have a good safety culture.

Safety culture covers those aspects of organizational culture that are relevant for safety – in safety-critical industries, in practice, organizational culture and safety culture refer to the same phenomenon but approach it from a different angle.

I use the term "culture for safety" in this book instead of "safety culture". By this I want to emphasize the word "culture". We cannot always know beforehand how the current culture affects safety, and thus we need to consider culture widely, without limiting ourselves to the obvious safety issues.

Edgar Schein was also critical of the safety culture concept and never accepted it as a scientific concept (Reiman & Rollenhagen, 2014) – although many writings on safety culture attribute his theory of culture as the underlying model of "safety culture". It is conceptually more precise to talk about the culture of a group and inspect how the group views technology and its hazards and makes trade-offs between safety and production (Reiman & Rollenhagen, 2014).

Silbey (2009, p. 359) proposes two main ways to further research on safety culture: 1) research that challenges hegemonic normalization – the identification of processes that successfully unsettle organizational routines to make the unthinkable cognizable and the invisible apparent (cf. Perin, 2005; Vogus et al., 2010; Weick, 1998); and 2) research that addresses power differentials and structured inequality in organizations, which might include topics such as the relation of financial risk and material hazards, the influence of stratification of societies on concerns about safety, and the issue of differentially situated interests. In her view, current research on safety culture seems to suggest that "responsibility for the consequences of complex technologies resides in a cultural ether, everywhere or nowhere" (Silbey, 2009, p. 362). One way forward is to see the dimensions depicted in Table 5.1 as interrelated but hierarchical: management and organizational issues are wider in scope and are higher-level dimensions than the group and individual dimensions. Even though they all influence one another, *interventions should primarily be made on the management and organizational level*, secondarily on the group level, and only in acute cases on the individual level.

The concept of safety culture emphasizes the fact that an absence of accidents is not enough to guarantee safety. The presence of capabilities, attitudes, and technical and organizational preconditions is needed, as is constant learning and improvement. Safety is not about reacting to events but constantly and proactively building organizational capacity for increased risk perception and risk management. The value of the safety culture concept is that it gives a direction for the organization and indications of the types of activities and ways of thinking that are typically needed for ensuring safety. The underlying assumption of the safety culture concept is that it is always possible to enhance safety, hence the motive for assessing and developing the culture regularly.

In addition to the structural normative criteria explicated in Table 5.1, there is a need for *process criteria* of a healthy safety culture. These process criteria are social and cultural mechanisms shaping how the system develops. Haslam et al. (2022) call for these kinds of mechanisms to be included in the models of High Reliability Organizations (see the end of this chapter). This book has proposed some ideas for process criteria in Chapters 3 and 4. Chapter 11 summarizes these criteria as warning signs of degrading culture for safety.

5.7 SUMMARY

The chapter discussed the concept of safety culture, its origins, development, and application across various industries. The concept originated in the nuclear industry following the Chernobyl nuclear accident in 1986. Despite extensive research, a universally accepted definition of safety culture remains elusive. Definitions vary, emphasizing attitudes, behavior, or knowledge of personnel, and include dimensions from individual to group to organizational levels. Challenges in defining safety culture arise from the generic nature of the concept, which neglects the varying demands of different industrial domains or functions. Many approaches to safety culture do not differ significantly from previous human error–focused traditions – the object of blame is just shifted from individual to the collective.

Industry models of safety culture focus on artifact-level characteristics, describing what good safety culture looks like in organizational processes and management. Characteristics of good safety culture can be divided into four domains: individual, group, management, and organization.

At the management and leadership level, prioritizing safety requires managers to be visibly committed to safety, show interest in safety issues, and allocate time to discuss safety, risks, and working conditions. Leaders must make their safety-related expectations and commitment visible, act as personal examples of safety-conscious behavior, promote safety behavior with incentives and rewards, and intervene in case of risky behavior. Integrating safety into operations ensures it is an integral part of daily work, connected with the core task and strategy of the organization.

At the organizational level, decision-making practices should be clear and documented, assessing the safety impacts of decisions and changes. Continuous improvement involves learning from successes and failures, collecting and implementing relevant lessons learned, and developing personnel competence. Competence management ensures personnel have suitable knowledge, skills, and abilities to carry out tasks safely. Problem identification and resolution involves recording and processing deviations, near-misses, and observations, and implementing corrective actions.

At the group level, cooperation and flow of information require well-defined roles and responsibilities, open communication, and information sharing. Employees and management must believe cooperation is necessary to achieve goals, and information should be shared openly. Openness and reporting involve actively raising safety and issues, listening to each other, and reporting mistakes without fear of consequences.

At the individual level, personal responsibility involves feeling obliged to do one's best to achieve goals, knowing roles and responsibilities, and demonstrating commitment to safety in actions. A questioning attitude involves vigilance and watchfulness, stopping when unsure, asking for advice, and offering different perspectives. A sense of control refers to the ability to perform work tasks well and carefully, without tasks being too demanding or stressful. Systems thinking recognizes the impact of all activities on safety, seeing mistakes as learning opportunities for the entire organization.

Safety culture has been considered a specific application of organizational culture. The concept covers aspects of organizational culture relevant for safety, emphasizing the importance of constant learning and improvement. The preventive nature of safety culture is an important insight, including the need for continuous improvement, since the absence of accidents is not enough to guarantee safety. Despite ambiguities, safety culture is useful for stimulating discussion about safety and paying attention to organizational issues necessary to ensure safety.

Organizational culture and safety culture are closely related concepts. Safety culture approaches culture from the perspective of how the given culture understands, appreciates, and manages safety, whereas organizational culture has a more neutral standpoint and looks at meanings, values, and assumptions in general. What safety culture as a concept could do is preserve these lessons we get from organizations – what aspects of culture support safety and what aspects do not. The approach in this book is that the best added value of the culture concept comes from this combination of seeking to understand the current culture and only after understanding it, making normative judgments about how the given culture affects safety (see Chapter 11).

Reflection Questions

How does your organization look from the perspective of the five leadership characteristics of safety culture? What are the strengths and weaknesses? How do those characteristics relate to organizational, group, and individual level dimensions?

How does your organization look from the perspective of the six organizational characteristics of safety culture? What are the strengths and weaknesses? How do those characteristics relate to leadership, group, and individual level dimensions?

How does your organization look from the perspective of the four group-level characteristics of safety culture? What are the strengths and weaknesses? How do those characteristics relate to leadership, organization, and individual level dimensions?

How does your organization look from the perspective of the five individual-level characteristics of safety culture? What are the strengths and weaknesses? How do those characteristics relate to leadership, organization, and group level dimensions?

Which of the safety culture dimensions needs most improvement in your organization? Can improvement or change in some other dimensions of safety culture facilitate also the weakest dimension?

How does resilience manifest in your organization at different levels (leadership, organization, group, individual)?

HIGH-RELIABILITY ORGANIZATIONS

Some organizations, even those that are complex and tightly coupled (see NAT in Chapter 3), have exceptionally good safety performance (at least so far). These so-called high-reliability organizations are organizations in which errors can have catastrophic consequences, but which consistently seem to avoid accidents.

The High-Reliability Organization (HRO) group formed after the Three Mile Island accident in 1979 at the University of Berkeley by Todd La Porte, Karlene Roberts, Paul Schulman and Gene Rochlin, and others has been influential in illustrating the organizational aspects of safety. They utilized an ethnographic approach and spent time studying how "nearly error free" operations were possible in nuclear powered aircraft carriers (La Porte & Consolini, 1991; Roberts, 1989; Roberts et al., 1994), nuclear power plants (Bourrier, 1996; Schulman, 1993), and the air-traffic control system (R. L. Klein et al., 1995). Karl Weick, together with Kathleen Sutcliffe and Karlene Roberts, significantly contributed to HRO theory (Weick, 1987; Weick & Roberts, 1993) and applied it to new contexts, including firefighting and healthcare. HRO is rooted in the social sciences, particularly social psychology, organizational psychology, sociology, and political science (Le Coze, 2019b).

According to La Porte (1996), well-performing organizations in high-risk domains shared some key traits: a unified mission balancing productivity and safety, skilled staff with a focus on technical competence, rigorous quality assurance, and adaptability to unexpected incidents. They employed flexible work processes, hierarchical structures that shift to skill-based authority during emergencies, and decentralized operational decisions while consulting experts for tactical choices. Decisions were swift, backed by comprehensive information, and feedback was actively collected to spot errors early. Their culture blended mission-driven norms with safety priorities, fostering professional pride and mutual support. Operators and leaders were able to make prompt, independent decisions in critical moments, with support from technical experts. These organizations had high competence and discipline to create viable schedules, and they also had transparent technical solutions, proactive self-assessment, stringent internal reviews that focus on actual operations, and clear distribution of responsibilities.

Further empirical findings emphasized the significance and possibilities of organizational culture, redundant organizational structures and functions, process orientation, decentralized decision making based upon the expertise needed in emergencies, training, and good organizational design and management, including the prioritization of safety as an overriding goal (Klein et al., 1995; La Porte & Consolini, 1991; Roberts, 1993; Weick et al., 1999; Weick & Sutcliffe, 2001). HRO researchers noted that organizations use approaches such as decentralization, redundant systems, and organizational structures to address the challenges of complexity and tight coupling, which Normal Accidents Theory (see Chapter 3) suggests would inevitably lead to accidents.

Although the original goal of HRO was to remain descriptive, Karl Weick, together with colleagues, summarized and elaborated the findings in a more normative way. Weick et al. (1999) introduced the concept of mindful organizing (originally called *collective mindfulness*) as the collective capability to discern discriminatory details about emerging issues and act swiftly in response to these details. Mindful organizing facilitates the discovery and correction of anomalies that could be combined with other anomalies and turn into a catastrophe unless acted upon (ibid.). Mindful organizing combines anticipation and resilience. It comprises five interrelated processes (also called "hallmarks of HROs"): preoccupation with failure, reluctance to simplify interpretations, sensitivity to operations, commitment to resilience, and deference to expertise (Weick & Sutcliffe, 2015).

(1) *Sensitivity to operations* means creating and maintaining an up-to-date, integrated understanding of organizational activities. By having an understanding of what is going on, anomalies are noticed while they are still tractable and can be isolated and dealt with. Sensitivity to operations is about the work itself, about seeing what the organization is actually doing regardless of intentions, designs, and plans. Sensitivity also means treating close calls as danger (in the guise of safety) rather than successes (safety in the guise of danger).
(2) *Preoccupation with failure* captures the need for continuous attention to anomalies that could be symptoms of larger problems in the system. Organizations that are preoccupied with failure systematically collect and analyze small, emerging failures – they report and learn from them. Preoccupation with failure directs attention to ways in which local activities can conceal or highlight such things as symptoms of system malfunction, small errors that could enlarge and spread, opportunities to speak up and be listened to, a gradual drift toward complacency, the need to pinpoint mistakes that need to be avoided, and respect for the day-to-day experience with its surprises.
(3) *Reluctance to simplify* means taking deliberate steps to create more complete and nuanced pictures of what the organization faces and who the organization is as they face it. It means actively questioning the advice received and operating assumptions in order to better uncover blind spots. Less simplification allows one to see system problems behind weak signals or events.
(4) *Commitment to resilience* involves developing individual, team, and organizational capabilities to adapt, improvise, and learn to recover better from unexpected events. The organization develops capabilities to detect, contain, and bounce back from the inevitable errors that are part of an indeterminate world. HROs are not error-free, but errors do not disable operations.
(5) *Deference to expertise* occurs when decisions migrate to those with the greatest expertise in the problem at hand, regardless of their formal rank. The organization cultivates diversity, not just because it helps the organization to notice more in complex environments, but also because it helps the organization to adapt to the complexities it notices.

According to Schulman (2020), a dominant value in high-reliability organizations is the primacy of safety over other values such as speed, efficiency, or even service. Reflecting this value, these organizations will suspend operations rather than operate outside of widely accepted margins of risk. This emphasis on safety supports the preoccupation with failure (see above) and leads to constant vigilance against complacency and hubris. This vigilance and skepticism show in questioning actions, plans, and assumptions. This approach forms the basis for detecting uncertainty and potential errors in measures, models, and planning assumptions (Schulman, 2020, p. 3).

Schulman (2020) notes that an additional element in HROs is the development of a "precursor" strategy related to errors and accidents. Significant effort and attention are directed toward identifying conditions or states that, while not failures or accidents themselves, can lead through chains of causations to more direct failure causes. These precursor conditions include not only physical conditions (e.g., excessive temperatures and pressures, loss of backup equipment) that exceed a bandwidth of acceptable operating conditions but also organizational conditions (e.g., cognitive load on operators, excessive noise in control rooms) that can degrade skills and invite error.

Schulman (2020) clarifies that a "deliberate bias in high reliability error management strategy is to accept possible errors of excess caution (even in shutting down a system) in order to suppress those of hubris and over-confidence". He further notes that HROs avoid errors connected to the systematic underestimation of risk, including both hazards and likelihoods of failure. In practice this means imagining the worst-case possibilities instead of probabilities.

Schulman (2020) also notes that a key element related to culture in HROs is the personal identification with safety by a significant number of individuals distributed across hierarchical levels, departments, and units and within individual work crews. These "reliability professionals" (Roe & Schulman, 2008) take responsibility for managing technical systems beyond their job descriptions. They often work in teams, gaining experience across different roles. Over time, they advocate for system-level risk perspectives. These individuals, and the culture that supports them, are crucial in preventing organizational drift as well as maintaining safety priorities despite competing organizational values such as efficiency or costs. Because of their long experience or technical knowledge and their ability to mix experiential and formal analytic perspectives, they may function as informal leaders in the organization (Schulman, 2020).

Several reviews have pointed out that even though the five HRO principles are more a matter of mindset than structure, practical solutions as well as studies have favored technical measures such as reporting systems and auditing tools (Cantu et al., 2021; Haslam et al., 2022). Thus, the original focus on social scientific aspects of organizations and organizing have been downplayed in the practical applications (Cantu et al., 2021). The five core principles should be understood and enacted as activities that are intrinsically motivated and identity-defining, rather than being merely the basis for rules, guidelines, and checklists (Haslam et al., 2020).

Cantu et al. (2021) points out in their literature review that "it is noteworthy that after 34 years of HRO publications and research, there is still very little published on implementation". Haslam et al. (2022) criticize HRO theory for underemphasizing the process focus of HROs, meaning the questions of how HROs are developed, and which activities lead to and support high reliability. They note that existing research can describe what an HRO looks like but is unable to explain how it forms (cf. Cantu et al., 2021). Recent reviews have also rediscovered the relevance of culture for HROs (Cantu et al., 2020), providing an interesting case of how the premises and foundations of theories and models may be "lost in translation" when models are implemented by practitioners and reinterpreted by new researchers (cf. Reiman & Viitanen, 2019b).

Another recent development has been made by Haslam et al. (2022), who proposed a social identity model of high-reliability organizations, which combines the idea of collective mind (Weick & Roberts, 1993) with shared social identity.

The five hallmarks of HROs have been investigated and some revisions have been proposed based on the characteristics of the specific operating environment. For example, Leveson et al. (2009) stress the importance of a system approach (see Chapter 1), and Saunders et al. (2016) propose that the ability to prosper in paradoxes is essential for high-reliability project organizing. Vogus and Welbourne (2003) propose that for more "prosaic" reliability-seeking organizations (in comparison to HROs), three of the five hallmarks are enough: reluctance to simplify, sensitivity to operations, and commitment to resilience. Rotteau et al. (2022) discuss the implementation of a hospital safety program based on the five high reliability organizing principles. Preoccupation with failure was difficult for the healthcare professionals to understand, as failure in terms of the death of patients is present in everyday work in healthcare.

Preoccupation with failure may be a principle that becomes important only when safety is already at a high level and there are few events or near-misses to maintain vigilance.

What Did We Learn about Safety in Complex Organizations?

The hallmarks of high-reliability organizations are a good addition to the features of a healthy culture for safety (see also the safety culture characteristics earlier in this Chapter).

The first hallmark, *sensitivity to operations*, highlights the importance of understanding the daily operations and the day-to-day workings of the organization. Safety management should prioritize continuous feedback loops, where operations and safety performance are constantly monitored, and employees have the tools to provide real-time feedback about potential safety issues.

The second hallmark, *preoccupation with failure*, reminds about the importance of keeping the possibility of failure always in mind, even when – especially when – things are going well. An active search for potential weaknesses and vulnerabilities and constant learning and development are needed.

Third, *reluctance to simplify* means acknowledging the inherent complexity of sociotechnical systems and avoiding "quick fixes", investing instead in thorough understanding and detailed problem solving. It also serves as a reminder of our natural tendencies to categorize issues and simplify reality and highlights the importance of adequate diversity and requisite variety in organizations.

Commitment to resilience is needed because surprises happen, and organizations need to be able to respond effectively to unexpected events and various mistakes. Errors should be contained as effectively as possible. Resilience Engineering has recently adopted the term *resilience* with a rather different interpretation. Views on Resilience Engineering are summarized at the end of Chapter 11.

Finally, *deference to expertise* means that decision-making authority is deferred to those with the most relevant expertise, rather than adhering strictly to hierarchical structures. This allows people closest to the situation to make decisions quickly and effectively in emergencies. High-reliability organizations are able to decentralize in a flexible manner in order to respond to various contingencies.

The way HRO research was carried out, in close collaboration with the case organizations and with an attempt to genuinely understand them, sets an important precedent for safety research as well as application.

NOTES

1. The following description is based on Mosey (2006), Higginbotham (2019) and Plokhy (2018).
2. RBMK (reaktor bolshoy moshchnosty kanalny, high-power channel reactor) is a Soviet-designed water-cooled reactor using graphite as its moderator. It was designed for plutonium production for military purposes as well as electrical generation.
3. This may also partly be an artifact of the questionnaire method, since questionnaires measure those issues that researchers choose to measure.
4. There are many other conceptual challenges in the concept of safety culture, which has led some academics to propose abandoning the term (Hopkins, 2018). One further source of confusion comes from the issue of whether all organizations have a safety culture, or whether safety culture is relevant only to those organizations that are committed to and value safety. Since the current book focuses on organizational culture, these debates do not concern us. However, my perspective is that all organizations have a culture for safety (safety culture, if you will) that is more or less healthy and adequate from the point of view of contributing to the safety of the organization.
5. More dynamic safety culture models include the European Railway Agency's (2020) safety culture model. However, being more dynamic, it is also more complex than the (apparently) simpler lists of key dimensions.

6 Structures and Organizing

This chapter deals with the organizing of complex organizations: What is needed in terms of organizational functions and processes. It looks at setting strategy and objectives, establishing structure, organizing work, making decisions, involving personnel, managing risk, and learning and continuous improvement. The chapter aims to demonstrate how the structures simultaneously reflect current culture and slowly change it, for better or worse. Structure creates culture but culture reflects back on the structures and how they manifest in practice. Safety management systems are an important structure that shapes culture but may also in some cases lead to discrepancy between structure and practice. The chapter advocates for mindful organizing, whereby the relationship between structure and culture is perceived and actively cultivated by leadership. The chapter ends with a different type of accident case: a comparison between the Fukushima and Onagawa nuclear power stations during the Great East Japan Earthquake in 2011.

6.1 SETTING STRATEGY AND OBJECTIVES

Le Coze (2021c) highlights the importance of considering strategy in safety management. He proposes the following definition of strategy and outlines an approach to consider strategy and safety:

> choices made by top managers over the main orientation of a business in its market and the ability of these orientations to be successful considering the human and technical capabilities involved.
>
> **(Le Coze, 2021c, p. 82)**

Safety needs to be clearly identified as a strategic goal and taken into account along with the other goals of the organization. Objectives for safety need to be set and communicated in the organization. These objectives need to be reflected in how the organization is structured and led. The organization should consider its strategy for ensuring safety. However, it should also be recognized that in complex systems, *all strategy is safety strategy.*

A typical goal for companies nowadays is "zero accidents", or "zero harm", that is, in practice the elimination of accidents that lead to absences from work. In safety science there has been a lot of debate about whether this is a sensible strategy, as it can lead to underreporting – especially when combined with financial incentives (Poole & Dekker, 2024). "Vision zero" works as a high-level vision for an organization, but care must be taken when breaking it down into lower-level strategies and actions. Actions must be such that they improve the work, the working conditions, and personnel's competences, not actions that measure the type of incidents in more detail and more frequently. The latter typically leads to suboptimizing and underreporting, as we will discuss in more detail below. It also easily leads to measures that target the most frequent and minor issues in the organization – and may miss the issues underlying serious injuries and major accidents.

I have noted that often both the actions and the measures are simplistic, and a brief good progress in optimizing the measures soon plateaus as the organization has picked the low-hanging fruit and done the most obvious things to manage the numbers. Many statistics also indicate that there has been more success over the years in decreasing minor injuries compared to serious injuries and fatalities (www.osha.gov).

As part of the strategy process, organizations typically define a set of official company values.

Surprisingly often, the official values seem not to be issues the organization is good at. Senior managers tend to select as their official values issues that are acute in their mind, issues they want their personnel to focus on, to step up and to improve. Chronic issues or issues that are currently

DOI: 10.1201/9781003491156-6

"solved" may not make the official lists. All this speaks about the importance of separating the deeper elements of culture from the surface features and espoused values (see Chapter 1). Official values always reveal something about the organization, but to know what, we need to dig deeper. Values should also always be contextualized:

- What do the values mean in practice for how we work and how we interact?
- What do the selected values mean for our organization, and how do the various elements of the organization (structure, practices, technology, our product) manifest them?
- How do the values relate to each other? How are they prioritized if they are in conflict?
- How do we know whether we live up to our values?

Strategy, including values, should show in how the organization is structured.

6.2 CREATING STRUCTURE

Structure includes the organizational structure and other physical structures such as technology, production facilities, and offices. Technology and its design are major factors in safety and accidents, as depicted in Chapter 3. Technology includes both the production technology and the tools used at work.

According to the meta-analysis by Dyreborg et al. (2022, p. 39), engineering approaches such as redesign or modernizations provided in general high reductions in injuries, also when compared to other safety prevention approaches. Especially passive engineering approaches that do not require active human interaction and adoption were found to be effective in reducing injuries (Dyreborg et al., 2022, pp. 39-40).

The basic technology differs between the various safety-critical fields, as discussed further in Chapter 10. This technology both constrains and creates possibilities for action. Simultaneously, it is the source of core task related hazards (see Chapter 4) that the organizational structure needs to manage.

Hopkins (2019) considers organizational structure as the communication and reporting lines, the structure of incentives, and the company priorities that are set by CEOs, boards, and remuneration systems. According to him (Ibid.), structure follows strategy and influences culture.

Deciding on an organizational structure involves choosing the extent to which the decision-making authority is dispersed or concentrated within the organization (Monteiro et al., 2020a). Centralized organizational structure and decentralized structure are the opposite ends of the organization design. Decentralized structure is typically considered to promote efficiency and innovation, but from a safety and engineering perspective, there are several challenges with decentralization when it comes to controlling major hazards (Hopkins, 2019; Monteiro et al., 2020a, 2020b). It needs to be ensured that the safety function has adequate authority, competence, and access to support and oversees the line activities. Organizing the safety function is elaborated in Chapter 8.

A safety management system (SMS) is a planned, documented, and verifiable method of managing hazards and associated risks in an organizational context. It is meant to be a systematic and continuously improving approach to organizational safety – a "vision of what the organization should be explicitly and intentionally doing for safety, under the leadership of its managers" (Pariès et al., 2019, p. 502).

The management system should provide management assurance that all significant hazards have been identified and are being adequately controlled and graded by their significance, using formal and documented methods (cf. Whittingham, 2008, p. 177). It includes the necessary organizational structures, accountabilities, policies, and procedures.

Even though it is formal, the SMS should be a living set of arrangements, developing as the organization does. It should be integrated into the business processes of the organization. It should not become a paper exercise developed solely for demonstrating compliance with the regulatory framework. However, the SMS is a good place to gather all requirements for managing safety in the organization and define and document how these requirements are to be met. Guldenmund (2010, p. 1477) notes that SMS should be a "framework for meaning and a reference for behavior".

Typically, safety management systems deal with at least the following issues:

- Leadership for safety and senior management commitment;
- Employee involvement in safety;
- Planning and organizing;
- Hazard identification and risk analysis;
- Risk prevention and control;
- Competence and training;
- Performance monitoring and assessment;
- Continuous improvement

The specific components of an SMS differ between industries. Furthermore, some industrial domains advocate integrated management systems, whereas in others there may be several management systems for different organizational goals (e.g. quality, occupational safety, environmental safety). For example, in the nuclear domain, the International Atomic Energy Agency has stated that an integrated management system shall be applied to achieve goals safely, to enhance safety, and to foster a strong safety culture by:

- Bringing together in a coherent manner all the necessary elements for safely managing the organization and its activities;
- Describing the arrangements made for management of the organization and its activities;
- Describing the planned and systematic actions necessary to provide confidence that all requirements are met;
- Ensuring that safety is taken into account in decision making and is not compromised by any decisions taken (IAEA, 2016a).

Sometimes the expectations put on the safety management systems are rather optimistic. The challenge is that, as was established in Chapter 4, safety is not a system and cannot be managed as an independent entity. Safety management systems thus need to be integrated into practice with the way work is managed (cf. Whittingham, 2008, p. 177).

An SMS easily becomes too burdensome and complex, resulting in processes that are incompatible with an organization's core activities and go against the common sense found in local practice (Accou & Reniers, 2020). An organization and its main technology form the system to be managed, the sociotechnical system. The SMS may end up sidelining safety if it makes safety a topic to be managed on its own. However, the safety management system can offer guidance for planning, implementing, monitoring, and continuously improving the appropriate controls needed to ensure safety in the organizational context.

An SMS is often perceived as pushing companies towards Work-as-Imagined (see Chapter 3) and a compliance-focused perspective. An SMS is typically based on a top-down control approach where information about the field is gathered and analyzed by the "controller", who then in turn plans, implements, and follows up on corrective actions (Reiman & Oedewald, 2007). They may end up being a paper-based system specifically developed for demonstrating compliance with the regulatory framework. Since safety management systems are designed to be verifiable, regulators also typically require companies to have one. This is because it helps the regulator; instead of verifying that the regulated organization is safe, the regulator only needs to verify the SMS is implemented (or at least that is the assumption). There is a potential for a vicious circle where regulators (or customers/other stakeholders) require verifiable and quantifiable safety performance, and the company builds a simplified SMS with simplistic indicators detached from the workplace reality but with easy-to-demonstrate safety indicators. These systems can then be audited and "verified". The regulatory focus also shifts from verifying technical issues, or the work itself, to verifying documented processes (Work-as -Imagined). Auditing as a safety management tool is described in more detail in Chapter 9.

Safety management systems and their emphasis on planning and standardization also risk alienating the workers from those that plan the work (cf. Almklov et al., 2014; Antonsen et al., 2008), further widening the gap between Work-as-Imagined and Work-as-Done. A detailed SMS can be seen as an attempt to reduce the autonomy of the workers (Pariès et al., 2019). Too strict planning and proceduralization can demotivate the people doing the work, and lead to employees finding creative ways of "enriching" their jobs, or shortcuts and rule violations to make the work easier or more efficient. These solutions usually remain informal parts of the employee work culture, often unknown to distant (safety) managers. Any endeavor to design the work for error-free performance and standardized outputs plants the seeds of its own failure by making the work (as imagined) rigid, inflexible, and predictable in an environment that is inherently dynamic and unpredictable (Hackman & Oldham, 1980; Hollnagel, 2009a; Snook, 2000).

Developing an SMS requires an organization to identify hazards and assess the associated risks that need to be controlled within the legal framework in which it operates. Additionally, the organization must have a clear understanding of what constitutes "good" performance and what behaviors are considered at-risk practices. A management system is structured to define the desired standards and behaviors, supporting employees in adhering to these through established processes. Without such a system, ensuring adequate controls and systematically improving organizational culture can be challenging.

It is important to recognize that management systems, by themselves, do not achieve results; it is the people who make the difference. Accident investigations have shown that an organization can have a very sophisticated safety management system on paper but fail to adequately manage the hazards in practice. Effective communication, interaction, and dialogue are essential for all employees, particularly managers, at every organizational level. The system merely offers a framework to facilitate these processes. Real compliance with SMS does not mean that the paperwork has been completed, but rather that the organization can demonstrate that it effectively, and as far as reasonably practicable, manages its hazards (Smith, 2024).

The culture and management system of a company affect each other. The culture supports or hinders operations in accordance with the management system, and the management system, in turn, supports or hinders the formation of the right kind of culture. The safety management system, when properly utilized, can also help in identifying desired trends in culture and contribute to its development.

6.3 PLANNING AND ORGANIZING OF WORK

6.3.1 Assessing and Controlling Risk

A critical element of any effective safety management system is a proactive, ongoing process to identify hazards and assess their risks. Risk management takes place at multiple organizational levels, from strategic risk management to operational.

Risk management must be dynamic, with constant re-evaluation of potential failures, especially in high-hazard industries. Risk management can be divided into hazard identification, risk evaluation, risk assessment, planning and implementing actions, and monitoring and review. These stages are complemented by communication and consultation as well as reporting and documentation. A Risk Management Process is described in ISO 31000:2018.

Identification of hazards is a prerequisite for functioning risk management as well as for an effective safety management system. There are various methods for identifying hazards during the system design and operation. During operation, hazards can also be identified by various means, such as reporting systems, assessments, audits, inspections, and incident investigations (see later in this chapter).

After the hazards have been identified, the evaluation (sometimes called risk analysis) stage aims at assigning them a probability of occurrence as well as the consequence, if they occur. This stage defines the risk that the hazard causes to the system and it is elaborated in Chapter 9.

Risk assessment refers to a process of deciding whether the analyzed risk is acceptable or not. Often various visualizations, such as risk matrices, are used at this stage. They can be valuable communication tools, but they need to be used carefully if they are utilized as a decision tool, as they may hide high-consequence low-probability risks. They also fail to consider the interaction between the various risks and may project a false sense of objectivity hiding the numerous assumptions and subjective judgments underlying the visualization.

Planning and implementing actions (sometimes called risk control) refers to deciding and applying actions to reduce, eliminate, or transfer risks. Monitoring and review refer to a regular checking of risks and risk controls to ensure their effectivencss.

Communication and consultation need to be conducted continuously with all stakeholders (including employees, managers, regulators, contractors, and customers). The aim is to make sure everyone understands the risks, their role in preventing them, and how the system of risk controls works.

Methods of hazard identification and risk analysis are complementary, and many methods can be used for both. However, identification of hazards is always the backbone of risk management and should be done systematically for different hazard types (see Chapter 4).

Methods of hazard identification include (e.g. Manuele, 2013, 2014; Leveson, 2023):

- Job hazard analysis;
- Task analysis;
- Process analysis;
- Functional Hazard Analysis (FHA);
- Preliminary Hazard Analysis (PHA);
- Hazard and Operability Study/Analysis (HAZOP);
- Brainstorming/analysis sessions;
- Audits and inspections (see Chapter 9),

Methods of risk analysis include (e.g. Manuele, 2013, 2014; Leveson, 2023):

- Failure Modes and Effects Analysis (FMEA), which lists possible failures and their effects, and ranks them by severity and likelihood;
- Fault Tree Analysis (FTA), a graphical method that maps out causes leading to a top event;
- Event Tree Analysis (ETA), which starts from an initial event and maps out possible outcomes and their probabilities;
- Quantitative Risk Analysis (QRA);
- Risk rating matrices, a visualization of risks by their probability and consequences.

Risk management should be the cornerstone of all work planning. Industries vary in their efforts devoted to risk management. In the nuclear industry, fault trees and failure trees are extensively utilized as part of probabilistic risk assessment, and these assessments contribute to grading systems, structures, and components based on their safety and operational significance, and to decision making and formation of procedures. PRA also informs plant modernization and redesign needs, and PRA results are used as a way to drive continuous improvement.

A hierarchy of controls can be applied to the process of eliminating the identified risks. This heuristic tool (see Chapter 2) provides guidance on how to treat hazards and how to systematically prioritize the various hazard control measures.

The hierarchy of controls according to Manuele (2013), from the most preferred to least preferred method, are:

1. Risk avoidance: Prevent entry of hazards into a workplace by selecting and incorporating appropriate technology and work methods criteria during the design processes.
2. Eliminate: Eliminate workplace and work methods risks that have been discovered.
3. Substitution: Reduce risks by substituting less hazardous methods or materials.
4. Engineering controls: Incorporate engineering controls/safety devices.
5. Warning: Provide warning systems.
6. Administrative controls: Apply administrative controls (the organization of work, training, scheduling, supervision, etc.).
7. Personal protective equipment: Provide personal protective equipment (PPE).

Risk assessment, like any human activity, always contains assumptions (about work, safety, environment, risks, and humans). The extent to which these assumptions are made explicit is critical for the success of risk management. One of the important assumptions concerns the nature of safety and the type of hazards that are considered. Another assumption concerns issues that are left out of the analysis as improbable or too minor to consider.

There are several tools for risk management in daily activities: pre-job briefs, post-job briefs, and mini risk assessments. These tools are considered in more detail in Chapter 9.

6.3.2 Division of Labor

An organizational structure aims to provide the members of the organization with a common view of the operations, division of duties, and supervisory relationships, including tasks, responsibilities, and work roles. Based on the structure, operations can be planned, organized, directed, and monitored. Structures create possibilities for and constraints on the functioning of work communities. They form the framework for how work should proceed (Work-as-Imagined). On the other hand, structures are interpreted and given meanings in the organization in question. Not all structures are followed as originally envisaged by the designer. Employees' perceptions of, for example, rules, training, or formal responsibilities affect how much these issues impact their ways of working. Local practices do not necessarily correspond to the standard operating procedures. Local practices arise because they are often necessary to make the work run smoothly within, or despite, the formal structure (see Chapter 3).

When organizing work, the focus should be on ensuring that employees can perform their tasks without interruptions and under suitable conditions, including appropriate lighting, posture, noise levels, and temperature. These issues have a significant impact on human performance, including fatigue.

It is advisable to conduct certain types of task analysis, such as cognitive task analysis (see Schraagen et al., 2014), especially for tasks that are critical to safety. Cognitive task analysis is an extension of traditional task analysis techniques that aims to yield information about the knowledge and mental processes that underlie observable task performance.

In safety-critical organizations, operations are typically divided into several specialized units and expert departments that assist, monitor, and supervise the line, and often each other. The basis of operations is usually the traditional line organization, where the responsibility for safety lies. However, many tasks require input from other parties as well, and in reality, the organizational structure and division of labor typically become quite complex. A complex organization places new demands on staff. The ways of coordinating work at the sharp end are often only partly described and remain the tacit knowledge of the supervisors and middle management. Especially supervisors face conflicting goals and need to fit the work of their subordinates with the work of the other units and functions while maintaining an overview of available resources and competencies as well as the working conditions.

In safety-critical organizations, areas of expertise are typically deep yet narrow. However, it is not unequivocal that reliability is best achieved by organizing work into narrow areas of specialization. Making the right decisions in a complex environment can be difficult if you don't know the big picture that your solution will affect in that situation. In practice, the question that often arises in organizations is which is the more sensible strategy: to deepen the competence of each of their own areas or educate multi-skilled professionals. A similar question applies to the organization of work; should the organization be divided into narrow areas of specialty, or should multidisciplinary teams and cross-functional processes be utilized?

As the complexity of technology grows, so does the need for specialized knowledge and competence. Specialists have their own concepts (language), and a specific way of looking at work. They understand each other, but not always the specialists in other areas. These subcultures may prefer to interact among themselves rather than risk communication with other subcultures. This can happen to safety departments too! It is much easier (in the short run) to isolate yourself and keep busy with each other than to expose yourself to "outside" criticism and potential disruptions to daily routines (see also Tett, 2015).

Specialization is needed, and to specialize, a possibility to focus on one area is needed. An organizational challenge is to balance specialization with the boundaries that it necessarily creates. Structures can enable collaboration or further facilitate specialization and silos. We typically interact more with people who are near us than those in separate floors, buildings, cities, or countries. Organizations are typically designed so that people who belong to the same function or department sit close to each other in the office, or work close by in the case of blue-collar workers. Social identity theory has shown how easily we form shared social identities with people whom we regularly interact with. Organizations should ensure adequate balance between specialization and generalist knowledge by, for example, facilitating interaction between the different specialist areas and creating processes that cross departmental boundaries. They should also create a common vocabulary of key terms and their meanings in this organization together with the specialists. There is likely considerable variation in the definitions of even basic terms among different professional groups. Moreover, the structural characteristics of the organization should be designed to facilitate interaction and joint learning among individuals. Information flow is key in this. Finally, the work of safety professionals needs to be organized as well. This is discussed in Chapter 8.

6.3.3 Information Flow

Organizations operate through a process of information reduction, where senior leaders receive data from their direct reports, who in turn have collected it from their own subordinates, and so forth. This approach is driven by the need for efficiency, as it is impractical for one individual (or even a group) to handle all the information. In this book there are many examples of information not reaching decision makers, or decision makers not believing the information that was delivered to them (see Chapter 3).

To counteract the effects of reduced upward reporting, organizations need to define adequate communication channels to ensure that safety-related information is exchanged among the different levels of the organization and with external parties, including contractors, partners, and suppliers. They need to ensure information reaches the people who need it, and that the information is understandable, accurate, consistent, and traceable. The needed communication channels should be specified in advance and promoted, encompassing top-down, bottom-up, and horizontal communication. The main thing is to acknowledge the effect of formal organizational structure on information flow.

Organizations must clearly define the type of safety-related information that needs to be communicated, determine the methods of communication, identify the recipients, and establish the conditions under which this communication will be initiated and processed.

Communication should also cover information regarding the safety policy, management expectations, current safety issues and risks, and safety goals and their fulfilment, as well as any changes in the organization, environment, or hazards.

Different reporting tools can serve as a method for bottom-up communication, and it is essential to ensure proper feedback from top down to maintain their effectiveness. Nevertheless, direct communication between supervisors and subordinates should always remain the primary channel of communication.

The organization that lifted the Apollo 13 space shuttle's oxygen tank from the maintenance module and accidentally dropped it (and unknowingly broke it) did not communicate the matter to the organization responsible for filling the container. When the other group encountered problems in filling the tank, they did not suspect that the tank was broken and improvised a way to fill the tank (Chiles, 2002). The oxygen tank exploded after two days of traveling while the shuttle was far from Earth on its way to the moon. The explosion damaged the service module and caused oxygen leakage and electrical interference. The mission turned into a rescue operation, where the improvisation of the crew saved them from death in space.

Organizational structures affect interaction and the subsequent flow of information. Viitanen et al. (2022) studied safety culture specialists' work and noted that the organizational positioning affected interactions and information flow, but that there were pros and cons for all organizational arrangements.

6.3.4 Rules and Procedures

Written rules and official procedures have a strong role in guiding the operations of safety critical organizations.[1] Rules are often considered to be a way to make the activities of humans and organizations more reliable. Procedures are often not considered to be a tool that operators can use to control the process but rather as something that controls the operator (cf. Schulman, 1996) – something that makes sure the operator does not make mistakes or otherwise endanger the system. However, routine noncompliance with written procedures and local adaptations are often the norm rather than the exception (Bourrier, 1996; Vaughan, 1999). What is more, these adaptations are typically necessary to get the work done and meet all the potentially conflicting goals. Only when things go wrong are these adaptations considered negative.

We need rules and standard operating procedures in order to save resources and pay attention to out-of-the-ordinary events. This is paradoxical since sometimes we can miss these events for the exact same reason, routines and the confirmation bias (see Chapter 2) created by the procedures. Studies of decision making have shown that the more we need to make decisions and choices about mundane daily issues, the less capacity we have for anticipating the future and reacting to various contingencies (Leisti & Poskiparta, 2022, p. 212). Procedures thus enable us to reserve cognitive capacity for handling more complex issues, unless the complex issue deals with the applicability of the procedure in the specific situation.

Organizations thus need flexible routines and ways to recognize situations that fall outside the standard operating procedures. They need to create a systematic approach to rules and their use – a rule on how to develop, interpret, and revise rules! Chapter 9 offers some guidance on this.

Safety critical organizations have a wide variety of rules. Hale and Swuste (1998) divide safety rules into three categories:

1. Goal rules that set objectives, such as
   a. Maximum value for toxic substances in the air;
   b. As low as reasonably achievable, ALARA.
2. Process rules that determine how decisions about operating methods and actions should be made, such as

   a. If normal instructions for use cannot be applied, the actions taken in the situation are determined jointly by the shift manager and the maintenance supervisor on duty.
   b. The calculation of risk limits must be based on certain emission and spread models.
3. Action rules that define concrete actions or required system states, such as
   a. Protective eyewear must always be used in laboratory facilities.
   b. Lifting appliances must be inspected at least once a year by a party competent for the task.
   c. Pressure vessels must have at least two independent pressure-reducing systems.
   d. Smoking is not allowed.

According to Hale and Swuste, action rules maximize both the benefits and disadvantages of rules. From the point of view of the rule follower, action rules save time and effort in familiar situations. They also clarify tasks and responsibilities, providing a basis for self-assessment and verification of compliance. On the other hand, action rules may make people blind to new or changing situations in which existing rules do not apply. Furthermore, detailed procedures may cause resentment (employees may feel they are being watched over or that they are not trusted). This, in turn, may increase rule violations. Detailed procedures may repress individual initiative and consequently can weaken learning in the organization. Maintaining rules and ensuring that they are followed also consumes resources. Hale and Swuste (1998, p. 169) suggest that rules should be treated as progressive restrictions on freedom of action. Goal rules define the objectives, process rules explain how to make decisions, and action rules impose actual restrictions on activities.

Procedures typically combine process and action rules by describing methods and specifying concrete steps, but they also often incorporate goal rules and serve as cognitive artifacts that guide, constrain, and document work.

It is important that organizations examine their rules and the foundations that they are based on, as well as determine the type of safety impacts that may result from bending rules or misinterpreting instructions. Furthermore, rule bending usually indicates system level problems and not only individual disobedience (Reason et al., 1998). Rules are bent because the inherent uncertainties of the system are not as apparent as pressures to be efficient, or because the rules do not consider the workplace reality (see Chapter 3). In Chapter 9 we will elaborate on rules as a safety management tool.

6.3.5 Contractors and Networked Organization

In addition to the complex structure and operating processes of one's own organization, the distribution of responsibilities among several companies poses a challenge. It has been noted in construction projects, for example, that generally the more complex the organizational structure of the project, including contracting, the greater the lack of clarity regarding safety responsibilities (Arditi & Chotibhongs, 2005). Assumptions, duplications, and gaps were found to be commonplace due to poor communication and lack of proactive management. Farrington-Darby et al. (2005) identified that the use of subcontractors greatly affected trust, negatively impacting on safety culture within a rail delivery unit.

Networks and contracting are part of the business model nowadays. This can be a challenge for safety management systems that traditionally focus on one organization. Traditional safety models such as the Swiss Cheese model or the High Reliability Organizations theory typically assume that work is conducted and safety is managed by one organization (Hayes & Tillement, 2022).

The networked organization, consisting of multiple independent companies, creates new risks and a culture of its own, which is difficult to consider with the typical safety methods of auditing, assessments, and management reviews. Further challenges are caused by each company relying on its own safety management system yet also trying to fulfil the requirements of the contract and the owner/main contractor's safety management system. Each company also has its own business strategy and core expertise, a limited responsibility of the whole, and its own role in relation to the authorities and other stakeholders, as well as its own culture.

Dechy and Largier (2022) noted, based on their study of contracting in nuclear power plant maintenance, that the power relations may block cooperation and learning since failures tend to be attributed to the contractor, overshadowing the customer contribution that must provide adequate conditions. Instead of separate organizations, in terms of learning, these organizations should be considered as one system. At the same time, it needs to be acknowledged that this system contains multiple subcultures that most likely identify more with their work tasks or organizations than an abstract network.

A lot of critique has been targeted at companies that always select the cheapest contractors, despite potential differences in quality. This has led to an increased focus on quality and safety issues during the bidding process. However, what it means in practice is all but clear. Smith (2024) has criticized the use of injury rate data as a criterion in contractor selection or performance monitoring. The critique is due to the inherent limitations of injury rates as a safety performance indicator as well as to their being easily manipulated. He further points out the challenges of including contractual requirements for safety, as a contractor will most likely agree to many requirements in the bidding phase, but these may end up being paper-based exercises aimed at meeting contractual requirements rather than improving safety. Of course, these challenges are not unique to contractor management but rather apply to safety management in general. In terms of safety management, supervision and monitoring of contractors' safety management and culture for safety may be more demanding (due to contractual reasons), thus aggravating the challenge.

Despite the challenges, safety issues need to be considered already during contracting. Later it is much more difficult, and expensive, to require certain practices or approaches if these issues have not already been explicitly mentioned in the contract. The contract should also give the owner the possibility to audit and monitor safety management and culture related issues in the supply chain. It is also better to avoid contractual requirements for quantitative measures for good "safety culture". Rather, emphasis should be put on contractors being aware of their own culture and its strengths and weaknesses. Contracts should also allow sharing of safety related information between contractors – and should specify the difference between safety and security information. In many cases, I have noted that contractors may be hesitant to share any information if they are not clear what counts as safety related information to be shared and what counts as security related confidential information, or other commercial information.

The management of contractors requires a clear definition of contractual arrangements and a clear specification of the job. Selection of the contractor needs to be done in such a manner that quality and safety issues are adequately considered as criteria in addition to cost and schedule considerations. Safety and quality requirements need to be clearly specified to the contractors before bidding is concluded, otherwise the contractor will be making the bid with insufficient understanding of the resources and competence required – and may need to cut corners later in order to make a profit. When the contractor has been selected, they need to be familiarized with the work, the work site(s) (if applicable), and the organization that they are working for (e.g. expectations for quality and safety and ways of working, reporting requirements, and safety policy).

Tillement and Leuridan (2022) present three lessons based on their case study of outsourcing in the nuclear industry:

- A correct assessment of the nature of an activity before outsourcing is essential.
- Hidden temporalities and embracing distant past and future are key in the contracting-out process.
- Contracting out is a dynamic process, which requires heedful interactions between principal and contractor in the long term.

In addition to this, the nature of the contract and the constraints and possibilities it creates for collaboration, joint learning, and management of safety need to be carefully considered beforehand. Additionally, when the contracted work is carried out in practice, adaptations need to be made (see Work-as-Imagined and Work-as-Done in Chapter 3). Ways of safely adapting to local conditions and ways of learning on a system level need to be considered. Contracting brings additional complexity that needs to be managed by other organizational arrangements.

Many contractors take an explicit part in the company's safety assurance, as certifiers, auditors, assessors or advisers. There is a conflict of interest there similar to credit-rating agencies or financial auditing companies as all of them are being paid by the company they are supposed to independently assess (Bazerman & Tenbrunsel, 2011). And with several potential contractors to choose from, with various levels of rigor and competence, companies can essentially choose the safety level, or safety awareness, they want to achieve. Furthermore, contractors are especially vulnerable to the unintended effects of setting specific numerical goals with ambiguous means to achieve them (see Section 6.6).

6.4 EMPLOYEE AND STAKEHOLDER INVOLVEMENT

As most safety theories (see e.g. descriptions of Man-Made Disasters, High Reliability Organizations, Resilience Engineering) as well as accident investigations have shown, typically someone had perceived the risks before they actualized, but this someone did not – for various reasons – get their message across to senior management. Involving employees is thus crucial for safety, not only because it motivates them but also because in complex organizations it is necessary for forming an overview of the risks and their management.

Organizations need to actively involve their own staff (or their representatives) as well as external interested parties in using and developing the safety management system to control risks over time. Involvement is also necessary because no single person has the information needed to manage safety – collaboration and sharing of information are required in complex organizations.

In some organizations, involvement can mean involving the customers (or patients) or contractors in the process of safety management. They can provide a diverse view of the risks and their current control measures, and in general can help in identifying both strengths and weaknesses in the organization.

External stakeholders include the labor unions. Their role, responsibilities and power differ between industries and countries. At best, labor unions can provide the management with a view into the workplace reality and work conditions and act as a safety resource for the organization. For personnel, labor unions can act as an important source of social identity and a strong interest group. The degree to which unions contribute to safety depends not only on their formal role, but also on the quality of trust and dialogue between the various stakeholders.

Safety committees are one way of involving personnel. Yet these committees are often set up without a clear idea of how they should influence safety in the organization. Typically, a safety committee is required by national law for companies of a certain size (e.g. in Finland, organizations that have over 20 employees need to establish a safety committee). However, employee involvement should be more daily than committee participation of selected representatives.

To involve employees in safety development, an organization needs to implement methods for employees to make recommendations and report observations or concerns. It is also necessary to ensure that two-way communications are in place and a rapid response is provided concerning the results of any safety-related reports and recommendations. Furthermore, easy, practical access to safety-related information through quick and reliable communication methods needs to be provided to all employees (Roughton et al., 2019, p. 103). Chapter 9 provides further details on reporting of observations.

Employees need to be involved in the writing of rules and procedures to ensure they are applicable in practice. In addition, active involvement in, for example, risk analyses, work development, and change planning is beneficial for safety.

Involving employees is one of the challenges of leadership and it is discussed in more detail in Chapter 7. Employee involvement is an issue that needs to be considered in all management, not only as a specific topic of its own. When it is made a specific topic or a goal, it risks the same disassociation from work as the concept of safety culture (Chapter 5). This can mean that employees are involved because it is a practice recommended by industry best practices or management standards, not because it is genuinely believed to be beneficial for the organization. As with other interventions that miss the actual work, these may only contribute to a wider gap between Work-as-Imagined and Work-as-Done (Chapter 3) as well as to loss of trust between management and the workers.

6.5 ORGANIZATIONAL DECISION-MAKING

Delegation of authority refers to the process of entrusting responsibility and decision-making power from one individual to another within an organization. Responsibility, accountability, and authority for safety-critical tasks need to be clearly defined and delegated.

Decision-making in organizations is made challenging by the complexity and the impossibility of ever completely anticipating all the (direct and indirect) consequences of our decisions. In decisions concerning safety, we typically try to gather all the facts and increase the certainty we feel about the consequences of the various potential choices. Analytical thinking is often seen as ideal in safety-critical contexts. As we discussed in Chapter 2, our everyday decision-making is intuitive rather than analytic. It is also strongly guided by our tendency to seek confirming rather than disconfirming evidence or not to seek evidence at all but settling for the first impression. Especially in situations where we are sure of ourselves, we seldom bother to analytically evaluate the alternative options or reflect on our decision-making premises. Analytical thinking is activated when there are conflicting goals or other obstacles in our path.

There is always both an intuitive and an analytical component to decision making in organizations. We should not aim to reduce either, as both are needed. Only relying on analytical thinking may slow us down considerably and even prevent us from making decisions as the uncertainty is experienced as too high. In some safety-critical domains this situation is described as "analysis paralysis", when the organization keeps on collecting more information and analyzing it in more detail, ruminating on it without being able to make a decision. There will never be a situation where all the information is collected and the decision is certain. In the end we always need to rely on our intuition (aka experience and tacit knowledge) to stop analyzing and decide. If we deny the intuitive and emotional components of decision making, we are ultimately unable to make any decisions. This need for adaptive action is one of the key requirements for leading complex systems (Chapter 7).

When making decisions about complex issues such as safety, it is good to remember that there are no optimal solutions to wicked problems (see Chapter 4), and compromises and sacrifices must always be made. It is also important to recognize that different goals are qualitatively different and cannot be measured with the same yardstick, for example safety and efficiency (see later in this chapter on indicators).

Trust and psychological safety are important prerequisites for good organizational decision making. Leadership plays a key role in creating an environment where good decision making is possible.

The decision-making heuristics, especially the confirmation bias discussed in Chapter 2, can be somewhat counteracted when decisions are made in collaboration with others. When making decisions with other people, we have to justify why our own arguments are better than those of others. When we say aloud our own thoughts, we can also evaluate them more critically (cf. Weick, 1995). With other people, we are also exposed to opinions and attitudes that differ from our own, which can help us become aware of our own preferences and prejudices (Leisti & Poskiparta, 2022, p. 174). However, if the group is not diverse enough in their attitudes and competence or lacks psychological safety for speaking up any ideas or concerns, there is a danger of groupthink (see Chapter 2), which typically leads to a decline in the quality of decisions.

Thus, to improve decision making in organizations, it is first important to recognize the decision-making situations – when a formal decision-making process is required and when intuitive decision-making is enough. In terms of learning, the decision-making process is more important than individual decisions. There are a multitude of variables that can affect the quality of a single decision, but if the decision-making process is sound, decisions will be better on average. If the decision has been made according to the process, and it is in line with the organization's values and strategy, its quality should not be judged merely by the (delayed and non-linear) outcomes. However, in case of repeat failures, it is important to inspect the process, including values and strategy. Similarly, in case of successful decisions, it is important not to overestimate the effectiveness of the decision-making process and acknowledge the inherent uncertainties of complex environments. Utilization of different tools such as SWOT-analysis or the premortem exercise (see Chapter 9) can help to improve the quality of decision making.

6.6 SAFETY ASSURANCE – MONITORING AND ASSESSMENT

Monitoring and evaluating performance is a vital part of an effective management system. Monitoring and evaluating includes the application and effectiveness of the management system and the sociotechnical system itself. These need to be planned and carried out in a systematic manner, including their documentation and follow-up.

Monitoring and improvement are closely related concepts that are typically treated separately in management system standards and guidelines. However, they are tightly related, and methods of improvement such as corrective action programs and incident investigations provide important information for monitoring purposes. Likewise, different types of assessments (internal and external assessments of the management system, culture and/or leadership) are needed both for monitoring and improvement.

There are several different ways of monitoring and assessment:

- Management oversight, including supervision and management reviews;
- Independent oversight, including external assessments;
- Self-assessments of the management system, culture and leadership;
- Safety performance indicators, data analyses of performance data, surveys;
- Auditing, internal and external – including auditing of processes, performance, and products;
- Incident investigations (as part of learning from experience);
- Operational data (if operations allow such data, e.g. process industries, aviation).

It is important that monitoring is able to give early warning to the organization about degradation of safety but also confirmation of the adequacy of the current measures. Furthermore, in case deviations or improvement needs are identified, there needs to be a clear mechanism for translating these findings into corrective actions.

Woods (2009, p. 499) highlights a lesson from the *Columbia* accident: "Organizations need mechanisms to assess the risk that the organization is operating nearer to safety boundaries than it realizes – a means to monitor for risks in how the organization monitors its risks relative to a changing environment."

Key Performance Indicators – What You Measure is Not What You Want To Get

The application of various safety performance indicators has become routine across all industries. Ideally, these indicators can provide safety professionals and top management with insights into the current challenges and advancements in safety. However, it is common for companies to measure easily quantifiable but less important aspects merely because they can be measured accurately. Senior management often focuses on, and manages, only what gets quantitatively measured. It's essential to remember that indicators are mere samples, proxy measures, of the company's safety level, and management should concentrate more on the underlying processes rather than the indicators themselves. Indicators should help leaders to manage the organization, instead of the indicators becoming objects to be managed (Poole & Dekker, 2024).

An indicator can be any quantitative or qualitative measure intended to produce information on a relevant issue (Reiman & Pietikäinen, 2012). Indicators are proxy measures for issues identified as important in underlying models of safety (Wreathall, 2009, p. 494). Typically, however, indicators focus on the negative. Measuring safety by its absence or setting a zero-injury goal has several drawbacks. Sharman (2019, p. 58) notes that striving for zero injuries builds a culture of intolerance and increases workers' fear of failure, affecting reporting and leading to underreporting or even falsification. *Zero accidents should be an outcome or a vision but not a specific numerical goal.*

Safety management literature makes a distinction between lagging indicators, which are reactive and often indicate negative outcomes that have already occurred (such as fatalities, the Lost Time Incident Frequency Rate (LTIFR), the Total Recordable Incidence Rate (TRIR), and process events), and leading indicators, which are proactive and include positive aspects observed and recorded before any adverse event occurs. The aim is to prevent adverse events from happening – those measured by lagging indicators. Leading indicators can consist of training, audits, near-misses, risk assessments, and various safety initiatives, often measured in straightforward terms like training hours per employee or the number of scheduled audits conducted.

The relevance of both types of indicators has been debated. Lag indicators have been criticized for focusing on past negative events. There is also limited evidence that lag indicators that measure minor incidents can be used in predicting major incidents – even when staying within the scope of a similar hazards, as in the construction industry study, which found no link between total recordable injury rate (TRIR) and fatalities (Hallowell et al., 2021).

Another challenge lies in understanding what safety means and how different safety types require varied measurement approaches. Hopkins (2000, 2019) has for a long time argued that the commonly used LTIFR can mislead management on major hazard risks (see also Smith, 2024). The two accidents at BP (Texas City and Deepwater Horizon) showed the consequences of confusing control of minor and major hazards (or local and core task related, see Chapter 4). Likewise, implementing corrective actions based on minor injuries may lead to the organization overlooking occupational risks that can cause fatalities (Poole & Dekker, 2024). This is a good reminder that local hazards also vary greatly in their potential severity.

A common solution to limitations of lag indicators during the past couple of decades has been the implementation of various leading indicators. However, Smith (2024, p. 6) criticizes leading indicators for focusing on either activity (how many times something is done) or bureaucratic compliance (whether a form has been completed) instead of the quality or effects of the activities. He also notes the danger that a metric becomes "corrupted" if the pursuit of the achievement of the metric becomes more important than the purpose of the metric. In those cases, the process becomes the purpose (ibid.). Poole and Dekker (2024, p. 2) agree and remind that "a measure with a target stops being a measure".

For example, some companies put targets on their safety indicators in ways that inadvertently encourage underreporting of minor incidents or near-misses. Similarly, organizations sometimes reward personnel based on the number of safety observations made, confusing the indicator (number of observations) with the goal (increasing the number of observations). Others have signs at the entrance to their site or even offices that indicate how many days they have had without anyone reporting a lost-time incident. This typically leads to underreporting or innovative ways of turning lost-time incidents into "replacement work", "light duty", or "at home accidents" (Busch, 2016; GAO, 2012; La Duke, 2020). The more non-reporting (or lack of incidents) is rewarded, the worse it gets (see below). This means that organizations in fact become dumber, not smarter, when relying on simplified indicators such as lost-time injuries (Poole & Dekker, 2024).

Smith (2024, p. 106) also emphasizes that organizations should not assume that their safety indicators are valid but rather should actively test their validity by questioning the information underpinning the metrics. Using measures which individuals have little control over[2] can be detrimental to their sense of control and sense of responsibility (see Chapter 5) and lead to decreased commitment to safety in the organization. The problem is not so much in the quantification per se, but rather the problem lies in what the organization does with the numbers. Specifically, the problem is in how the same numbers are used simultaneously to both drive and measure/monitor performance (Reiman & Pietikäinen, 2012).

More than a decade ago, we differentiated among outcome indicators, monitor indicators, and drive indicators (Reiman & Pietikäinen, 2012) partly due to the same reasons Smith notes above. Outcome indicators, typically lagging, measure system outputs. Monitor indicators gauge activities, processes, and capabilities within the system. Drive indicators involve issues deemed significant by senior management and set as organizational goals. As soon as these goals are promoted and rewarded, they stop being mere monitoring or outcome indicators and instead become focal points of interest. This shift changes what is measured, transforming the indicator into a new area of interest whereby effective activities will reflect in the monitor and outcome indicators.

Normalization of deviance (Vaughan, 1996), which was introduced in Chapter 3 (see also the accident case at the end of Chapter 2) impacts safety indicators' usability. Normalization causes potential danger signals to be disregarded or misinterpreted as normal, and thus not reported. Traditional outcome indicators struggle to capture normalized signals since they involve practices, variations, actions, and events not yet resulting in adverse outcomes. Qualitative auditing or organizational safety (culture) assessments might detect incremental norm departures if they are conducted thoroughly. Addressing normalization includes measuring and improving related organizational aspects like resource scarcity, production pressures, unruly technology, structural secrecy, intolerance of dissenting opinions, and distant information patterns (Dekker, 2011).

One consequence of misunderstanding different types of indicators is the conflation of prediction and influence. When an indicator is chosen as critical, personnel and management focus on optimizing the indicator, sometimes afterwards using the same indicator to monitor performance. If we accept that indicators sample some underlying phenomena, such as system safety, efforts to boost one indicator must not be presumed to automatically affect the underlying phenomena unless this effect is shown by other unbiased indicators. The same issue applies when optimizing single questions on employee surveys (see Chapter 11). Furthermore, what an indicator tells about the underlying phenomena changes as the organization and the way it produces safety changes – for example, when there are a lot of unmanaged hazards and people start reporting near-misses, these reports can at first actually act as a proxy measure of safety (combined with indicators of lost-time incidents, etc). However, as there are less easily perceived hazards, the observations either decrease or, as often happens, under management pressure to make more observations, personnel report more and more minor and insignificant issues.

Using health as a metaphor helps illustrate safety measurement challenges. Just as individual health encompasses more than the absence of illness, safety requires active pursuit – nutrition, exercise, vitality, hobbies, social network, and so on. People often take health for granted until negative signs appear; similarly, safety involves constant activities, processes, and mental states contributing to organizational safety levels. Noticing a lack of incidents or declining trends isn't enough; understanding why things are safe and evaluating current safety management processes are crucial. Direct measurement of safety is impossible because safety itself is not an outcome that can be captured by indicators. Safety is a dynamic non-event (Weick, 1987), and as such, non-events cannot be quantified or characterized.

We need to examine the "dynamic" aspect to understand how non-events are created, recognizing that we will never actually capture the non-event itself (Reiman & Pietikäinen, 2012). This means that indicators alone are never a sufficient method for monitoring the level of safety. They can provide one source of information on how safety is managed, but other, more qualitative measures are also needed. Safety management should centre on managing the sociotechnical systems, not just optimizing specific indicators. Safety assurance relies on continuous attention to past deficiencies via lagging indicators, monitoring of current organizational capabilities, and processes promoting safety through leading drive indicators.

Safety Case

A *safety case* refers to a systematic assessment and documentation of major hazards and their management in a sociotechnical system. In essence, it is the presentation of evidence that the particular system is safe, or that risks to the people and the environment are as low as reasonably practicable (the so-called ALARP principle).

A safety case is typically required for chemical and petrochemical installations, for example, as well as railways and aviation. In the nuclear industry, the safety case is typically documented as the Final Safety Analysis Report (IAEA, 2021a) and it has been in use since the 1960s. After the Three Mile Island nuclear accident (see Chapter 2), requirements for probabilistic risk assessments and demonstration of defense-in-depth were added.

The accident at offshore Piper Alpha in 1988 was a turning point in the oil and gas industry in terms of formal safety management requirements, including the safety case (Cullen, 1990). In the railways, the Clapham Junction rail crash in 1988, when three trains collided in Clapham, London, was a similar turning point (Hidden, 1989). For major hazard installations within the European Union, the Seveso II directive, adopted in 1996, required a structured safety management system and a safety report, their equivalent of a safety case.

The safety case runs the same risk of becoming a paper exercise as the safety management system (cf. Smith, 2024). The document may end up as regulatory delivery that reflects wishful thinking rather than reality. Modern safety cases include management and culture aspects, which is a good in principle but in practice runs the risk of separating the Work-as-Done from the Work-as-Imagined (see Chapter 3). Subsequently, it runs the risk of separating "safety culture" from the "company culture" (see Chapter 5).

For example, the IAEA Safety Guide for the Final Safety Analysis Report notes that "the information provided … should cover establishing, assessing, sustaining and continuously improving effective leadership and management for safety". In addition, several specific issues need to be addressed, including:

- The different management processes aimed at ensuring the overall objectives for safety, and how these processes are established, controlled, monitored, and reviewed, while ensuring that safety is given the highest priority;
- How goals, strategies, plans, and objectives are established by the organization, consistent with the organization's safety policy;
- The overall accountability for the management system and the assignment of individuals for the coordination, development, application, and maintenance of the management system;
- How the effectiveness of the management system will be monitored and assessed, including all processes and arrangements made to ensure continuous improvement;
- How the management system establishes the framework to foster and sustain a culture for safety (IAEA, 2021a, pp. 116–118).

Documenting these issues can be a good exercise for the organization. However, rather than describing in detail the actual management processes and working practices, it is good to refer to the management system documentation for the most up-to-date practices. Otherwise there is a risk that the safety case hampers continuous improvement since it is more time consuming to change, and changes require regulatory approval.

The safety case and the safety management system are closely related. The safety case is produced when the system is designed and commissioned, and the safety management system serves as a way to monitor that the safety case remains valid. The safety case also needs to be regularly updated, based on safety management system inputs, system changes, and periodic reviews.

6.7 ORGANIZATIONAL LEARNING AND IMPROVEMENT

As we have already established, learning and improvement in complex organizations are both essential and challenging. At an organizational level, learning may be embedded within:

- Organizational beliefs and assumptions: culturally accepted views about the system (what hazards are present, what risks are important, what is normal, what is taken for granted, what should be ignored);
- Organizational routines, procedures, and rules;
- Organizational structure and relationships within organizations;
- The design of equipment and implementation of technologies within the sociotechnical system;
- The knowledge of people working within or interacting with the sociotechnical system.

To learn from failure, we must first identify that we have in fact failed. This is difficult in complex organizations where feedback is delayed and in many cases the success or failure of our actions depends on countless other actions and their interactions. We succeed or fail by adapting to the environment. But we often learn based on the immediate consequences of our actions, and not on their overall adaptive effectiveness on the entire system (see Chapters 2 and 3 on the challenges of learning). Without systematic organizational learning mechanisms, our learning is typically local.

Very often, formal learning efforts in organizations are driven by the severity of the accident – the more severe, the more resources and attention are given to the investigation and the learning process. However, in complex systems, outcomes seldom match the effort one-to-one. There is always the element of chance, involving external factors as well as the non-linearity of the interactions that may cause surprises. Based on the outcome alone, it is impossible to know how much there is to learn, if the aim is to understand and improve the organization. Furthermore, in complex organizations, the exact same accident will never happen again. When we strive to learn from accidents in complex systems, we need to think wider instead of deeper, and instead of decomposing the system into its constituent parts, we need to take a step up and see connections and similarities (cf. Dekker, 2015).

Learning in safety-critical contexts cannot be driven only by errors and mistakes. More proactive measures aimed at improving the ability of the organization to recognize the boundaries of safe performance, and to succeed in the future, are needed. At the same time, errors and near-misses should be considered learning opportunities, and human performance issues should be tackled in incident investigations. Organizations must learn to anticipate potential surprises and recognize how vulnerabilities evolve.

Weick and Sutcliffe (2007, p. 72) point out that learning from unexpected events should not lead to more elaborate defenses, such as new rules and restrictions, but more elaborate response capabilities. Recovery from events does not mean the organization returns to the previous status quo, but rather to a position where the capability for successful recovery in future is improved.

Chris Argyris and Donald Schön developed the idea of single and double-loop learning (Argyris, 1999). *Single-loop learning* involves detecting and correcting errors within a set framework, leading to incremental changes. It focuses on fixing problems without changing the underlying structures and processes, aiming for small adaptations to complete tasks. This approach often seeks "quick fixes", which can overlook the need for double-loop learning, potentially resulting in repeatedly addressing symptoms instead of causes. Single-loop learning can be compared to a thermostat adjusting temperature based on feedback. It is fast and sufficient for many immediate needs, and often organizations excel at it, offering flexibility as long as conditions remain consistent. Different reporting systems frequently facilitate single-loop learning by implementing mechanisms to track corrective actions for individual observations. As we pointed out in Chapter 3, organizations need to balance between closing individual issues and gathering evidence and analyzing to identify wider patterns. This means that in addition to assessing the significance of individual issues, organizations need to analyze trends and significance of several similar (or similarly caused) issues concatenating.

In *double-loop learning*, improvements extend beyond the specific situation. The values, assumptions, and policies that initially prompted actions are also examined. This involves altering the understanding of how processes function, what influences decisions, and the assumptions underlying actions. The benefit of double-loop learning is its potential to prevent future errors by addressing systemic issues and so-called root causes. This requires more in-depth investigation of events and typically trending of multiple minor observations than a single major observation.

The goal of organizational learning should be to increase the organization's potential to succeed under varying conditions, as well as to improve adaptive capacity, including the ability to learn from future successes and failures. To achieve this, organizations must strike a balance between implementing quick solutions (or enforcing existing solutions such as rules and instructions) and fostering a deeper, more deliberate comprehension of events with double-loop learning. Both single-loop and double-loop learning are applicable to successes as well as failures. The challenge for organizational learning is to learn the right lessons, whether the outcome has been a success or a failure. We cannot escape the effect of hindsight, whether we analyze success or failure (see Chapter 3).

For example, BP's own investigation of the Texas City refinery explosion (see Chapter 4) is a good example of how the focus of analysis affects the identified root causes and subsequent recommendations. BP identified worker fatigue as one of the root causes of the accident. However, according to Silbey (2009, p. 353), BP failed to notice that worker fatigue was associated with management policies that relied on routine overtime to meet production needs instead of hiring additional employees. The subsequent Baker Panel investigation noted this in their analysis (Baker, III, 2007). It is also emblematic that the BP Report used the term "safety culture" more than 390 times in the investigation report but in its analysis and recommendations focused on individual actions rather than on systemic processes (Silbey, 2009, p. 352).

In addition to learning from one's own successes and failures, safety-critical organizations need to learn from external events. This means the successes and failures of other safety-critical organizations, not only in the exact same industry but also in other safety-critical domains. Very often organizations limit their learning efforts to other same-technology companies. This makes sense when the lessons are technical, but learning to improve safety includes organizational and cultural lessons that are more generalizable than technical lessons.

Learning from external events is made more difficult by the biases and tendencies discussed in Chapters 2 and 3. Organizations need to actively avoid simplifying their external reality and considering themselves so unique that they cannot learn anything from others.

Benchmarking has other challenges too. All organizations are unique in the sense that all complex adaptive systems are (see Chapter 1), with their own history and culture. Best practices successfully implemented by one organization do not necessarily translate to the environment of another organization. In fact, what is needed is translation, not copying.

Continuous improvement is a term that is used in many safety-critical industries to denote the idea that safety needs to be actively and continuously enhanced; otherwise there is a risk of degradation. This improvement includes technological investments as well as organizational and human-related improvements. One of the areas that needs continuous improvement is competence. Training as competence management tool is described in Chapter 9. Continuous improvement needs to be an organizational goal, and it requires active effort from organizations. Sometimes it also requires organizational changes.

6.8 ORGANIZATIONAL CHANGE

Kurt Lewin proposed his three-stage model of organizational change during the 1950s: unfreeze, change, freeze. A key insight from Lewin was that a change to something is always a change from something. A change is always a departure from a previous state or condition. This means that in order to change something, you first need to understand the current state. Lewin famously introduced the idea that one needs to unfreeze the current state in order to transition to the new state.

In complex organizations, the question is not only about unfreezing them in order to change them but also about freezing them to assess and understand them (see Chapter 11). The balance leaders need to tackle is between the frozen image of what the organization was and the unfrozen, constantly changing organization of what is, and what will be.

There are numerous change models and approaches available, with different assumptions about what change is, or what an organization is (Weick & Quinn, 1999). A typical differentiation is between models of intentional, episodic change (*a planning approach*) and emergent, continuous change (*a process approach*) (Alvesson & Sveningsson, 2024). The planning approach has produced various n-step models of change. These are typically top-down-driven models starting with a vision and motivation for change; involvement of employees and implementation of the change, including communication and learning of new skills and practices; and finally some type of consolidation and evaluation (Palmer et al., 2022). The term *n-step models* refers to the tendency of the models to introduce a various number of often linear steps to be taken to implement the change.

One of the problems of the planned approach is that it typically underemphasizes changes that take place all the time in social systems. According to the planning approach, organizational changes are episodic and planned in advance. They approach organizations from a mechanical perspective and often consider the change that is planned as the only change taking place, and that this change can be controlled in a top-down way by management (cf. Weick & Quinn, 1999). Thus, change models that are based on the idea of change as planned and episodic usually emphasize the planning phase more than the implementation phase (Alvesson & Sveningsson, 2024). This means that change projects driven by these models may be unprepared for the surprises and adaptation needs that happen during the implementation phase – when Work-as-Imagined meets Work-as-Done.

The process approach strives to approach organization change as an emergent, processual, continuous, and unpredictable process. Sensemaking (Chapter 2) is a key element affecting change. The process approach seeks a more thorough understanding of the politics, context, and substance of change by acknowledging local and emergent interpretations, experiences, and sensemaking (Alvesson & Sveningsson, 2024, p. 33). The process approach is more in line with the cultural perspective endorsed in this book.

From the safety perspective, organizational changes are relevant for two reasons: First, changes that aim to improve productivity, cut costs, or accomplish some other organizational goal may have unintended safety effects. Safety thus needs to be considered in all change efforts. Second, improving safety is more and more often an explicit goal of organizational change projects. An unfortunate feature of these change efforts is that very often they start (and end) with quantified lagging indicators (see Chapter 9) such as lost-time incidents as both the problem to be solved and the solution. Furthermore, these change programs often approach change from the episodic, or planned, perspective, missing the changes taking place all the time in the organization. What they miss is the *nature of change in social systems* (see Chapter 12).

Organizational change may require utilization of any of the eight leadership principles that will be depicted in Chapter 7. What change means in a given system depends on the current state of the system. A young and innovative company may need to structure their activities and constrain them in order to be more predictable. Another type of change may be required in a highly complex system that has operated for many years and developed subcultures and local rules threatening the system level goals. More typical changes aim to increase the organization's innovation capabilities or effectiveness. Whatever the explicit goals of the change, organizational change should always consider the organizational core task and the hazards (see Chapter 10). Hopkins (2012, p. 101) points out that when BP reorganized its activities into a streamlined, decentralized business model during the 1990s, "the change was a commercial success, but the seeds were sown for both the Texas City and the Macondo disasters". This was due to the fact that the change left the refinery managers on their own, without clear process safety expectations, responsibilities, or accountabilities (see Chapter 4).

Often organizational changes are made in order to improve information flow and thus the efficiency (sometimes even safety) of the activities. The organization may have identified that the interfaces between departments or work units are where the problem lies (and where the information gets buried) and decides to introduce a new structure to compensate. These attempts typically change the information flow, but they also move the manifestation of the problem elsewhere, to the new boundaries. Structural secrecy and the division of labor always affect information flow, and just attempting to "fix a broken flow" risks misunderstanding the dynamics of the system.

Ramanujam wrote more than twenty years ago that the "current explanations of the organizational origins of accidents understate, or even ignore, the role of organisational change" (Ramanujam, 2003, p. 614). This is still true today. A consistent approach to risk assessment and risk control is needed before and after an organizational change is made. The change management process must include an analysis of the effects of proposed changes on partners, suppliers, and others with which the organization has an interface. Organizations need to balance between making the change process excessively cumbersome, which can lead to tolerating deficient practices to avoid bureaucracy or extra effort required by the process, and making the change process so minimal in analysis that potential risks are overlooked.

All organizational change and development is cultural, intentionally or not. This is one of the reasons so many development projects and change initiatives fail: They neglect the cultural dimension, the importance of human interaction and our need to belong, be heard, and be involved, and the effect of beliefs and assumptions on our behavior.

6.9 SUMMARY – TOWARDS MINDFUL ORGANIZING

This chapter delved into various aspects of organizational structures and management practices. Safety objectives and performance monitoring are established based on organizational strategy. Organizational structure, including communication and reporting lines, significantly influences organizational culture.

A Safety Management System (SMS) is a formal, documented, and verifiable method for managing hazards and risks. It should be integrated into business processes and continuously improved. SMS includes issues such as employee involvement, planning, hazard identification, risk control, competence management, performance monitoring, and continuous improvement.

Organizational structure aims to provide a common view of operations, division of labor, and supervisory relationships, but local practices often diverge from standard operating procedures due to the necessity of smooth workflow. These need to be considered from the perspective of Work-as-Imagined and Work-as-Done.

Specialization versus multi-skilled professionals is a key consideration in safety-critical organizations. Equally important is communication. Effective information flow requires defining adequate communication channels, specifying safety-related information, and ensuring proper feedback mechanisms.

Rules and procedures guide operations in safety-critical organizations. They can save resources but may also lead to rigidity and non-compliance. Organizations need flexible routines to handle situations outside standard procedures.

Managing contractors involves clear contractual arrangements and monitoring performance. Complex organizational structures can lead to unclear safety responsibility issues. Effective contractor management requires collaboration and clear communication. Managing contractors requires clear definitions of job specifications, safety requirements, and monitoring performance, with an emphasis on avoiding purely paper-based compliance. Contracts should allow for safety audits and information sharing.

Active involvement of employees and external parties in safety management systems is essential for effective risk control and organizational safety. Methods include safety committees, reporting systems, and participation in rule-making and risk analyses.

Decision making involves balancing intuitive and analytical thinking. Trust and psychological safety are essential for good decisions. Collaborative decision making can counteract biases but requires diverse and safe environments. Learning from failures and successes involves single and double-loop learning. Organizations must identify systemic issues and adapt to varying conditions. Organizational learning should focus on increasing adaptive capacity and understanding the broader system, rather than just addressing specific incidents. Learning should focus on both successes and failures. Learning from external events is also important.

Monitoring and evaluating performance is vital. This includes management oversight, independent assessments, self-assessments, and safety performance indicators. Indicators should be used to manage underlying processes, not just as targets. Organizations should avoid measures transforming into targets, for example by setting financial incentives for lost-time incidents or amount of safety observations. Indicators can be lagging (reactive) or leading (proactive). Both types have limitations, and organizations must ensure indicators reflect actual safety performance. Overreliance on quantitative measures can lead to underreporting and misinterpretation.

The flow of information does not guarantee that sense is made of it; rather, the organizational processes should aim at promoting the seeking out of both positive as well as negative information at all levels in the organization. Organizations must continuously work to override the typical human tendencies to seek confirmation and avoid disconfirmation (Weick & Sutcliffe, 2007). Mindful organizing and mindful organizational change are key in this.

Reflection Questions

How does your organization's management system support the shared value of safety?

How does your management system and organizational structure support awareness of and solving competing priorities?

How do your management system and organizational structure support the core task? What aspects of the system are experienced as difficult, demanding, or frustrating by employees? Why is that so?

How do you balance the formalized approach of the management system and the informal competence and adaptive capacity in the field?

How does your organization structure support learning and change?

What cultural assumptions should your organization's structures and management system reflect?

What cultural assumptions about control, trust, and uncertainty are embedded in your current monitoring practices?

How does your organization structure support understanding of current operational realities at all organizational levels? How can employees raise concerns and get their voices heard?

How does your organization structure support constructive challenge from colleagues, supervisors, or other functions/departments?

FUKUSHIMA AND ONAGAWA NUCLEAR POWER PLANTS

On 11 March 2011, a massive 9.0-magnitude earthquake was followed by a tsunami (the Great East Japan Earthquake and Tsunami) off the east coast of Japan. Three nuclear power stations in the region were affected, Fukushima Daiichi, Fukushima Daini, and Onagawa, with Onagawa being the closest to the proximity of the earthquake, about 60 kilometers nearer to the epicenter than Fukushima Daiichi. Despite this, the worst accident happened at Fukushima Daiichi nuclear power station.

The Fukushima Daiichi accident is regarded as the worst nuclear accident since the Chernobyl accident in 1986 in present day Ukraine. Both are rated seven (the highest score) on the International Nuclear Event Scale (INES). The Tokyo Electric Power Company (TEPCO) operated Fukushima Daiichi and Daini, and Tōhoku Electric Power operated the Onagawa nuclear power station.

The Fukushima Daiichi nuclear power plant was designed to withstand both earthquakes and tsunamis. However, the plant was designed to handle tsunamis up to 5.7 meters (19 feet) in height. The tsunami that struck the plant on 11 March was about 14 meters (46 feet) high and it thus exceeded the plant's design basis. In the nuclear industry, this is called an out-of-design-basis accident. Originally the plant was built to withstand a 4.2-meter-high tsunami, but it was later modified to withstand a 5.7-meter tsunami. The design basis for earthquakes was a maximum of 8.9 on the Richter scale. The earthquake was thus beyond the besign basis of the power plant.

The earthquake knocked out the external power supply, and the tsunami flooded the plant's electrical systems and backup generators, which were critical for maintaining power to cool the reactors. Immediately after the earthquake, the units that were in operation at the time (units 1–3) automatically shut down. Due to the decay heat generated by the nuclear reaction even when shut down, nuclear power plants require electricity for safety systems that cool the reactor. With both external power lines as well as the emergency diesel generators out of operation, the reactors' cooling systems couldn't remove the heat generated by the nuclear fuel. This full station blackout situation did not occur at Fukushima Daini or Onagawa. The inability to cool the fuel led to a gradual increase in temperature in the reactor buildings, causing a meltdown of the nuclear fuel in three of the six reactors (Units 1, 3, and 4).[3] Hydrogen-air explosions took place in Units 1, 3, and 4 (Filburn & Bullard, 2016; NAS, 2014).

The backup diesel generators were in vulnerable areas due to choices made during the construction phase. Original design drawings from the US GE were used without considering the local conditions. These plans indicated that the diesel generators were to be placed in the basement or the lowest level of turbine building. Due to Fukushima's location on the coastline, this was a poor position for safety-critical equipment, as they flooded during the tsunami and became non-operational. TEPCO utilized also in many other areas plans taken directly from the United States without carrying out its own critical analysis of whether the plans applied as such to the plant's specific conditions.

The world's attention was on the accident and aftermath of the Fukushima disaster. The causes of the Fukushima nuclear disaster have been largely attributed to cultural factors (weak safety culture and regulatory capture). However, even more interesting but much less familiar is an example of a nuclear power plant that manifested good "safety culture" during the tsunami, namely, the Onagawa nuclear power plant operated by Tohoku Electric Power Company (EPC).[4]

The Onagawa nuclear power plant has three boiling water reactors (BWR Mark I) commissioned in 1984, 1995, and 2002. The Onagawa nuclear power plant was the closest nuclear power plant to the March 2011 earthquake epicenter. The plant experienced "the strongest shaking that any nuclear power plant has ever experienced from an earthquake" (IAEA, 2012) but was able to shut down safely all its three reactors and take care of the residual heat. No radioactive material was released to the environment. The plant units suffered only minor damage, with some internal flooding in unit 2 and turbine damage in all units.

Ibrion et al. (2020) note that

> lack of seismic damages at Onagawa NPS can be attributed to continuous seismic improvements throughout the life cycle stages of design, construction and operation. The safety culture during operation can be credited to both a continuous upkeep and upgrade of systems.
>
> **(Ibrion et al., 2020, p. 9)**

Following the tsunami, many local residents took refuge in the plant's gymnasium, as the reactor complex was the only safe area in the vicinity to evacuate to. This indicates the level of trust the power company had established with the local community, as during the crisis, the locals run towards the nuclear power plant and not away from it.

It has been argued that the safe shutdown of Onagawa in many respects can be attributed to the plant's high level of safety culture (Ryu & Meshkati, 2014). Especially conservative decision-making, a questioning attitude, continuous improvement, and management commitment (see Chapter 5) were clearly present at Onagawa NPS.

Tōhoku Electric clearly practiced conservative decision-making and continuous learning (see Chapter 5). Before beginning construction in 1968, Tōhoku Electric reviewed literature and interviewed local people about past tsunamis. For example, they took into account the massive Jōgan earthquake (estimated magnitude between 8.6 and 9.0) and tsunami of the year 869.

These initial predictions showed that tsunamis in the region historically had an average height of about 3 meters. Based on that information, the company positioned its plant at 14.8 meters above sea level, which was almost five times the estimated height of an average tsunami.[5]

New studies conducted by the power company raised the estimate first to 9.1 meters in 1987 when preparing the license application for unit 2 and then, as a result of in separate study in 2002, to 13.6 meters. Several improvements were made based on the new estimates. Further, Tōhoku Electric learned from past earthquakes and tsunamis – including one in Chile on February 28, 2010 – and continuously improved its countermeasures.

Tohoku EPC was continuously involved in conducting surveys, studies, and simulations for estimating tsunami levels and ensuring preparedness at Onagawa (Ibrion et al., 2020). They also actively sought lessons from earthquakes around the world. Continuous seismic evaluations, regular seismic re-evaluations, and seismic improvement works significantly increased the seismic margins at Onagawa power station. For example, they had reinforced the old administration building with braces and were in the process of building a new seismic isolation building to house an emergency response center. Fortunately, the old administration building survived the quake and acted as a response center.

Tohoku Electric's emergency response was well organized, collaborative, and controlled (Ibrion et al., 2020). Protocols were in place for emergencies, including tsunami, and operators had been trained to cope with a variety of simulated emergencies, including the loss of off-site power. Tohoku Electric established emergency disaster response centers both at the Onagawa plant and at company headquarters.

There was good support from the headquarters throughout the disaster. Supervisors and chief engineers were dispatched to the main control rooms of the damaged reactors to make decisions, and information was sent in a timely manner to all levels of the response team.

Finally, the successful shutdown of the plant can be attributed to top management commitment to safety and interaction with all stakeholders in matters of nuclear safety. Yanosuke

Hirai, vice president of Tōhoku EPC from 1960 to 1975 – a time that preceded the 1980 groundbreaking at Onagawa – was adamant about safety protocols and became a member of the Coastal Institution Research Association in 1963 because of his concern about the importance of protecting against natural disasters. With a senior employee in upper management advocating forcefully for safety, a strong safety culture formed within the company. Representatives of Tōhoku Electric participated in seminars and panel discussions about earthquake and tsunami disaster prevention held by the Japan Nuclear Energy Safety Organization. The company implemented strict protocols for disaster response, and all workers were familiar with the steps to be taken when a tsunami was approaching.

The experts themselves from Tōhoku EPC conclude in their article:

> The key points for safe cold shutdown were first, the plant site grade was higher than the maximum tsunami height and, second, an emergency diesel generator for each unit and one of the off-site electrical power lines remained available. In addition to these factors, preparedness (such as seismic reinforcements for all units, updating of tsunami predictions where appropriate, and regular fire drills and simulator training for loss of off-site-power) contributed greatly.
>
> **(Obonai et al., 2014)**

One of the writers of the above-mentioned article was Takao Watanabe, who was superintendent of the Onagawa NPS at the time of the accident. WANO presented a Nuclear Excellence Award to Watanabe in 2012. They cited the following three reasons for the award.

1. For providing for the safety of many residents of the local community following the earthquake and tsunami;
2. For his inspirational leadership enabling the safe shutdown of three reactors during the largest earthquake and tsunami encountered by any nuclear plant in the world;
3. For his superb leadership in preparing the Onagawa nuclear plant and staff to be ready for emergencies.

In terms of culture development, I believe a crucial aspect noted by WANO was point 3, "superb leadership in *preparing* the … plant". Watanabe was not only a hero at the moment, but he was a hero for a long time when contributing to a healthy culture for safety.

Takao Watanabe declared that, prior to and after the 2011 earthquake and tsunami, the mindset of the Onagawa nuclear power station was to "Handle normal times with emergencies in mind so that you are able to handle emergencies like normal times".

Reactor unit 2 of the Onagawa nuclear power station restarted on October 29, 2024, following significant safety improvements. The restart of unit 3 is pending. Unit 1, the oldest reactor, is to be decommissioned.

What Did We Learn about Safety in Complex Organizations?

The accident at Fukushima Daiichi is often attributed to characteristics of the Japanese culture. Although there were some features of the Japanese nuclear culture in particular that contributed to the accident, the example of Onagawa shows the strength of organizational culture when identifying reasons for success or failure.

The story of Onagawa demonstrated how safety of an organization is a chronic issue that cannot ever be completely solved. Onagawa's success was attributed to their preparation for emergencies. They had a mindset where, even though in many aspects they already had made more safety improvements than, for example, Fukushima Daiichi, unexpected events may still happen, and continuous improvement of safety is always needed.

It is not enough to conduct risk analyses; these analyses need to lead to actions. TEPCO did analyze the tsunami risk but did not initiate adequate corrective actions. The Tōhoku Electric and Onagawa stations not only continuously conducted new studies and analyses but initiated several safety improvements based on the results.

The safety of an organization is affected by decisions taken when the system is designed, manufactured, and/or constructed. Critical safety-related issues are decided when sociotechnical systems are designed, which are difficult, time-consuming, and expensive to change when the system is in operation. Some of these decisions can make life easier during the pre-operational phase (such as the removal of a cliff face to ease moving heavy equipment at Fukushima) but affect safety much later, when the system is in operation. At this stage it may be impossible, or very costly, to upgrade the design.

In safety-critical organizations, workers, including safety professionals, need to work for their successors. They need to leave them with a system that is sustainable in the long term. This requires that personnel have the long-term consequences of their actions in mind. For example, Yanosuke Hirai faced tough decisions with the management board in the 1960s and early 1970s about raising the elevation of the site at Onagawa due to the risk of tsunamis. This incurred a rise in the construction costs, but Hirai was adamant that it was the right thing to do. He was proven right by the 2011 Tōhoku earthquake and tsunami. Hirai himself did not live to see that day, as he passed away in 1986 at the age of 83. One could easily argue that it was his legacy that saved Onagawa.

NOTES

1. Rules and procedures are close concepts and often used interchangeably. Here, rules is the broader term, referring to both written prescriptions and unwritten norms, whereas procedures refers to formal, documented instructions that specify methods and steps.
2. A frequent complaint I have heard from shop-floor workers concerning "x days since last incident" cards or rewarding if the quarterly or annual goal of zero incidents is reached that "it does not matter how well and carefully I do my work, since it can be the guy from the other team that gets hurt". And this does not even count the fact that even hard and careful work is not always enough to avoid incidents.
3. The fuel damage at unit 4 actually took place in the spent fuel pool where the unit had just loaded the fuel from its previous operating period.
4. The 2011 Great East Japan Earthquake is also called the Tōhoku Earthquake due to the location of the epicenter right of the coast of the Tōhoku region.
5. This was the complete opposite of the strategy at Fukushima Daaichi, where a steep cliff face rising to an elevation of 35 meters above sea level was present at the site. After a detailed cost–benefit analysis, TEPCO decided to remove the cliff before the plant was constructed. Filburn and Bullard (2016, pp. 78–80) note that there were three main reasons for the costly leveling of the cliff. First, TEPCO wanted to anchor the reactor and turbine buildings to the bedrock to better withstand earthquakes. In Japan's coastal region, the bedrock was relatively deep, and covered with sediments. Second, as the plant would use sea water for cooling the steam condensers, pumping water to a higher elevation would cost much more than pumping it to sea level. Third, had the cliff remained, it would have been very difficult during construction to get the large components, manufactured elsewhere and transported to the site, into place.

7 Leadership and Culture

In this chapter I will discuss the role of leadership in safety. I will first look at the differentiation between leadership and management and then illustrate the various styles of leadership and their relation to safety. After that I will present a framework of adaptive safety management. The framework emphasizes the need for leaders to balance various tensions, and adapt their style to situational requirements. I will present in detail the eight principles of adaptive management and leadership for safety, consisting of issues such as promoting system goals, supporting local actions, prioritizing and deciding, and facilitating novelty and exploration. After that, the specifics of leadership in teams are considered. The chapter ends by presenting two typologies of organizational culture: the Competing Values Framework and the Multi-Focus Model.

7.1 LEADERSHIP AND MANAGEMENT IN COMPLEX ORGANIZATIONS

Leadership is hard to pin down, much like culture and safety. Commonly, a distinction is drawn between managers and leaders. Managers are part of the organizational hierarchy and wield formal authority over their team members. Leadership, on the other hand, is a voluntary act whereby an individual influences colleagues socially to achieve collective goals and drive positive change within the organization. While leaders shape and transform the culture, managers operate within the boundaries and potential of the current culture.

There are a plethora of definitions and approaches to leadership in organizations. To give a few examples, Bass (1990) emphasized that leadership is about interaction between two or more members of a group, and this interaction typically involves the structuring or restructuring of the situation and the perceptions and expectations of the members. Kotter (1996, p. 25) has noted that "leadership defines what the future should look like, aligns people with that vision, and inspires them to make it happen", whereas Chemers (1997) defined leadership as "a process of social influence in which one person is able to enlist the aid and support of others in the accomplishment of a common task".

Similar to how the understanding of the human contribution to safety has developed from individual "bad apple" theories to more systems-oriented and positive conceptualizations, leadership has also recently been viewed from a systems perspective as a process rather than an individual. Also, positive psychology has emphasized the energizing and motivating aspects of leadership (Cameron, 2021; Quinn, 2015). All these developments highlight the increasing understanding of the interplay between culture and leadership (mediated with power). We can thus define leadership as an activity affecting culture by either transforming, strengthening, or weakening some of its structures or changing its dynamics.

At its core, leadership is the process of influencing others to think, feel, believe, and act in a certain way. It is a process of defining what is meaningful and what is not. A leader is anyone who facilitates meaning, motive, and structure in a given situation (Smircich & Morgan, 1982). Leaders frame and shape the context of action in such a way that others can use the meaning thus created as a point of reference for their own action and understanding of the situation (Smircich & Morgan, 1982). Leadership thus shapes the perceived possibilities and constraints of organizational performance.

Leadership is not a formal position in an organization, even though formal management should also act as leaders especially when it comes to safety issues. Recently more and more emphasis has been put on the shared and processual aspects of leadership. Identity leadership refers to a process where leaders work not only to build a sense of shared identity in the teams they lead, but also to

DOI: 10.1201/9781003491156-7

shape the identity-based direction and purpose of those teams (Haslam, 2004; Steffens et al., 2014). According to identity leadership theory, leadership should be seen as a process of group influence that centers on a sense of shared social identity (ibid.). A key to influencing culture is in the leader being seen as "one of us".

When we look at leadership from a cultural perspective, the traditional differentiation between leadership and management gains a new meaning. They are no longer distinct types of activities, but rather both influence and are influenced by culture. However, leadership requires transcending and potentially transforming the current culture, whereas management takes place within the context of the current culture. Both deal with power, but leadership is more about ideological influence, whereas management is about authority (see Chapter 1).

Whatever leaders do and say, employees always interpret it according to their more or less shared frame of reference, including culture and individual experiences. In strong cultures, leaders' behavior is typically interpreted similarly by all employees, if (s)he acts according to the cultural values and assumptions (or directly in opposition to them). In less strong cultures, there is more variation in terms of interpretation of the leaders' intentions and meanings, and more ambiguity in how to respond. This does not mean that strong cultures always have good leaders, since, as I have argued in this book, when it comes to culture, strong does not always equal good.

7.2 LEADERSHIP STYLES AND SAFETY

There is no universally superior style of leadership applicable to any setting or circumstance. The effectiveness of any leadership style depends on many factors, including the risks and hazards associated with the organization's activities. When it comes to leadership in a safety-critical context, a specific form of leadership, safety leadership, is typically highlighted. Similar to generic leadership, there has been a lot of research on the question of which style of safety leadership is most effective. Quite a few different leadership styles have been shown to relate positively to safety, and the conclusions are very similar to those of generic leadership studies: effectiveness of a particular leadership style depends on many factors, including the context/environment (minor versus major hazards, production pressures), the specific situation (normal, abnormal, or emergency), and the followers (their skills, abilities, and leadership-related expectations). A leader should adapt their style based on the demands of the situation, the requirements of those involved, and the challenges facing the company (see Chapter 10 on contextual influences on organizations).

Creating a distinct division between safety leadership and general leadership can negatively impact safety, similar to the adverse effects observed when safety culture is segregated from organizational culture (refer to Chapter 5). Despite this, leaders need from time to time to specifically highlight safety issues and safety as a value. It has also been noted that whatever the leadership style, it makes a difference whether leaders specifically emphasize the importance of safety or not (Mattson Molnar et al., 2019). I argue that it makes sense to focus specifically on safety issues in leadership because of the complexity of organizations and the difficulty of perceiving all risks. Studies have shown that safety-specific leadership is needed especially when the physical risks are low (Nguyen et al., 2023), but I argue that leadership always needs to balance between integrating safety into everyday work and keeping the discussion on safety alive as a specific topic of interest. In complex organizations, risks may be perceived to be low due to complexity and delayed effects, making leadership necessary for maintaining adequate risk awareness. Also, safety leadership should aim not only to change things but also to maintain good practices, safety barriers, and other controls, and to verify their adequacy. Leadership is often considered as being about change and increased efficiency, but leading safety requires also leading for stability, as was discussed in Chapter 4.

Leadership literature differentiates several types of leadership styles. In safety research, the following leadership styles have been used most frequently: transactional, transformative, empowering, authentic, and leader–member exchange (LMX) style. Empirical studies of leadership and safety have typically focused on transactional–transformative leadership or leader–member exchange (Pilbeam et al., 2016).

Bass and Avolio (Bass & Avolio, 1994, see also Bass, 1985a) developed the *Transformational Leadership Theory* by capitalizing on the broad distinction between task-oriented and relationship-oriented leadership. They distinguish between a transformational and transactional approach to leadership[1]. According to them, transformational leadership focuses on the needs and capabilities of individual subordinates, engendering both a motivation to contribute and a challenge to pursue excellence. With transformational leadership, the followers feel trust, admiration, loyalty, and respect toward the leader, and they are motivated to do more than they originally expected to do (Bass, 1985a). *Transactional leadership* in turn involves the application of rewards that are contingent on performance and the management of deviance from the accepted norm (Bass & Avolio, 1994).

Transactional leadership is characterized by two primary factors: contingent rewards and management-by-exception. Management-by-exception means intervening only when subordinates do not meet acceptable performance levels or when corrective action is required to improve performance. Active management by exception refers to leaders actively monitoring the performance of their subordinates, identifying deviations and taking corrective action to prevent mistakes. Passive management by exception involves leaders stepping in only when standards are not met or performance does not meet expectations. Contingent rewards involve rewarding subordinates for their efforts and achievements, thereby motivating them to maintain or improve their performance.

According to Bass (1985a), transformational leadership increases follower motivation and performance more than transactional leadership, but effective leaders use a combination of both types of leadership. The subcomponents of transformational leadership are as follows:

- *Idealized influence* is based on trust and occurs when leaders demonstrate high standards of moral conduct in their own behavior, becoming role models for their subordinates.
- *Inspirational motivation* occurs when leaders provide clarity, communicate a positive value-based vision for the future, and challenge employees to go beyond their personal interests and focus their attention on the goals of the collective.
- *Intellectual stimulation* means that leaders encourage employees to share their perspectives on issues, to challenge organizational norms, to question assumptions, and to think creatively. Leaders draw on a variety of opinions to make decisions.
- *Individual consideration* refers to leaders recognizing the unique needs and abilities of the followers and adapting their approach by seeking to coach or mentor them in order that they might reach their full potential (Bass, 1985a; Bass & Avolio, 1994).

Leader–member exchange (LMX) reflects the degree to which a supervisor–employee dyad relationship is characterized by mutual obligation, respect, and trust, and therefore can be defined as a high-quality exchange relationship (Graen & Uhl-Bien, 1995). High-quality exchange relationships share similarities with transformational leadership. However, LMX focuses on individual relationships, and the quality of these relationships can vary among team members, even when they have the same manager.

The *authentic leadership* model complements and enhances the transformational leadership model by emphasizing authentic behaviors, role modeling, and value-based leadership, which create an inclusive, ethical, and caring climate. Authentic leadership focuses on values and building trust in the organization through transparent relationships, which aligns stakeholders and ultimately influences the organizational culture (Avolio et al., 2004). Gardner et al. (2005) integrated authentic leader perspectives into a self-based model of authentic leader and follower development. Three of the model's distinguishing features are balanced processing, relational transparency, and an internal moral compass. Since authentic leaders are driven by an internal moral compass, they are not easily influenced by external pressures (Walumbwa et al., 2008).

The *empowering leadership* model by Arnold et al. (2000) argues that the main function of a leader is to increase a team's potential for self-management. The empowering leadership model distinguishes five dimensions corresponding to different behaviors that empowering leaders must show: leading by example, participative decision-making, coaching, informing, and showing concern/interacting with employees. Empowering leadership is closely associated with participative as well as transformative leadership, and some studies treat these as identical (Clarke, 2013).

Research comparing the transactional leadership style and the transformational leadership style has demonstrated that both compliance-enforcing leadership (an aspect of transactional leadership style) as well as leadership that is more focused on inspiring employees to exceed expectations, question assumptions, think creatively, and improve performance (aspects of transformational leadership) are needed in safety-critical organizations. However, the findings also imply that transactional leadership and especially management by exception have a positive effect only if there already is trust present in the working group (Clarke, 2013; Hoffmeister et al., 2014; Kim & Gausdal, 2020; Lyubykh et al., 2022; Mullen & Kelloway, 2009).

Humble leadership and humble inquiry (Schein, 2013; Schein & Schein, 2018) emphasizes the importance of humility in leadership and how it fosters trust, collaboration, and effective team dynamics. In complex environments, dialogue and joint learning are needed to develop safety. These require skills and an attitude of humility and mutual respect. A humble leader recognizes that they do not have all the answers and actively seeks feedback and learning opportunities. Humble inquiry refers to "the fine art of drawing someone out, of asking questions to which you do not know the answer, of building a relationship based on curiosity and interest in the other person" (Schein, 2013). In addition, it is an "attitude of listening more deeply to how others respond to our inquiry, responding appropriately, and revealing more of ourselves in the relationship building process" (Schein & Schein, 2021, p. 3). Schein argues that:

> In an increasingly complex, interdependent, and culturally diverse world, we cannot hope to understand and work with people from different occupational, professional, and national cultures if we do not know how to ask questions and build relationships that are based on mutual respect and the recognition that others know things that we may need to know in order to get a job done.
>
> **(Schein, 2013, pp. 1–2)**

Humble inquiry means approaching the other person with curiosity, care, and an intent to learn about the other (Schein & Schein, 2021). Humble inquiry means engaging in dialogue with the other person instead of an argument or a debate. Humility and modesty are strengths that acknowledge the importance of everyone's contributions. Building and promoting psychological safety (Chapter 2) is an important task for the humble leader. Schein and Schein (2021, p. 5) argue that it is especially our failure to ask humbly and with the right attitude that has led to so many toxic work climates. Humble inquiry reminds managers that before telling people what to do, they should ask them what they need and listen to what they answer.

Humble inquiry is about how we treat each other in the workplace. In this way, we are all leaders in workplaces, for better or worse; the way in which we treat each other, the way in which we ask questions, the way we listen, and the way we tell others what to do and what not to do all influence trust and psychological safety of the group and consequently the willingness of our co-workers to speak up and contribute to our joint core task.

Kim Cameron (2021) and Robert Quinn (2015), the authors behind the Competing Values Framework presented at the end of this chapter, have recently advocated positive leadership and positively energizing aspects of leadership. The positive and energizing aspects of leadership are important to consider also from a safety perspective, as employee motivation and wellbeing are closely related to safety issues. Similar to recent safety theories such as Safety-II and Human and Organizational Performance (HOP), positive psychology builds on the strengths of people and acknowledges the crucial role people have in success. Trust, psychological safety, employee well-being and engagement are in focus when approaching work from positive psychology perspective.

7.3 ADAPTIVE LEADERS

There are several contradictions and paradoxes in leadership that make it a complex activity. For example, leaders need to balance the following tensions:

- Listening to people versus telling them what to do;
- Telling why versus telling how;
- Deliberating on issues versus acting swiftly;
- Appreciating uncertainty versus emanating certainty;
- Meeting long-term goals versus meeting short-term goals;
- Establishing own standards versus meeting others' expectations;
- Diversifying practices and thinking versus simplifying operations;
- Learning deeply versus optimizing;
- Sharing leadership versus keeping authority to oneself;
- Including others versus excluding others.

A key feature of good leadership is the ability to adapt the management style to the requirements of the situation. This is not easy, as many leaders have one established style that is applied in almost every situation. The leadership style is partly tied to the personality, but the style can also be changed and developed with feedback and practice.

Adaptive leadership[2] builds on the collective intelligence and capabilities of the group. An adaptive leader is aware of their own leadership style and is also able to change their style depending on the situation.

Many of the existing models of safety leadership do not take adequately into account the complexity of our organizations. They focus more on the problem of implementing the management's safety vision and getting others committed to it. This is important too, but in the age of complexity, organizations need more than a top-down implementation of someone else's ideas. Organizations need critical reflection, vigilance, joint learning, intellectual humility, and the ability to, as Edgar Schein put it, "step outside the culture".

Leaders demonstrate a *personal example* by their behavior. Leaders with integrity and honesty say exactly what they mean and are willing to admit mistakes (Donovan et al., 2016). To promote safety, leaders need to be consistent in what they say and do. They also need to be honest and truthful, including when saying unpopular things. Leaders set high performance standards and a good example by the way they behave (Donovan et al., 2016). In order to promote safety and safety commitment among personnel, leaders need to visibly demonstrate their own commitment to safety (Donovan et al., 2016; Stiles et al., 2018; Zohar, 2002). These correspond to idealized influence as described in the model of transformational leadership (see above).

Values show in actions, not in words. Thus, employees interpret the leader's real values based on how they behave and what kind of decisions they make. Leaders need to integrate the value of safety into their personal value system.

Leaders need to accept that they are partly responsible for problems that are faced. They should not pass the buck or shift pressure that belongs to their role. They need to make themselves accountable for their actions and require others to do the same.

A leader needs awareness of their skills and abilities, including the limitations. Admitting one's own imperfections requires *mindfulness*. As a general principle, leaders need to seek to disconfirm their hypotheses rather than try to confirm them (see confirmation bias in Chapter 2). They need to actively seek negative information ("bad news") and solicit views that challenge their deeply held positions and carefully listen to different points of view before coming to conclusions (Donovan et al., 2016). Otherwise, they easily only look for information that supports their preconceptions. Leaders need to adjust their thinking in light of new information.

Leaders have to continuously reflect on the current safety level and wonder "is there something we have missed?" This orientation of expecting the worst and preparing for even worse has been called, with small differences in emphasis, "productive paranoia" (Collins & Hansen, 2011), "chronic unease" (Reason, 1997), or "preoccupation with failure" (Weick & Sutcliffe, 2015). Fruhen et al. (2014) note that managers need to channel their unease into flexible thinking, approach safety-related issues from many angles, think about them critically, and question assumptions. In this way, chronic uneasiness contributes positively to safety instead of paralyzing the leader. In order to question assumptions, leaders first need to be able to recognize them. This requires reflection and introspection, but also genuine dialogue with others whereby assumptions can be surfaced and safely challenged.

Leadership requires ability to "step outside" the current culture. There are two main reasons why leaders need to transcend cultural beliefs and assumptions. First, as we have stated previously, culture always includes blind spots, and a "way of seeing is always a way of unseeing" (Turner, 1978). A heightened awareness of these blind spots is essential for leadership that aims to develop the organization. Second, cultural assumptions include the group's conception of who they are (who are "we"), and how they are to be led. Haslam et al. (2020) argue that "leader prototypicality" has a major influence on whom we consider leaders. The problem is that what we want from our leaders is not always the same as the leadership we need (e.g. to develop, change, or to ensure safety). For a leader, an awareness of what personnel expect is important, as well as an awareness of what the personnel (and the entire organization) needs.

Another important characteristic is *systems thinking*, wherein the focus is on relationships rather than individuals and processes rather than outcomes (see Chapter 1). This helps to avoid an overly individualistic approach to failures or successes. Systems thinking is also important for the leader's sense of control, since in complex systems an individual's ability to control the system is limited, and the effects of one's actions can sometimes be surprising for everyone. Without systems thinking, the leader can easily learn the wrong lessons or burn out due to unrealistic expectations about what the leader needs to (and can) control in complex organizations.

Leaders are not external observers or controllers of the organization, but rather an integral part of the system. Leaders need to realize that they are affected by the system as much as they affect it. Systems thinking combined with intellectual humility mean that the leader reflects on their own leadership style and how it affects others, and whether the style is suitable for a given situation. They are aware of their emotions and how emotions affect cognition and behavior.

An integral part of systems thinking is understanding the nature of safety. Safety as an emergent property of sociotechnical systems means that it cannot ever be solved permanently. Like all wicked problems, it can only be temporarily tamed (see Chapter 4). Safety is a chronic issue, and as such it goes against the grain of modern "tick the box" thinking prevalent in much of safety management today (Poole & Dekker, 2024; Provan et al., 2020; Smith, 2024). When it comes to long-term safety, leaders should avoid the temptation of looking for quick fixes in places they cannot be found. Yet, leaders need to act, even when all consequences cannot be known and uncertainties remain. This requires *adaptive action,* where the leader acts while remaining sensitive to the various intended and unintended consequences of one's actions.

The complexity of our organizations means that no-one can understand all the interconnections and all the consequences of one's own actions, or the actions of others. This applies to leaders as well as anyone else in the organization. This means that there is always a need for adaptive action based on the situational constraints and possibilities. Adaptive action is an action that probes the environment, trying to identify leverage points while being mindful of unintended consequences. It pays attention to the fact that the system constantly evolves. What was known about the system a while ago may no longer apply. In complex systems, two big challenges in terms of action are over- and underreacting to issues. Adaptive action means that the leader needs to probe the environment and be prepared to revert decisions if there are signs that actions are leading the organization in the wrong direction (cf. Kurtz & Snowden, 2003).

There are situations where action is needed fast, and there is no time to conduct a thorough analysis of all action possibilities and plan them accordingly. Amalberti (2013, p. 39) notes that the demand for "perfect" situation awareness may even be dangerous in quickly evolving situations where the timeframe for intervention is limited.

Adaptive action also means an ability to recognize so called "high influencing moments". A high influencing moment means an event, a condition, or a situation whereby leader's actions have a wide impact on the entire organization in terms of how the actions will be made sense of by the employees, and how they will subsequently influence the assumptions. Mistakes and failures, as well as organizational crises, are such events, as are successes. *A good leader sees mistakes as an opportunity for the organization to learn, and as an opportunity for the leader to build culture.* The same principle applies to successes, as we also discussed in Chapter 3.

The values that guide a leader's actions in a time of crisis will quickly become inputs to the stories and rationalizations that employees make of the event as well as its outcomes. Regardless of the so-called objective outcomes of the event, the way the leader conducts themselves in the crises has a huge effect on what the group learns and, subsequently, how the entire organization develops.

7.4 PRINCIPLES OF ADAPTIVE SAFETY MANAGEMENT

Figure 7.1 presents the principles of adaptive safety management and leadership. They have been developed based on research in complex sociotechnical systems (Reiman et al., 2015), originally inspired by the Competing Values Framework (Cameron & Quinn, 1999) and elaborated based on my practical experience of using the safety management principles in my safety work. There are four main domains– *facilitate, create*, *control*, and *direct* – and eight principles associated with them. In addition, in the middle of Figure 7.1 are shown four general characteristics required from an adaptive leader, based on lessons concerning the nature of complex organizations and safety especially in Chapters 4–5.

The appropriate balance between the principles depends on at least the following issues (see also Chapter 10 on contextual factors to consider):

1. Requirements of the task (e.g. technology used, hazards, strategy);
2. Situation (normal, emergency, project, change project, crisis);
3. Personnel (abilities, knowledge, skills, habits, culture);
4. Current performance of the group or the organization.

Some leaders are more skilled in some types of leadership than others. Leaders who are adaptive and able to change their style based on context and the situational requirements are more likely to be effective leaders.

The leadership style that is needed also depends on what kind of a culture the organization wants to create. When defining this ideal culture, the organization should consider the nature of technology and hazards as well as the environment in general: How much can be standardized, how much uncertainty is there in the main technology, how about environmental uncertainty? In addition, the current level of safety and culture needs to be taken into account (cf. safety culture maturity models in Chapter 5). Leadership should utilize and maintain the organization's strengths and change its weaknesses. The trick is in recognizing which are which. We will illustrate the influence of the contextual features on ideal culture in Chapter 10 and elaborate on assessment of culture in Chapter 11.

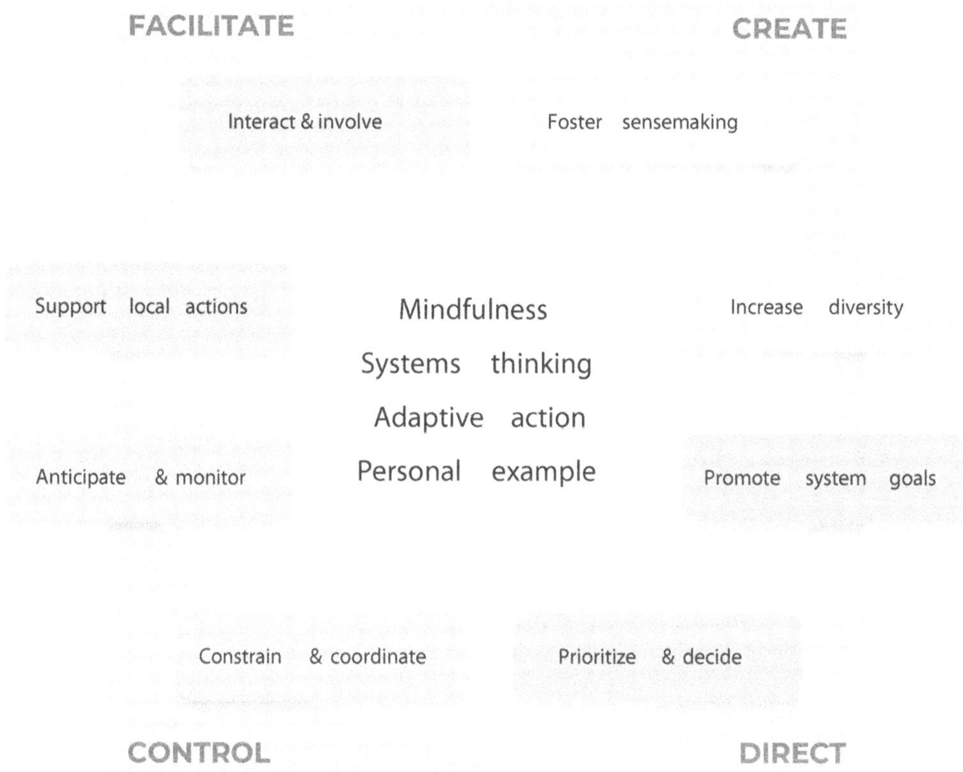

FIGURE 7.1 Principles of adaptive safety management

DIRECT – Promote System Goals

Why: An organization requires a clear core task to provide clarity and overall direction for everyday activities. There must be certain shared guiding principles that inform situational decision making. A system where decisions are made based solely on personal priorities would not be safe. Safety should be promoted as a shared and guiding principle (Reiman et al., 2015). It should be prominently reflected in both goals and expectations and integrated as a fundamental part of the organization's overall objectives.

What: System goals should be connected to the organizational culture as they influence how the organization operates, highlight priorities, and underscore the importance of safety. Goals and expectations must align with the organization's strategy. Safety should be integrated into the organization's strategy and vision from the beginning. This strategy needs to be communicated to personnel, and management expectations should be clearly stated. At different organizational levels and for different roles, it is important to define which activities are permitted and which are prohibited. The role and importance of safety as part of the organizational core task should be clearly defined and communicated to personnel. Additionally, any distinctions between different types of safety should be clarified.

Transformational leadership includes the idea of inspirational motivation (see above), where the leader provides a vision for the future. This vision needs to include safety considerations as a clear organizational objective and a natural part of the core task.

System goals are promoted by leaders constantly keeping safety on the agenda, reminding colleagues of the importance of safety, and most importantly, by showing in their own behavior and decisions the priority of safety issues. When promoting system goals, it is better done by explaining the overall purpose of the tasks (the why), not only by telling personnel what needs to be done and how.

Before system goals can be promoted, they need to be defined. When defining and setting goals for the organization, it is important that leaders motivate with the content of goals (why), not numbers (what, how much). The challenges associated with numerical indicators were discussed in Chapter 6.

It is also good practice to regularly discuss and remind people about the expectations and the shared core task – why we are here and why we are doing the things we are doing. Performance expectations need to be consistently communicated. Leaders need to clearly define where the line should be drawn between acceptable and unacceptable actions, and between genuine errors and intentional violations (Reason, 1997).

When: Promotion of system goals is needed constantly but it is especially important in times of organizational changes, or when the organization is facing external pressure, for example, competition or critique. Furthermore, when decisions are made, their connections to the organizational core task should be clearly stated. System goals should also be clarified when new personnel join the organization.

In complex organizations, there is a need for priorities and "simple rules" for situational decision making. Personnel are only able to successfully and safely adapt locally if they share a core purpose and simple underlying guiding principles on an organizational level (McMillan, 2008, p. 198).

DIRECT – Prioritize and Decide

Why: Organizations need to focus on the most significant issues at any given time. Simultaneously, they need to take action even in ambiguous situations and make decisions that are not favorable to everyone, even inside the organization. This requires prioritization and strategic and operational choices.

What: Leaders need to select areas where they will focus their efforts and to emphasize some connections and some people over others, depending on their potential contribution to organizational goals. This means that not everyone's wishes can be fulfilled and not everything can be a priority at the same time. In this role, the manager defines what is important and what is not important by making decisions about activities and resources, giving feedback and recognition.

Feedback and recognition can be used as opportunities to motivate staff and positively reinforce desired behavior (Caldwell, 2018). Feedback needs to be consistent, transparent, and fair. Positive feedback on performance should be given whenever possible. Sometimes there is a need for negative feedback. In those cases, it is important to be as specific and concrete as possible and to focus on performance, not on the character. The feedback should convey an appreciation of the person while acknowledging that there are aspects of that person's performance that need to change. This type of feedback is best given in private. However, it is important that the leader does not publicly give the impression that they approve of the behavior in question.

It is important to praise and reward initiatives and process (effort) rather than, or in addition to, outcomes. In complex systems, outcomes depend on many issues that are outside the control of the employees. On the other hand, if outcomes are measured and rewarded, it is typical that the work is steered so that the measured issues are increased. This does not mean that the outcomes improve, only that the instances of the measured issues are increased (see Chapter 6). By giving feedback and rewarding the process, leaders can promote certain ways of behaving (based on the organizational core principles and values) rather than imply what numbers the leaders would like to see. Giving feedback and communicating expectations facilitates the wellbeing of the subordinates (Tawfik et al., 2023b) and inspires both individual and team confidence (Caldwell, 2018).

Leadership also has a major role in detecting and countering toxic behavior, including toxic leadership in organizations. Especially leaders in senior positions need to monitor their organization for signs of toxic behavior and, if necessary, intervene in a constructive manner. Effective workforce planning, recruitment, and promotion are other mechanisms to both develop leadership capabilities and prevent people who lack the ability and willingness to lead from becoming managers. It is essential that the organization has multiple career paths, and that employees have a possibility to advance in their careers (more demanding tasks, larger responsibilities, higher salary, better title) without necessarily seeking a management position.

One area where leaders, sometimes implicitly, prioritize organizational objectives and values is recruitment and promotion. What type of people are recruited and promoted, those that challenge or abide by the current culture? Is safety an advantage for career development? How have recruitment consultants and HR been instructed on the criteria for a good employee? We have a tendency to prefer others who are similar to us and who share the same worldview (Chapter 2). Leaders need to resist this tendency and consider what the team/organization needs to cover the needed roles and ensure adequate requisite variety of opinions and competence. Prioritization needs to be balanced with the principle of *Increase Diversity* (see below).

Another method of prioritization is development discussions and performance appraisals. It is important to keep safety issues and culture on the agenda of development discussions. The main point is not to evaluate performance but to discuss safety aspects in performance as well as to have a dialogue about the meaning of safety, the current preconditions for it at work, and the employee's role in affecting safety.

Prioritizing and deciding resembles in many respects the ideas put forward in the model of transactional leadership, especially concerning rewarding and management of deviations (Bass & Avolio, 1994).

When: Not everyone's wishes can be fulfilled and not everything can take precedence at the same time. Prioritization and decision making can therefore conflict with the experience of inclusion and fairness. Communication of the priorities and their reasons is especially important to create an understanding and commitment among personnel. Prioritization and deciding is needed in organizational changes, and whenever top-down influence is needed to steer the organization.

Prioritization is also needed in emergency situations and other high-pressure situations. Studies have shown that a directive leadership style is appropriate in, for example, trauma resuscitation teams when trauma severity is high or when the team was inexperienced (Yun et al., 2005). However, when the team expertise was high, empowering leadership was more effective even in high trauma cases. Their results also demonstrated that an empowering leader provided team members with more learning opportunities than a directive leader did (ibid.). However, directive leadership is typically not the best course of action when someone makes a mistake. A strong immediate response may deteriorate psychological safety in the group. It is better to approach the situation by enabling and encouraging people to make sense of what happened, rather than telling your perspective immediately. Strong top-down decision-making also inhibits people from speaking up (Marquet, 2020, p. 25). Directive leadership should thus be balanced with a more empowering leadership style.

FACILITATE – Support Local Actions

Why: Local conditions and the realities in the field need to be taken into account when managing safety. Centralized solutions, generic system goals, and standard operating procedures need to be adapted to specific circumstances in a manner that takes local conditions and the organizational core task into account. The challenges faced by groups and individuals need to be considered.

What: Daily adjustments must be made to match the work as imagined (by designers, planners and writers of procedures) and the actual conditions under which the work is carried out (Hollnagel,

2009a). Local conditions include personnel, resources, time, competence, other actors involved, and the various interdependencies of the work with other tasks going on in the organization (or in the wider system).

Supporting local actions relies on the ideas of empowering leadership (Arnold et al., 2000) as well as individual consideration subcomponent of the transformational leadership model (Bass, 1985b). In addition, understanding the workplace reality and its ongoing construction is critical (cf. ERA, 2020). Leaders need to create a psychologically safe environment for subordinates to speak up on any issue that is on their mind. Trust between the leader and the subordinate is an important requirement for psychological safety in a working group (Carrillo, 2020).

Supporting local actions can also mean adapting tools and technology to the conditions in the field and making sure they are applicable in real-life situations.

It is important for employee wellbeing that the leader arranges time for their subordinates and is regularly available for them (Tawfik et al., 2023b). In time management and leadership in general, equality and respect for everyone are important. However, supervisors must also listen to the concerns of individuals and take into account the individual needs and wishes of their subordinates as far as possible. The leader takes the time to discuss each team member's concerns patiently and shows concern for members' wellbeing. In this role, leaders may be able to identify and mitigate psychosocial hazards (see Chapter 4) among their subordinates.

A supportive leader acts more like a coach than a supervisor. They try to find ways to develop and grow their subordinates, to give them more responsibility, if they want it. They give positive feedback for safety-oriented behavior (Sleiman et al., 2020). They trust their team members so that they can trust the leader in return. For example, in situations where staff raise safety concerns or stop working due to perceived risks, the manager should trust that the employee made a decision that made sense to them at the given time. The first reaction should not be to question the employee or doubt the choice, because this will increase hesitation among the personnel to raise safety concerns next time.

An important part of supporting local actions is discussing the daily challenges of the employee at the workplace. These challenges can relate to combining the various goals and reconciling contradicting requirements of work, to workload and fatigue, or to social issues of fitting in with the group. Here the leader needs to balance the short-term requirements of the tasks at hand with the requirements of the organizational core task. An important question to ask is "what do you need from me in order to succeed in your work?" Supporting local actions also means ensuring resources and adequate tools are available and the working conditions support safe work. This requires being mindful of the various external pressures to finish the job, whether these come from the employees themselves, from management, or from the customers.

When: Supporting local actions is a continuous work of any leader, but it is essential for supervisors. Many occupational safety hazards can only be perceived and controlled close to the source, so supporting local actions is even more important for local hazards than core task–related hazards (see Chapter 4). When a leader supports local actions, the leader also becomes a member of the team and contributes to creating a shared social identity. This conception of a leader as "one of us" is important for building a culture for safety and for intervening in case of risky behavior (see below).

However, when failures and successes take place, these should not be treated only locally, as often is the case (Weick et al., 1999, p. 39). As discussed in Chapter 6, organizations (or individuals) are typically not very good at learning objectively or at a system level. Leaders need to encourage reporting of events and sharing of safety related information upwards in the organization for trending and learning purposes – see *fostering sensemaking* below.

FACILITATE – Interact and Involve

Why: Interaction between employees at all levels, both horizontally and vertically, is needed to guarantee organizational cohesiveness, information flow, and enough structure for the system to act in a coherent manner during normal situations and to organize in a decentralized manner during abnormal situations (Reiman et al., 2015; Weick et al., 2005, p. 413). Employee involvement is needed to build ownership and trust at all organizational levels and ensure the local knowledge and understanding can be fully utilized in the organization.

What: When leaders interact and create connections between the various actors in the organization, the system becomes more adaptable because information can be shared, and employees are able and more willing to ask for help. They also enable the organization to make better use of existing knowledge and create new understanding together.

Leaders need to communicate both formally and informally across organizational units. They need to share knowledge with others. Leaders have to talk about their values and beliefs of the importance of safety, providing continuous encouragement to others to do their jobs safely (Donovan et al., 2016). Communication is also important to show that the leader understands the work and the challenges of the organization (Carrillo, 2020, p. 34). In networked organizations, leaders need to communicate and interact with the other organizations to ensure that information necessary to work is shared.

Leaders need to learn to trust others, especially their subordinates and colleagues. Carrillo (2020) notes that genuine communication happens when there is a foundation of trust between individuals.[3] When communicating, it is important to listen to understand. We usually listen to respond, not to learn and understand (Schein, 2013). It is important that the leader cares about the other person and their wellbeing. Leaders give employees a feeling of being seen and heard, of being taken seriously, of being respected. These ideas resemble those of authentic leadership (Avolio et al., 2004).

Leaders can promote employees to speak up on safety concerns by using inclusive and encouraging language. This can be done by either using implicit or explicit inclusive language. Weiss et al. (2018) showed that implicit inclusive language, meaning language where the leader refers to the group with first-person plural pronouns (i.e., *we, us, our*) works when the leader and the team already share a social identity (see Chapter 2) by belonging to the same occupational group. Leaders also need to use explicit inclusive leader language such as invitations (e.g., "What do you think?") and appreciations (e.g., "Thank you for this idea!") for team member input especially when team members belong to a different professional group (have a different social identity) than the leader (Weiss et al., 2018).

Employees need to be involved in the development of the organization and safety, and they have to have a genuine opportunity to make a difference. Leaders create engagement by giving employees autonomy to make decisions in their work and responsibility for the outcomes of their work. They involve employees in decisions that influence their work and encourage staff to use their knowledge and experience to identify and resolve problems. Engaged employees actively contribute to their work and the organization, strive for excellence, solve problems, and motivate others to do the same. Leaders foster collaboration between team members, encourage exchange of information, and facilitate group problem solving.

In this role, the leader builds on and widens the social identity of the group, creates more awareness of the outgroup members (including their roles and responsibilities), and introduces new ingroup members (see Chapter 2). This is important for future collaboration and information sharing as well as responding to unexpected events.

Facilitating interaction necessitates going out to the "field", talking and listening to employees at various positions and organizational levels. This applies to leaders at every level of the organization. Senior leaders need to visit the workplace frequently and stay in contact with the front-line staff, including both workers and supervisors. During visits, it is important for leaders to engage in dialogue with the personnel and not be normative or judgmental.

Facilitating interaction also means helping others to interact and communicate. This may require that structures such as seating arrangements or work facilities need to be rearranged or redesigned to allow for better interaction, but often it is enough to just introduce people to one another or at least make it easy for employees to find out who is who, and who is an expert in which issue. It is good to remember that especially for office workers, the feeling that they have control and autonomy in their workspace is associated with job satisfaction and work wellbeing (Knight & Haslam, 2010). Thus, when changes in the workspace arrangement are needed, it is good to take local conditions into account and involve the personnel in the planning (see below).

Leaders should also encourage cross-departmental communications to prevent unwitting structural secrecy from building up (cf. Dekker, 2018) An important part of facilitating interaction is getting to know one's team and learning not only about the team members' knowledge, skills, and abilities, but also their personalities, motives, and wishes. Listening and dialogue are needed for this. A leader develops effective working relations with team members, recognizing and developing followers' potential. The leader demonstrates confidence in team members' capabilities (Donovan et al., 2016).

Leaders delegate tasks and responsibility, and the necessary authority to carry out the tasks. They open a two-way communication channel that enables subordinates to suggest ways to improve safety in their organization and voice their safety-related concerns.

When: When engaging and involving people, the leader needs to be careful and not to always turn to the same active people for opinions or volunteers. The leader needs to be inclusive and give everyone an opportunity to contribute. This requires diversity in the potential ways in which employees can take part in developing the organization.

CONTROL – Constrain and Coordinate

Why: Organizations need structure for people to work in a coordinated manner. Roles and responsibilities must be clearly defined to organize work efficiently. Clear expectations on working practices enhance coordination of activities and improve their predictability. Hazards and risks need to be assessed and risk mitigation measures implemented, including standard operating procedures. Standard responses increase the system's capacity to handle predefined events.

What: Works needs to be organized in a way that employees know their role and responsibilities, understand the typical hazards associated with their tasks, and are familiar with the risk mitigation methods. Clear roles and responsibilities are a key feature enabling the coordination of activities. Of the various working practices, this is the one most associated with management instead of leadership. Despite this, it is highly important for safety in complex organizations.

For effective coordination, it is important that the expectations concerning working practices are as clear as possible. This requires certain shared decision-making principles that can be embedded in the operating procedures. Organizations also need to define ways to identify human errors and violations and mitigate their consequences in case they cannot be prevented.

Contractors and other external personnel need to be incorporated into the organization, and their work needs to be coordinated with other ongoing work.

Coordination include issues such as setting and managing organizational structures, planning shifts and resources, organizing meeting structures, planning for work procedures, and managing concerted actions. Grill and Nielsen (2019) call these *indirect safety leadership behaviors* and note that these behaviors are also compatible with organizational performance goals other than safety performance, for example, productivity. Coordination also includes taking safety into account in situations where there are competing goals or time pressure. Effective communication channels have to be established for the dissemination of safety-related information in all directions.

Establishing a management system is important for the systematic development of the organization. Internal governance processes need to be established and documented in the management system. The processes include risk management, internal auditing, compliance, delegation of authority, and performance monitoring and assessment. Safety must be an integral part of all governance processes. It is also important that safety responsibilities are clearly allocated to the line organization, with adequate support and supervision from a safety function or department. The management system acts as an upper-level structure for organizational activities, including roles and responsibilities. One way to establish clear roles and responsibilities is to utilize RACI charts (Pilbeam et al., 2025). RACI refers to Responsible, Accountable, Consulted and Informed, and it can be used in process modeling. Management systems and their role in safety are discussed in Chapter 6.

When: All safety-critical organizations require clarity in structure, including the roles and responsibilities for various work-related issues. Rules and procedures are also needed, as discussed in Chapter 6. The role of rules is further elaborated in Chapter 9 as one of the tools of safety management.

CONTROL – Anticipate and Monitor

Why: The current performance and culture of the organization has to be understood in order to define the needed activities or, in some cases, interventions. Monitoring activities and culture aim to maintain a view of the current performance, whereas anticipating refers to creating standard operating procedures and ensuring that the system stays within the prescribed limits and conditions. This principle aims at ensuring the adequate uniformity of practices and identify deviations from agreed standards and procedures.

What: Leaders need to ensure that the organization complies with the set rules and procedures. Leaders must also intervene if they see activity that is not in line with organizational values or characteristics of good safety culture. These can include harassment, shooting down of ideas, risk taking, or bullying. Further, if leaders perceive risky activities, they need to intervene immediately. The proper ways to intervene must be considered in advance. Intervention works best in an open and blame-free atmosphere – on the other hand, the wrong kind of intervention can deteriorate the climate.

In addition to organizational compliance, leaders – especially supervisors – need to monitor local hazards (Chapter 4), identify new hazards, keep track of old hazards and their countermeasures, and consider how work can be organized or reorganized so as to minimize human contact with the hazards or mitigate and guard against their consequences.

Individual underperformance must also be addressed. Underperformance can be caused by many things, including work planning, organizing, and management. Therefore, intervention must always be carried out in such a way that the factors affecting the situation are taken into account extensively.

Paradoxically, monitoring system activities and boundaries requires that the organization defines what can be ignored and what information is not worth analyzing (cf. Weick, 1998). Monitoring always requires simplifying and attending to those signals that are considered relevant for safety.

Auditing is one of the basic tools for monitoring safety-critical systems. Another often-used method is the utilization of various safety performance indicators. These were introduced in Chapter 6 and are elaborated in Chapter 9. Leaders need to set up effective safety management arrangements, including procedures for monitoring for compliance and deviations. Leaders need to acknowledge that quantitative safety targets can hinder transparency and drive incidents underground (de Boer, 2021). Paradoxically, safety performance indicators are thus not the best means of monitoring the organization's safety performance.

The competence of employees needs to be monitored and continuously improved. It is also important to monitor whether competence in the organization is actually utilized in practice.

When: Monitoring can also mean paying attention to the conditions of equipment or tools, as well as general working conditions. In addition, monitoring of personnel motivation, fatigue, and stress is important. Monitoring is needed especially in highly standardized settings with high levels of routinization, where the danger of slips and errors of omission is increased. In these cases, monitoring is best organized in close collaboration with the safety department. Monitoring needs to encompass the organizational capacities, not only individual performance. Further, and most importantly, monitoring should encompass why things go right in addition to why they sometimes go wrong (Hollnagel, 2014).

CREATE – Foster Sensemaking

Why: Fostering sensemaking refers to improving the organization's capability for situational responses and ability to self-organize activities, structures, and mindsets based on the current or anticipated situational requirements. It also aims to create generic learning capacity for the organization. Sensemaking is a critical social process in organizations (Chapter 2), and leaders have a major role in fostering it. Fostering sensemaking in the organization is about teaching the members of the organization to view their problems in different light and giving them analytical tools to make sense of and learn from the daily work.

What: Fostering sensemaking requires from leaders an ambition to break up typical categorizations. This approach focuses on increasing variability within the organization rather than constraining it. Instead of standard responses and conventional interpretations, a wider variety of options are given to personnel. They are also encouraged to create new categorizations and interpretations of events and needed actions. Fostering sensemaking entails promoting equality and engaging a diverse group of individuals.

Leaders need to be reluctant to simplify (Weick & Sutcliffe, 2015). They take deliberate steps to create more complete and nuanced pictures of what the organization faces and who they are as they face it. Leaders actively question assumptions in order to better uncover blind spots. They need to encourage curiosity and a questioning attitude among employees. These are needed to avoid normalization and drift (see Chapter 3).

Fostering sensemaking includes giving employees permission to redefine boundaries and roles, as well as adjusting and interpreting rules and standard operating practices according to situational requirements. A good understanding of the core task as well as the technology and the hazards that need to be taken care of is critical for successful adaptations (Reiman & Oedewald, 2007). Adaptive sensemaking allows both honoring and rejecting past beliefs and practices while interpreting an abundance of data into actionable knowledge (Weick et al., 2005, p. 412). It requires an open environment that supports learning and development. Leaders need to foster a culture of openness and sharing, where people communicate, share information actively, and help each other out.

Leaders must encourage others to listen to one another and underscore the importance of listening through their own words and actions. Further, leaders need to consider the team roles so that the teams have sufficient diversity, but at the same time they have a sufficiently shared understanding of goals and ways of working, and they share mutual respect.

Leaders need to reflect and make explicit the underlying decision principles, norms, and assumptions that guide the decision making and way of working in their groups. In this manner, they are open to alternative courses of action and can avoid committing themselves to the most obvious ones. By becoming conscious of and challenging commonly shared assumptions, the leader becomes more aware of the culture of organization (cf. Eoyang & Holladay, 2013, p. 193).

Organizations need to enable reporting of safety-related observations, worries, and incidents and make sure there is feedback from the reporting. Leaders need to encourage people to make suggestions, raise issues, and actively participate in resolution; people are encouraged to speak openly and honestly, voicing also issues that may not be popular (Caldwell, 2018). Leaders need to promote a sense of candor and allow people to share mistakes and concerns without the fear of retribution (de Boer, 2021). Intellectual stimulation mentioned in the model of transformation leadership (Bass, 1985a) resembles many of these principles.

Fostering sensemaking is needed in situations where the team is stuck, demotivated, or discouraged. In these situations, leaders need to help the team find a new perspective or a fresh look to the situation at hand with, for example, new metaphors or questioning of assumptions. Fostering sensemaking among personnel also contributes to them avoiding overlearning routines and developing mindless habits. For example, emergency exercises and drills run the risk of becoming shows where the organization demonstrates how good they are, not as actual learning and development opportunities. Leaders have a role in facilitating sensemaking before, during, and after these exercises.

Fostering sensemaking and encouraging personnel to think and learn are achieved through several activities. By sharing information about their own decisions, challenges, and uncertainties, leaders show an example of openness. Actively seeking input from employees shows that leaders value their contribution. It is also important that leaders are willing to listen to concerns without an immediate attempt to solve them, and without requiring the person who brings up the issue to propose a solution, as this may lead to self-censorship. Leaders need to have an open and curious mindset and avoid trying to solve all problems for their employees (Acosta, 2024, p. 79).[4] A check-the-box mentality seldom fosters sensemaking.

Leaders also need to be mindful of the type of questions they ask, since one of the most powerful influencing tools leaders have at their disposal is the effective use of questions (Lloyd, 2020, p. 51). It is also important for leaders to reflect on one's own communication style and what kind of a message the style delivers, and whether it conveys trust or fear (Lloyd, 2020). This does not mean that leaders should never propose solutions or solve problems, but rather that they should be mindful of their influence on learning and thinking in the group. Leaders should suggest new ways of working while encouraging team members to also express their ideas. Leaders who foster thinking in the group listen to suggestions and ideas and use them to inform decisions that affect the work group (Donovan et al., 2016).They create space where employees have the possibility to pause and speak up. They do this by controlling the clock instead of obeying the clock, and calling for a pause and reflection (Marquet, 2020, p. 82). Often employees do not speak up because they worry about the schedule. Leaders need to make sure they are not unintentionally adding to the pressure when they ask if someone has worries or wishes to speak up.

Leaders need to continuously update their understanding of how the work is currently done and what needs to be developed. De Boer (de Boer, 2021, p. 12) points out that the challenge for many managers is to be curious without being normative. He notes that many leaders find it difficult to discuss how work is done in practice, without referring to a baseline of how the task "should" be done. They should let the employees explain and tell their rationale for working the way they are working without judging or assessing compliance with the present criteria.

When: Discussions, dialogue, and conflicts are a constant presence within organizations. Leaders have both the capacity and the responsibility to influence the direction of these dialogues by actively participating in them. Carrillo (2020) emphasizes that leaders can influence the direction of conversations by participating in them, listening to what the various participants wish to say, and then articulating a new level of understanding from what everyone has said. This means that leaders do not provide concrete solutions to problems but rather create the psychological atmosphere that allows for a higher quality of problem solving (ibid., p. 29).

Effective leaders encourage their subordinates to think critically and solve problems independently, rather than providing them with ready-made solutions. Situational requirements need to be considered as well. Leaders should assess the demands of the situation and the team's capability to address the problem without external assistance. In situations that require prompt action, a more direct approach from the leader may be necessary.

When discussing work without judging, the leader has a much better chance of learning something new about the worker, the work itself, the work conditions, or how the leadership can better support safe work. Humble inquiry (Schein, 2013, see above) is useful when fostering sensemaking as well. To foster sensemaking, the leader needs to be competent in systems thinking and mindful of their own heuristics and ways of making sense (see above in this chapter).

CREATE – Increase Diversity and Explore

Why: Safety in complex organizations is a continuous accomplishment, and as the system changes, ways to ensure safety may also change. In complex systems, development of activities is needed even to maintain the status quo. Novelty, variance, and diversity are the key ingredients of development, and these need to be nurtured in the organization.

What: Facilitating novelty means active interaction with diverse others, both internal and external to the organization. It means bringing new ideas and insights, as well as exploring the boundaries of the system, including its cultural assumptions.

A key part of leadership is promoting intercultural dialogue within the organization and with external stakeholders, including regulators, educational institutions, and government. Even organizations with a single nationality have subcultures based on education, profession, and work location. Managers must possess "cultural humility" (Schein, 2010, p. 263) to integrate these subcultures towards common goals. Respecting and leveraging these differences can turn them into positive forces.

It is important that leaders seek information on matters and do not merely wait for the information to find them. They should proactively commission audits and assessments to identify any organizational weaknesses and question audit findings that only deliver good news. Leaders seek outside perspectives and support candid, thorough assessments of their operations. Leaders openly communicate the results of monitoring and assessment throughout the organization (Caldwell, 2018).

One of the areas where leaders need to challenge the organization is the metrics associated with measuring safety. Leaders should not be passive recipients of safety information, but rather actively challenge and question the various indicators of safety, both leading and lagging (Smith, 2024).They need to be mindful of the potential side-effects of safety objectives and indicators (see Chapters 6 and 9).

Leaders should use incidents and accidents that happen in other organizations as opportunities to check whether similar issues and potential interactions exist in their organization. In a similar fashion, they need to share safety-related lessons learned with other organizations as well as with the regulators.

Leaders need to pay continuous attention to anomalies that could be symptoms of larger problems in the system. They value and reward the reporting of near-misses and errors because they are viewed as learning opportunities and a means of obtaining a realistic picture of operations (Weick & Sutcliffe, 2015).

Exploring can also mean looking at complex problems through a variety of different lenses (Jackson, 2024) or trying different solutions with the aim of learning how the system really behaves. Sometimes this requires dropping old tools, whether physical or mental (Weick, 1993, 1996), and seeing the problem from a new perspective (see Chapter 3). Leaders need to actively avoid the biases described in Chapter 2 by intentionally connecting with people who are different in terms of their experience, values, ideas, and competence, and try to learn how the world looks from their point of view. This can broaden the leader's perspective by either revealing blind spots in their own thinking or by helping them to better understand others' viewpoints.

When: Fostering sensemaking, the first principle associated with "create", aims at changing the existing mindset of the employees and facilitating how they learn. Increasing diversity and exploring aims at the same thing, but here the leader is actively contributing to learning by bringing in new ideas rather than helping others think. Both are needed, and a balance between them depends on, for example, how well the employees can currently make sense of the situation and how much help they need. Leaders also need to balance exploration with stability and predictability. Ways of safely exploring the system include various simulations and scenarios (see Chapter 9).

7.5 LEADERSHIP IN TEAMS

At team level, the competence of the leader in the specific domain of the team is more important for leadership than it is in the higher levels of the organization (De Meuse et al., 2011; Künzle et al., 2010b). The context knowledge is not enough, however. In addition to task-related issues, team leaders need to manage relations (the team) and change (Künzle et al., 2010b; Yukl & Gardner, 2020).

A team leader needs to balance the requirements of the organizational core task with the role and tasks of the team. The leader also needs to create structure for the team to perform while encouraging the team members to develop those structures further. The leader has a role in monitoring performance and keeping it within the agreed range while extending the range and introducing diversity and change. Finally, the leader needs to balance directive style and prioritization of actions with participation and trust.

In the healthcare domain, Tawfik et al. (2023a) identified that the local leadership behaviors of being predictably available, setting expectations, and providing feedback that is frequent, useful, and positive were associated with lower emotional exhaustion (burnout), better safety climate, better teamwork climate, lower perceptions of workload, and reduced intentions to leave.

Leadership that focuses on support, development, and encouragement of the team members has a positive effect on team performance (Künzle et al., 2010b). Team leaders have a major role in creating psychological safety in the team, which is needed for both development as well as trouble shooting (see Chapter 2). The previously discussed humble inquiry is essential for team leaders as they build a team where every member feels appreciated and safe enough to speak their mind on any issue. Further, as discussed above, it is not enough if only the team leader contributes to psychological safety; the team members must do so likewise. The team leader is in a key position to remind the team members about how to interact with each other and to monitor for potential toxic behavior in their team.

Team leaders need to structure and direct the team members' activities, including the distribution of tasks and resource planning. Communication and coordination are key principles here. In addition, the leader needs to make sure that the team has the prerequisites to perform, including resources, competencies, and adequate working conditions (lighting and heat, work posture, no unnecessary interruptions during safety-critical tasks, etc.).

Different situations require a different leadership style. Leaders' ability to analyze situations and adapt their leadership behaviors to the situation are critical for team effectiveness. Various studies show that in times of high stress and time pressure, a more directive leadership style is suitable, especially if the leader is also competent in the domain area (Cooper & Wakelam, 1999; Yun et al., 2005). In emergencies and other high workload situations, the leader needs to initiate a structure within their team, to take command and give direction, guidance, and assistance. The leader needs to be flexible, but at the same time engender trust and respect from his team (Cooper & Wakelam, 1999). Clarity of communication and assertiveness of the leader are important in times of stress. Furthermore, control of oneself is an essential feature of providing effective leadership in stressful situations (Wetzel et al., 2006).

The preferred style of the team leader depends on both the competence and experience of the team as well as the particular situation. Leadership in teams has been extensively studied in healthcare. Klein et al. (2006) have shown that leaders of medical emergency teams switch between more directive and more shared forms of leadership depending on the complexity of the emergency. They call this *dynamic delegation*, where the active leadership role is delegated up and down the leadership hierarchy based on the characteristics of the task and the composition of the team.

Yun et al. (2005) found in their study of leadership in trauma resuscitation teams that directive leadership was more effective when trauma severity was high, whereas empowering leadership was more effective when trauma severity was low. In addition, when team experience was high, empowering leadership was more effective. Similar results have been found concerning workload; in situations of high workload, more directive and active influential leadership is required (Künzle et al., 2010a, p. 9). Finally, in critical care teams, leaders who prioritize, give direction, distribute tasks explicitly, delegate, and communicate contribute to successful team performance (Künzle et al., 2010a, p. 10). Leadership needs to be unambiguous, otherwise there is a risk of confusion in the team if there are contradictory orders from different people (Cooper & Wakelam, 1999).

This means that in high-stakes situations, leaders need to rely mostly on two principles: 1) constrain and coordinate, and 2) prioritize and decide (see Figure 7.1). Supporting local actions is also needed as high-level guidance needs to be applied to the particular situation. Team leaders also have a role in monitoring performance and detecting and correcting errors. Another important task for team leaders is to give feedback on performance (Künzle et al., 2010b). Debriefs are an important tool for team learning as well as for psychological recovery after stressful events (see Chapter 9).

Grinerud et al. (2021) note that in the road transport industry, leaders find it difficult to observe workers' performance because drivers work outside the physical boundaries of their organizations. This limited visibility makes executing leadership challenging in these organizations (Zohar & Luria, 2005) because management must lead workers and implement any needed changes from afar. It can be argued that almost all safety effects are limited in their visibility, more so than effects on efficiency, which are easier to measure in quantitative terms (e.g. profits or savings, schedules kept, amounts of products produced).

7.6 SUMMARY

Leadership is viewed as an interaction between group members, involving the structuring of situations and perceptions. It defines future visions, aligns people with these visions, and inspires action. Leadership is about influencing others to define what is meaningful and to take action.

Leadership is not confined to formal positions. Leadership includes fostering an environment where all team members participate in both thinking and doing in a climate of psychological safety. Recent research has focused on shared and process-based elements of leadership, describing how leaders contribute to the development of a shared identity and guide teams' direction and objectives. Leadership is often characterized as a group influence process based on a collective social identity.

The effectiveness of different leadership styles depends on task requirements, situations, personnel abilities, and team performance. Adaptive leaders who can change their style based on situational requirements are more effective. Effective safety leadership styles depend on context, situation, and followers' skills.

Adaptive safety management relies on eight principles:

Promote system goals: Leaders should integrate safety into the organization's core task and strategy, constantly keeping safety on the agenda and explaining the purpose of tasks.

Prioritize and decide: Leaders must focus on significant issues, give consistent feedback, and manage deviations. Prioritization is also crucial in emergency situations.

Support local actions: Leaders need to understand the local conditions, support the local adaptation of organizational procedures and tools, create a psychologically safe environment, and be available for their subordinates.

Interact and involve: Leaders need to communicate, listen, and involve employees in decision making.

Constrain and coordinate: Leaders need to set up effective safety management arrangements with clear roles and responsibilities.

Anticipate and monitor: Leaders need to monitor activities and culture, and ensure the organization complies with the set rules and procedures.

Foster sensemaking: Leaders should encourage curiosity, question assumptions, and create an open environment for learning and development.
Increase diversity and explore: Leaders should promote interdepartmental, interorganizational, and intercultural dialogue, seek information, and bring new ideas.

For people to actively develop safety, speak up on concerns, and report them, there needs to be trust in the management that they will treat the information fairly and act on it wisely, and psychological safety in the work group so that individuals are willing to risk speaking up. Additionally, there needs to be enough diversity and reflection in the group so that they can avoid groupthink and normalization of deviations and risky conditions (see Chapter 3) combined with shared goals and values and respect for each other's expertise to avoid power plays and conflicts of interest.

At the team level, leaders need to balance organizational core tasks with team roles, create structure, monitor performance, and adapt leadership styles based on situational requirements. Effective team leadership involves setting expectations, providing feedback, and fostering psychological safety.

Adaptive leadership can shift from one principle to another based on the situational circumstances at hand. Static leadership is stuck in one role and is unable to adapt even when circumstances change. An adaptive leader is able to balance between different management principles and thus may overcome some of the blind spots created by the organization's dominant culture. An adaptive leader is able to "step outside the culture", as Edgar Schein put it.

Reflection Questions

Consider a critical event in your company, positive or negative.

- How did you as a leader/your leaders react to the event?
- What issues did your reaction emphasize? What cultural values, beliefs, or assumptions did you want to change, maintain, or reinforce? How conscious was your response?
- How well did you succeed? What could have been done better? What did you learn?

What is good leadership in your organization?
How do you receive feedback about your leadership style and leadership skills?
What kind of feedback would benefit your work as a leader?
How have you tried to change leadership expectations in your organization?
Which leadership principles require most development in your organization? What evidence do you base your assessment on?
What principles do you find most difficult to implement in your work? And what principles come to you naturally?
Can you find adequate coverage of the various leadership principles from your management team?
How do you show in practice your commitment to safety?

TYPOLOGIES OF CULTURE: COMPETING VALUES FRAMEWORK AND MULTI-FOCUS MODEL

Competing Values Framework

In the early 1980s, a Competing Values Framework (CVF) was developed to integrate criteria of organizational effectiveness. Quinn and Rohrbaugh (1983) conducted a two-part study, where they first asked seven academic experts to evaluate which of Campbell's (1977) thirty effectiveness criteria were relevant for organizational effectiveness. They then analyzed the responses with multidimensional scaling and uncovered three underlying dimensions, which were referred to as focus, structure, and means–ends.

The first dimension is related to organizational focus, from an internal, micro emphasis on the wellbeing and development of people in the organization to an external, macro emphasis on the wellbeing and development of the organization itself. The second dimension is related to organizational structure, with, at the one end, emphasis on stability and, at the other end, emphasis on flexibility. The third dimension is related to means–end relationships, for example, planning and goal setting vs. outcomes (productivity). Cameron and Quinn (1999) developed these findings into a Competing Values Framework (CVF) for assessing and profiling the dominant cultures of organizations.

According to CVF, organizations and associated organizational values can be viewed along two dimensions; i) internal versus external processes and ii) control versus flexibility. In CVF, the two core dimensions form four quadrants, each representing a distinct cluster of criteria representing what is seen as good, right, and appropriate, that is, the fundamental values that exist in the organization.

The upper left quadrant (Collaborate) identifies value creation and performance criteria that emphasize an internal, organic focus. It is typified as a friendly place to work where people share a lot of themselves, like an extended family with best friends at work. Success is defined in terms of internal climate and concern for people. The organization places a premium on teamwork, participation, and consensus. The dominant organizational values are commitment, communication, and development.

The lower right quadrant (Compete) identifies value creation and performance criteria that emphasize an external, control focus. The glue that holds the organization together is an emphasis on winning. Success is defined in terms of market share and market penetration. Goal achievement, profitability, and market share are valued.

The upper right quadrant (Create) identifies value creation and performance criteria that emphasize an external, organic focus. The glue that holds the organization together is a commitment to experimentation and innovation. The emphasis is on being at the leading edge of new knowledge, products, and/or services. Success means producing unique and original products and services. Innovative outputs, transformation, and agility are valued.

The lower left quadrant (Control) emphasizes internal, control value creation and performance criteria. The long-term concerns of the organization are stability, predictability, and efficiency. Formal rules and policies hold the organization together. Efficiency, timeliness, consistency, and uniformity are the dominant values.

When naming the quadrants, Cameron and Quinn built on Ouchi's (1980) separation of markets, bureaucracies, and clans (see also Chapter 2 and discussion on trust).

What is notable about these four quadrants is that they represent opposite or competing assumptions. Each continuum highlights value creation and key performance criteria that are opposite from the value creation and performance criteria on the other end of the continuum, that is, flexibility versus stability, internal focus versus external focus. The dimensions, therefore, produce quadrants that are also contradictory or competing on the diagonal (Cameron & Quinn, 1999). These competing elements in each quadrant give rise to one of the most important features of the CVF, the presence and necessity of paradox. CVF emphasizes that successful managers need to work simultaneously with several contradictory logics and shift their dominant value sets when circumstances require. Most organizations develop a dominant culture style, yet effective organizations need capabilities that allow them to succeed in each of the four quadrants (Cameron & Quinn, 2011; Quinn, 1988).

The competing values framework identifies two general tensions: internal focus and integration vs. external focus and differentiation, and flexibility and discretion vs. stability and control (Cameron & Quinn, 2011). Cameron and Quinn (2011) describe the resulting quadrants as follows: hierarchy culture (clear authorities, standardized rules and procedures, high control and accountability); market culture (customer-orientation, focus on competition and results); clan culture (shared values and goals, cohesion, participation, focus on teamwork); and adhocracy culture (temporary, specialized, dynamic, focus on innovation and reconfiguration). The CVF suggests that successful organizations integrate elements from all four quadrants into their culture in a balanced manner.

Cameron and Quinn (2011) present the Organizational Culture Assessment Instrument (OCAI) for creating an overall organizational culture profile by self-assessing six dimensions. The OCAI identifies how the current organizational culture is perceived, as well as what the organization's preferred or future culture is like. The six dimensions are: Dominant Characteristics, Organizational Leadership, Management of Employees, Organization Glue, Strategic Emphases, and Criteria of Success. These dimensions are assessed with four questions each (representing the four quadrants), and the scores are calculated for each of the four quadrants.

Cameron and Quinn (2011) also present the Management Skills Assessment Instrument (MSAI), which helps managers identify the key competencies they will need to develop or improve in order to foster organizational culture change. In a discussion of competencies and leadership roles in the competing values framework, Quinn (1988) suggests eight roles: innovator, broker, producer, director, coordinator, monitor, facilitator, and mentor.

Multi-Focus Model based on Hofstede's Theory of Culture

Hofstede, who developed the model of national culture that will be further investigated in Chapter 10, developed his theory of culture to be applicable also to organizations (Hofstede et al., 2010). This model was further developed conceptually and technically by Waisfisz, Schram, and Schachner working under The Hofstede Centre (nowadays called The Culture Factor Group) during 2009–2015.

The approach, called the Multi-Focus Model™, consists of six independent dimensions.[5] Organizations' scores on these dimensions are not inherently good or bad; it rather depends on the type of business as well as the chosen strategy. The following characterizations are from: www.theculturefactor.com/organisational-culture.

The first dimension differentiates between *means-oriented versus goal-oriented* cultures. Means-oriented cultures focus on the way the work needs to be carried out, whereas goal-oriented cultures pay attention to the end results. This is a difference between "how" and "what". The dimension is associated with risk taking, since in an extremely means-oriented culture, risks are strongly avoided, whereas in extremely goal-oriented cultures, risks are actively sought. For safety-critical organizations, a means-oriented score can be very functional, but generally goal-oriented cultures are more dynamic, innovative, and development oriented.

The second dimension differentiates, at its extremes, *internally driven versus externally driven* cultures. This customer orientation dimension refers to whether the employees emphasize customer requirements (externally driven) or their own ideals and ethics (internally driven). Internally driven cultures emphasize consistency and adherence to standards or procedures. Externally driven cultures in turn focus on meeting customer needs and flexibility.

The third dimension, *easy-going versus strict work discipline*,[6] refers to the amount of internal structuring and control. Easy-going cultures have much looser internal structures and less predictability in behavior than cultures with strict work discipline. Tight, or strict, work discipline shows up as a higher number of instructions and closer supervision by boss, or self-control.

The fourth dimension is called *local versus professional*, and it measures where the employee loyalty and identification lie – in the work unit and the boss, or in their profession and work contents. Extremely local[7] cultures are internally focused, short-term directed, and cohesive, whereas extremely professional cultures have a longer time perspective and an external focus. Local cultures are driven by strong norms which extent outside of work as well – social control applies at work as well as home. The various 24-hour safety culture approaches can be seen as attempts to create a local culture. This dimension has connections to the concept of social identity discussed in Chapter 2.

The fifth dimension separates *open systems versus closed systems*. This dimension does not refer to systems theory (see Chapter 1) but rather to how open the organization is to, for example, newcomers or outsiders. Safety issues would benefit from an open culture, but security-related concerns typically push the organization towards a more closed system.

The sixth and final dimension is *employee oriented versus work oriented*. In employee-oriented cultures, the focus is on people and their welfare and concerns, whereas work-oriented cultures emphasize task completion. Moreover, employee-oriented cultures are more collective than the individual-centric work-oriented cultures. In an extremely work-oriented culture, there can be heavy pressure to complete tasks despite the toll on people.

The model also includes internal normative windows. This means that each dimension has a window where these aspects of culture are dysfunctional. Dysfunctional areas concern a too high means orientation, being too highly internally driven, a too easy-going work discipline, a too local culture, a too closed system, and finally, a too work-oriented culture.

A highly local culture resembles the extreme form of clan culture in the CVF, with strong norms preventing speaking up on issues that would threaten the identity of the group. The Multi-Focus Model emphasizes the existence and identification of subcultures in organizations more than the CVF does. This also means that averaging survey scores hides important cultural characteristics (see Chapter 11 on assessment of culture).

NOTES

1. The distinction between transactional and transforming leadership was originally developed by James MacGregor Burns in his 1978 book Leadership, where he drew on the ideas of e.g. Max Weber and Chester Barnard. Bass (1985b) extended and operationalized many of Burns (1978) original ideas.
2. I use the terms *adaptive management* and *adaptive leadership* interchangeably. The framework is called *adaptive safety management* but it includes elements of both leadership and management.
3. She (2020, p. 29) notes that groupthink results from fear of exclusion. She emphasizes that thinking in groups is what humans do, and inclusive leadership is needed to combat the negative effects of groups and amplify the positive effects.
4. I have noticed that some organizations have an informal policy that if you bring up a problem you also need to propose a solution. This approach is detrimental to openness and silences people on all complex, vague, and uncertain concerns and issues – exactly those that the leaders should want to know!
5. Description of the dimensions is based on Hofstede et al. (2010) and www.theculturefactor.com.
6. Hofstede et al. (2010) call this *loose versus tight control*. This definition should not be confused with Perrow's loose and tight couplings, which are inherent properties of the sociotechnical system and not cultural features (see end of Chapter 3). However, loose or tight control can be the organization's response to either loose or tight couplings (and complex or simple interactions), with various consequences depending on the type of system. See Chapter 10 for elaboration.
7. Hofstede et al. (2010) use the term "parochial" instead of "local".

8 Safety Departments and Safety Professionals

In this chapter I focus on the role of safety professionals and safety departments in creating and maintaining a culture of safety. I will utilize the model introduced in Chapter 7 to illustrate how safety professionals are leaders when it comes to safety but have different responsibilities compared to line managers. Safety is always a line responsibility, but safety professionals are needed to support, monitor, ensure, and challenge the line activities. The organizing of the safety departments should take this into account and promote both involvement and independence in safety professionals' activities. The chapter ends with an accident case of a Boeing 737 MAX aircraft that experienced two crashes in 2018 and 2019. The accident demonstrates the challenges of independent safety oversight, whether it is internal or external to the organization.

8.1 SAFETY PROFESSIONALS IN COMPLEX ORGANIZATIONS

The role and responsibilities of safety professionals vary significantly between different industries and regulatory contexts. The role has also changed during the past years as both the profession has developed and the world has changed. We will take a look at some of these more global changes in Chapter 10 and focus in this chapter on the role of safety professionals from the perspective of developing, monitoring, and maintaining safety in complex organizations.

Safety professionals have multiple roles in organizations, ranging from experts and analysts to coordinators and facilitators. Their potential roles correspond with the leadership principles presented in Chapter 7, with some important distinctions that we introduce below. Safety as a profession is multidisciplinary, combining engineering, management, psychology, and sociology as well as, depending on the field of application, physics, chemistry, law, statistics, and/or medicine.

Safety professionals are leaders in safety issues, but they should lead in a manner whereby the line responsibility is always emphasized. They should make it possible for the line to lead safety. The role is demanding because very often the task that is left to safety professionals is one of reminding, warning, and "slowing down work". This too is often needed, but it should not be the defining feature of safety professionals' work.

Many recent studies have illustrated the challenges and the multifaceted nature of safety professionals' roles (Carroll et al., 2022; Pilbeam et al., 2025; Provan et al., 2017, 2020). The goal conflict is strongly present in the work of safety professionals (Provan et al., 2017). It is also often left to the safety professionals themselves to find their own role and identity in organizations. This can create anxiety (Le Coze, 2024). One of the reasons for this is the many competing paradigms of safety management that are based on different views on what safety is and how it should be improved (Chapter 4). Safety practice has been described by both academics and practitioners as divided as well as disconnected from, or at least only loosely coupled with, the science of safety (Busch, 2016; La Duke, 2020; Le Coze, 2024; Maldonado, 2019, Marriott, 2018).

Safety professionals struggle with the same tensions that managers typically struggle with in complex organizations (see Chapter 4). Pilbeam et al. (2025) note in their study of safety professionals' working practices that experienced safety professionals recognized the importance of rule-based practices for safety but also considered adaptive, more flexible principle-based practices to be even more necessary. At the same time, the safety professionals considered these principle-based practices to be harder to implement. Pilbeam et al. (2025) suggest that rule-based and principle-based practices act as a duality, rather than being in opposition to each other. As Reiman et al.

DOI: 10.1201/9781003491156-8

(2015) noted, in complex safety-critical systems, various safety management principles that are contradictory to each other need to be utilized in an adaptive manner to manage safety.

Much like any professional, safety professionals tend to stick to the methods they know and would rather widen the field of their application than look for new methods and ways of working. As we illustrated in Chapter 4, safety is too often underspecified, and what works for one type of safety in a particular organization operating in a particular context may not work in another setting. Safety professionals need to be adaptive in their working practices and localize and contextualize the generic safety models.

Safety professionals occupy an awkward position between management expectations and the realities of the work and the working conditions. Sometimes, when local adaptations fail, the safety professionals get blamed (Le Coze, 2024; Provan et al., 2017), indicating senior management's misunderstanding of the role of safety professionals in managing safety. At the same time, the management wishes for simple answers and clear views on an issue that cannot, and should not, be simplified. However, safety professionals lack tools to adequately express this complexity and often find it easier to succumb to the pressures for easily quantifiable measures – all this until something happens, and safety professionals find themselves faced with a new conundrum; should they try to question the premises of their safety work, including the numbers that management loves to glance at, or should they continue to push forward with stronger measures (the same but more)? This is the individual response to the challenge of organizational learning and a balance between single-loop and double-loop learning discussed in Chapter 6.

In November 1987, a fire broke out on the old wooden escalators at London's King's Cross underground station due to a cigarette dropped by a passenger. Metro station fires were common between 1958 and 1987, when a total of 400 fires occurred at underground stations. The metro management preferred to call them "smoldering" instead of fires so that they wouldn't sound so scary. The common belief was that they could not be prevented, but luckily, they did not spread but merely smoldered away.

This time, the fire did not smolder out. About 15 minutes after the fire was detected, it suddenly spread up to the ticket hall, accompanied by black smoke. Thirty-one people died and several were injured by the flames and toxic smoke. The staff had not been trained in case of fires. The metro company had been focusing their main attention on operating trains at the expense of station maintenance. They were considered secondary goals. The wooden escalator also seems to have disappeared from the London Underground organogram. One consequence was that the escalator received inadequate maintenance. James Reason (Reason, 2016, p. 30) *argues that the organization and the reorganization of the London Underground in the 1980s contributed to the accident. In that reorganization, responsibility for safety was scattered, with no-one having an overall safety responsibility.*

The accident investigation report (Fennell, 1988) was full of recommendations related to fire safety, all of which were implemented. The end result was a bureaucratic jungle and a loss of focus from ensuring security to ensuring that the papers were up to date. No benefit–cost analysis was carried out on the recommendations, and the London Underground almost went bankrupt when it implemented recommendations that ultimately had a marginal impact on overall safety. The recommendations and corrective measures only concerned fire safety. Over the next few years, metro passengers had to endure multiple evacuations from the slightest smoke. On a hot day, during an evacuation due to a suspicious bag found on a train, train traffic was suspended for five hours, and several people died from heat, claustrophobia, and crowding (Maidment, 1997, p. 71). Subsequently, the authorities commissioned an external investigation that found that Fennel's report had led to disproportionate expenditure on firefighting and emergency operations without taking into account other safety risks – such as crowd control (Maidment, 1997).

Safety practitioners, and especially managers, simultaneously lead and influence the system and act in the system. This means that they need to balance not only between different safety leadership principles (see Chapter 7) but also between actions targeted towards creating preconditions for others to act in a certain manner, and actions that manifest this type of wanted behavior – sometimes going against the dominant values of the culture in question. For example, on the one hand, a manager can define system boundaries and create rules and standard operating procedures for the personnel to follow, and, on the other hand, the manager themselves obeys the rules and boundaries of the organization. On the other hand, a safety manager who perceives their organization as too proceduralized and dependent on written instructions and guidelines may decide to introduce some variance in the system by provoking discussion on hazards, potential blind spots as well as the role of rules in general. Changing cultural values is slow and usually requires change in other elements of the sociotechnical system as well. The safety manager needs to work with all the elements of the sociotechnical system, including cultural assumptions.

Safety professionals rely on the different sources of power depicted in Chapter 1. Their formal authority depends very much on their place in the organizational structure (see below) but typically they rely on other ways of influencing instead of or in addition to formal authority (Daudigeos, 2013; Madigan et al., 2021; Provan et al., 2017).

Traditionally, safety professionals have focused on ensuring compliance with safety regulations. This has contributed to the bureaucratization of safety (Dekker, 2014; Provan et al., 2017) along with a general rise in the self-regulation of organizations (Dekker, 2021). This is inspected in more detail in Chapter 10. Safety professionals need to widen their role from compliance to culture development (Manuele, 2015). They are in a key role when integrating safety into the daily business of the organization. It is important that safety professionals both realize the influence they have on culture and build on that to facilitate a culture for safety.

The role and practices of safety professionals should depend on the types of hazards to be managed in the organization (see Chapter 4). The more complex the risks, the more support employees need in identifying them and taking them into account in daily work.

8.2 SAFETY PROFESSIONALS' PRACTICES

In Chapter 7, the adaptive safety management framework was introduced. Based on my practical experience as a safety manager, the model was first elaborated into a model of nuclear safety professionals' practices (Reiman et al., 2023) and further into the general model of safety professionals' practices for this book.

Similar to the model of adaptive safety management applied to line managers, the model for safety professionals has four main domains and eight principles associated with them (see Figure 8.1). The domain areas differ slightly; instead of controlling work, safety professionals *regulate* it, and instead of creating change, safety professionals *challenge* existing practices and assumptions. These are small nuances to the framework but they exemplify the fact that safety professionals need to make sure the line organization leads and owns safety and the safety function ensures that safety limits are defined, understood, and met, provides support, monitors safety performance, and challenges the line, if needed.

Similar to leaders (Chapter 7), the prioritization of safety management principles depends on many factors: requirements of the task (e.g. technology used, hazards), situation (normal, emergency, project, change project, crisis), personnel (abilities, knowledge, skills, habits, culture), and current performance of the team.

Systems thinking is important for safety professionals. For example, when participating in or leading incident investigations, safety professionals should always consider the context and acknowledge that there almost always are multiple contributing factors to any event. The safety professional needs to listen to people's stories and acknowledge the local rationality of people's actions and seek multiple perspectives on the event under investigation. The approach should not be overly judgmental concerning variability and deviations. The safety professional needs to identify and understand what a normal and necessary performance variability/adjustment in the given situation is, and what unwanted variability is. When learning from events, the safety professional needs to go "up and out" instead of "down and in" (Dekker, 2015). They need to look at the system more widely and consider the conditions and interactions – instead of digging into the "root cause". Finally, safety

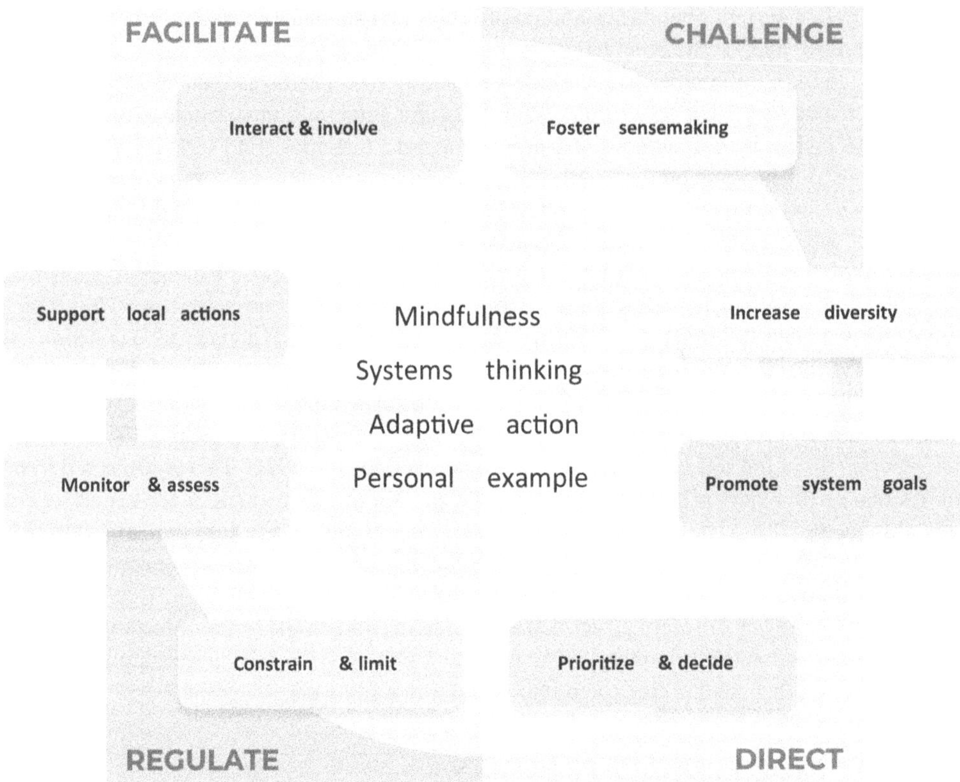

FIGURE 8.1 Principles of adaptive safety management for safety professionals

professionals should use events to learn about normal work, not only about the current incident or failures.

Adaptive action for safety professionals means a balance between two extremes: taking the "trial-and-error" approach that gives little a priori consideration to the consequences of actions, seeing them as opportunities for learning, or being paralyzed by the perceived complexity and unpredictability of sociotechnical systems and avoiding action altogether (Madigan et al., 2021). Safety professionals need to be able to anticipate how the system can fail but also identify how it currently functions and how to facilitate its improvement. Safety professionals create adaptive capacity to the organization by being adaptive themselves (Poole & Dekker, 2024, p. 109).

Setting a *personal example* is important for safety professionals. Nothing demotivates employees nearly as much as a safety professional who does not follow the safety rules (senior management who does not follow the same rules comes second). The safety professional should also be able to explain why the rule needs to be followed, demonstrating not only compliance but also genuine commitment to safety and risk awareness. The importance of personal example extends further, as safety professionals must also consider their interactions with people. They should aim to foster a culture of trust rather than one of fear.

Mindfulness is required from safety professionals when considering risks as well as when reflecting how their own work and way of interacting with personnel affects culture and, ultimately, safety. Safety professionals need to reflect on their competence and assumptions about safety and accidents. Safety professionals are subject to the same dangers of mindless repetition of routines and nonreflective application of old tools as any occupation.

DIRECT – Promote System Goals

Why: As mentioned in Chapter 7, an organization needs to have shared guiding principles that inform situational decision making, with safety being one of these guiding principles. Safety needs to be clearly connected to the other goals of the organization and not seen as something extra to the core task.

What: Safety professionals communicate the importance of safety as an integral part of the organizational core task and a key objective for the organization. They define together with management what safety-related activities are mandatory and what kind of behavior is not tolerated. The role and importance of safety as part of the organization's strategy should be clearly defined and communicated to personnel. Additionally, any distinctions between different types of safety should be clarified (see Chapter 4).

Safety professionals need to facilitate an increased awareness of hazards and risk as well as an understanding of the priority of safety as a guiding principle in uncertain situations. They need to facilitate risk awareness of workers as well as management.

Safety professionals need to remind the line organization about their safety responsibilities, and personnel and management to remain mindful about risks. Safety professionals need to get line leadership engaged and active in leading safety. They also regularly communicate safety-related expectations to personnel.

When: Promotion of system goals is important, especially when there are pressures to perform the work faster or cheaper, or when there have been changes in organization and there may be unclarities about the current priorities of the senior management. In addition to promoting system goals in the sharp end, safety professionals need to promote safety as a system goal to senior management, who are similarly prone to emphasize short-term targets or may lose sight of the long-term and "chronic" issues such as safety when dealing with acute cost and performance pressures.

DIRECT – Prioritize and Decide

Why: Organizations need to focus on the most significant issues at any given time. Simultaneously, they need to take action even in ambiguous situations and make decisions that are not favorable to everyone; they need to reward some and reprimand others. This requires prioritization and strategic and operational choices. Safety professionals need to ensure safety is prioritized and decisions take safety into account.

What: In this role, safety professionals make different types of decisions: First, they prioritize the resources they have and select focus areas and objectives for regular safety work. Second, they facilitate the priority of safety in management decisions and objective setting on an organizational level. Third, they provide recognition and positive as well as negative feedback based on the safety performance of the organization and the various work units. Sometimes this may require stopping work or intervening to stop dangerous events from progressing. Safety professionals, or the safety function, can also act as the waiver authority for deviations from technical specifications. They may reject proposals, set limiting conditions, or require additional analyses/safety barriers.

Safety professionals need to facilitate prioritization of safety in organizational goals and actions taken at all organizational levels. They do this by helping managers manage work instead of optimizing measures and by helping them to assess the safety significance of various issues and conditions. They further need to help the management understand the potential safety consequences of different decisions and strategies and explain why certain actions or decisions are not recommended. Sometimes there is a need to challenge management decisions that do not adequately consider safety aspects. Safety professionals need to use a graded approach to prioritize issues having the highest safety significance and make sure these issues are considered when organizational goals are defined.

In some cases, safety professionals need to adopt the role of "whistle-blowers" and disclose inappropriate behavior to senior management (Hale, 1995), or in some cases even to the regulator. More often than blowing the whistle, safety professionals prioritize by defining objectives for safety improvement, selecting focus areas and themes, preparing an annual plan of activities, and selecting control measures and corrective actions based on safety significance. In addition, they recognize safe performance and reinforce and reward it. The other types of more acute prioritizations come up when safety professionals observe unsafe practices or leadership that is not in line with organizational values. Intervention and speaking up is required in those cases.

When: Prioritization and decisive action are needed regularly in safety professionals' activities. They need to be constantly prepared to change from participating and facilitative role to a more authoritative role where decisive action is taken. However, care must be taken that this action is carried out in an appreciative manner. This role is challenging since it requires courage to speak up and challenge the line or the management as well as confidence in one's observations and judgment.

FACILITATE – Support Local Actions

Why: Local conditions and the realities in the field need to be taken into account when managing safety in complex organizations. Centralized solutions, generic system goals, and standard operating procedures need to be adapted to specific circumstances in a manner that takes local conditions and the organizational core task into account. Safety professionals, working closely together with supervisors and middle management, are in a key position to facilitate this.

What: Safety professionals need to acknowledge the daily realities at the workplace and the various conflicting goals that employees face. Daily adjustments must be made to match the work as imagined (by designers, planners, and writers of instructors) and the actual conditions under which the work is carried out (Hollnagel, 2009a). Safety professionals should help employees to make sacrificing decisions and combine local goals and pressures with global organizational level requirements. They also should make sure the local system–specific safety expertise of employees (Almklov et al., 2014) is considered when safety approaches are applied.

Safety professionals contribute to the psychological safety of the work group. By being open to concerns and worries without unduly apportioning blame, they contribute to a climate where it is easy to speak up.

Safety professionals need to participate in various line activities and be familiar with the work conditions and requirements of the tasks. When changes or other projects are carried out, safety professionals need to be involved early. For example, they need to participate in the early stages of contractor approval to ensure adequate coverage of safety issues.

How: Safety professionals work together with supervisors to adapt procedures and processes to local conditions. This requires knowledge of available resources, tools, and competencies as well as requirements of the task and the characteristics of the object of work (e.g., in industrial settings, the wear and tear on the particular machine). In addition, several environmental characteristics need to be considered, such as weather and the possibilities and constraints created by other actors.

It requires making oneself available for interaction with employees by regularly visiting the field, having an open-door policy, or otherwise making it clear that one is available for discussions and willing to have a dialogue with the workers. Humble inquiry (see Chapter 7) is needed to connect with line personnel and gain understanding of their perspective.

It is important that the safety professional uses terminology that is understood by the line organization. In fact, when supporting local actions, it is more important to understand how the line speaks about and understands risks, safety, and hazards than what the official definitions are. Likewise for management priorities, safety professionals need to translate them to the discipline's culture and language – for example, what "safety first" really means in the particular context.

When: In this role, the safety professional is better able to become a member of the work community than in other roles. This has its pros and cons. On the one hand, as discussed in Chapter 2 on social identity, the recommendations of the safety professional are more easily accepted by the group if they are considered to originate from "one of us" rather than an external group ("experts" or "headquarters", etc.). On the other hand, I have observed that many safety professionals seem to at first enjoy this role as they are able to visibly help the workers, but they may also gradually drift into a position where they are either fighting fires in the field or defending the workers' position toward management, even in cases where they should take the management side. Thus, it is important that safety professionals balance the support given to local actions with actions aiming at change at higher levels of the organization or the prioritization of issues associated with system goals. Otherwise, they risk drifting into a role that Graeber characterized as "duct taper", where they exist to "solve a problem that ought not to exist" (Graeber, 2018, p. 40) or what Rollenhagen (2010) characterized as an excuse for not addressing technology or design issues.

The independence–involvement tension (see below) refers to, on the one hand, safety professionals' need for freedom from the line management's excess influence on safety professionals' decisions (independence) and, on the other hand, the need for the safety professionals to understand the daily realities of the work and be involved early in various decisions and projects. Supporting local actions should aim to balance this by being involved in activities but considering how the system goals are taken into account in the particular context and situation.

FACILITATE – Interact and Involve

Why: Safety professionals have a significant role in facilitating discussions concerning hazards, risks, and safety between different work groups and organizational levels. In complex organizations, risk awareness requires adequate information flow between the different work units as well as an employee involvement in risk analysis.

What: Safety professionals should promote open dialogue with personnel and senior management. They should coach employees to perceive hazards and risks in their workplace and foster an environment in which speaking up on risks and safety concerns is normal practice. Safety walks and other methods (see Chapter 9) should be used to connect with personnel, to listen to their concerns, and to have a dialogue with them concerning their work and its risks.

As discussed in Chapter 7, employees need to be involved in the development of the organization and safety, and they have to have a genuine opportunity to make a difference. Safety professionals have a role in empowering employees by having a dialogue with them about their work and how they perceive the hazards as well as the mitigating measures.

Work teams and departments all have a localized view of the risks, which may not adequately consider all other stakeholders. Safety professionals need to facilitate cross-functional collaboration to ensure that safety is not siloed but integrated into all business processes.

Trust is an important factor in how safety professionals are able to influence the line organization. Trust can also differ between safety professionals and the various groups inside the organization, for example, line workers, line management, and senior management. Humble inquiry (see Chapter 7) helps to build connections between safety professionals and other stakeholders.

Facilitating interaction necessitates going out to the "field", talking and listening to employees at various positions and organizational levels in their "natural habitat", whether that is an office, laboratory, workshop, or something else. It means contacting people and discussing safety and hazards with them. Safety professionals can act as facilitators, bringing different views on safety (and the people holding them) inside the organization together and initiating dialogue on the differences.

Additionally, this principle requires organizing seminars, workshops, or other types of gatherings where individuals can gather to meet and engage in discussions.

Safety professionals need to foster an inclusive environment that values and shares diverse perspectives and enables collaboration and information sharing (Pilbeam et al., 2025).

When: Facilitating interaction is important for all other principles, as the safety professional builds relationships and trust, within and outside the organization that are critical for implementing various safety practices (Pilbeam et al., 2025, p. 10) as well as carrying out the other principles described here.

REGULATE – Constrain and Limit

Why: Organizations need constraints for people to work in a safe manner. Limits need to be set, including safety limits, operating conditions, and technical specifications. Safety rules and instructions are required for many tasks.

What: Safety professionals need to make sure there are adequate constraints on performance in terms of safety limits, technical safety barriers, and standard operating procedures. These limits need to be set in close collaboration with the line organization. Safety professionals need to coordinate risk assessments and analyses, coordinate writing of safety instructions, follow up on corrective actions, and organize training on safety issues. In this role, safety professionals develop and maintain technical standards and specifications.

In this role, safety professionals arrange safety messages and implement warnings and signs (see Chapter 9). It also involves employing a variety of communication mediums and tailoring them to the specific context and audience (Pilbeam et al., 2025). Various subcultures in an organization may require different communication methods depending, for example, on their access to computers, language skills or some specific features of the work. Safety professionals need to ensure that important information concerning safety, including lessons learned from incidents, are shared across the organization and integrated into future practices.

The utilization of safety rules is described in more detail in Chapter 9. An integrated management system or a safety management system is one method of safety management that was discussed in Chapter 6. Whether the system is an integrated management system incorporating all the organizational goals (quality, safety, environmental, etc.) or a self-standing safety management system (see Chapter 6), it is important that the safety professional utilizes standard terminology and creates a transparent system that ensures consistent practices which can be audited and verified (Pilbeam et al., 2025).

In industrial settings, constraining and setting limits involves creating clear design specifications, risk registers, and standards for equipment and system design as well as optimizing maintenance schedules based on experience feedback (Pilbeam et al., 2025).

Objectives for safety and culture improvement need to be coordinated with other non-technical development programs in the organization and together with top management (and other interest groups when applicable) to ensure a unified approach (Viitanen et al., 2022).

When: Safety professionals need to balance between relying on structure (including standard operating procedures and predefined roles) to control behavior and relying on personnel to self-organize based on their situational judgment and expertise. As discussed previously, excess rules and procedures can hinder employees' autonomy and their sense of responsibility and agency. Rules can never prescribe work precisely. Rather, they always need to be interpreted and applied in context. Further, as work and the environment change, rules need to be revised also. This is especially challenging for safety rules, many of which are typically written to be applicable to the entire organization – meaning in practice that all work teams need to interpret them by themselves based on their local understanding and realities.

The answer to the problem of generalized safety rules is not to make more detailed rules. In fact, if this principle is taken to an extreme, it may lead to what Rae et al. (2018) call the creation of "safety clutter" and Dekker (2014) the "bureaucratization of safety". The principle of coordination and organization works best when the system is simple, or complicated, at most. This means that it works for hazards that can be understood and managed locally. The paradox is that by introducing rules and mitigating measures, safety professionals can accidentally increase the complexity of the workplace so that they in effect sabotage their own efforts at controlling the hazards. This is especially so if the rationale of the various mitigation measures is not explained to the workers, if the mitigation measures make the hazards more opaque, or if they instill a false sense of security among the employees.

REGULATE – Monitor and Assess

Why: The current performance and culture of the organization has to be understood in order to define the needed activities or, in some cases, interventions. Safety professionals need to monitor organizational performance and anticipate potential problems.

What: Safety professionals monitor organizational performance and verify compliance with systematic methods that include various types of assessments, indicators, safety walks, and observations as well as investigations and inspections.

Safety professionals need to ensure that the organization complies with the safety rules and procedures that have been implemented.

Organizations need to enable reporting of safety related observations, worries, and incidents and make sure there is feedback from the reporting.

In practice, this principle means, for example, conducting audits, assessments, and evaluations and formulating corrective actions; assessing the safety effects of organizational and technical changes; and conducting safety walks and field visits. Auditing is a regular tool for monitoring safety-critical organizations. Another often used method is the utilization of various safety performance indicators. These are described in more detail in Chapter 9. As discussed in Chapter 6, quantitative safety targets can limit transparency in the organization and can drive incidents underground (de Boer, 2021).

After development needs are identified, safety professionals can be involved in the design and implementation of corrective actions as advisors, organizers of the process, or, in some cases, as problem solvers themselves (if the identified development need is a safety assurance process, for example). They also always have a role in the follow-up of the corrective actions and analysis of their effectiveness.

It is also important to adequately involve safety professionals in document review, whether the documents concern management system procedures, work instructions, or designs. Safety professionals must be able to reject or require modifications to documents with safety deficiencies. Involving safety professionals early in design modifications can also save a lot of time later, as making modifications in design phase is easier and cheaper than during construction/implementation or during use.

Safety professionals also monitor the use and effectiveness of safety requirements and procedures in everyday work and update them when necessary. When something happens, the lessons learned are incorporated into the procedures.

When: Safety professionals need to make sure their work does not drift into what Graeber calls "box tickers", where their role is to "allow an organization to claim that it is doing something that it is, in fact, not doing" (Graeber, 2018, p. 45). Another term that was recently introduced by Rae et al. (2018) is "safety clutter". Safety clutter, according to Rae et al. (2018), is "the accumulation of safety procedures, documents, roles, and activities that are performed in the name of safety, but do not contribute to the safety of operations". Poole and Dekker describe it as the "relentless addition of rules, procedures, systems, and processes surrounding the capture, reporting, tabulation, tracking, ranking, storage, and analysis of all the data" (Poole & Dekker, 2024, p. 11).

Many safety professionals seem to devote large amounts of time to counting and categorizing various incidents and accidents to be presented to the management. In addition, they may count the amount of safety walks, audits, safety training, and so on as proof of how safety has been taken care of during the past reporting period. Some of this may be beneficial for organizations (depending on what data is collected and how it is used), but taken to an extreme, it alienates the safety professionals from the realities of the field and the real challenges of safety.

CHALLENGE – Foster Sensemaking

Why: Fostering sensemaking by safety professionals happens by questioning practices and assumptions. This is important for developing safety in complex organizations. Constructive challenge also improves the organization's ability to make sense of issues and events, contributing to their ability to learn and develop – contributing to their culture.

What: Safety professionals need to challenge the line to think about their performance and their assumptions. Safety professionals can also encourage people to make suggestions, raise issues, and actively participate in their resolution. They also need to make sure people can comment on the various safety rules and propose modifications based on the realities of the field.

Safety professionals need to promote a sense of candor and allow people to share mistakes and concerns without the fear of retribution – to do that, they have to be aware of their communication style and make sure it conveys trust rather than fear (de Boer, 2021).

Questioning assumptions is a type of ideological power (see Chapter 1) whereby the safety professional tries to shape how employees understand the world, how they assign causality and rationalize events.

The management system needs to be challenged from time to time from a safety perspective, and regular reviewing and revising of the procedures are necessary to ensure that the system corresponds to practice.

Safety professionals must promote the importance of leadership by providing training in leadership skills, fostering visible leadership in the organization, and facilitating leaders' understanding of the importance of leadership to culture and safety (Pilbeam et al., 2025).

It is important that the safety professional discusses issues related to work and work conditions with personnel without judgment or preconceptions about the "correct" working practices. Avoiding this normative stance is difficult for safety professionals, yet it is imperative if they wish to bridge the gap between Work-as-Imagined and Work-as-Done (de Boer, 2021). They may also learn something new about the worker, the work itself, the work conditions, or about how safety function can better support work. This requires humble inquiry (Schein, 2013), which we discussed in Chapter 7.

In industrial settings, fostering sensemaking involves developing a shared understanding of the current configuration of technical systems, of the importance of managing the configuration during technical changes and maintaining the systems with good quality maintenance programs (cf. Pilbeam et al., 2025).

When: Safety professionals are necessarily normative when auditing, making inspections, or verifying activities, and this may create tension between these two roles. The safety professional must also avoid the impression among personnel that if they say something negative, the safety professional (or their colleague) will come back the following day to audit the same issue and issue a non-conformity or a new safety rule. Paradoxically, questioning and challenging the line activities may be best done without a normative approach but rather with a shared intent to learn.

CHALLENGE – Increase Diversity and Explore

Why: Safety in complex systems is a continuous accomplishment. As the system changes, the methods of ensuring safety must also change. In complex systems, development of activities is needed even to maintain the status quo. Novelty, variance, and diversity are the key ingredients of development, and safety professionals need to nurture these in their organization.

What: Safety professionals need to challenge the organization in how they see safety and how they measure it. They should not just passively deliver to the management the numbers they think they need, but rather actively facilitate a questioning attitude towards safety and an awareness of the uncertainties associated with knowing how safe the system is. Safety professionals should challenge cultural assumptions that the line organization and senior management holds about safety and hazards. They should challenge the line organization to drop their old tools (Weick, 1993, 1996) and try new ones, whether the tools are physical or mental. This may help to see issues from a different perspective and bring in needed variety.

Safety professionals need to identify new risks and opportunities. They construct risk scenarios to support senior management decision making based on personal and collected lessons as well as sociotechnical safety theories. They amplify voices from the field and translate them into a language and concepts that management and the given discipline understands.

Provan (2021, pp. 282–283) presents three types of "everyday work explorations" that a safety professional can utilize to improve the organization's understanding of how safety is created:

- Normative explorations are used to identify specific improvements;
- Descriptive explorations are used to increase understanding of the work;
- Formative explorations are used to look for and decide alternative approaches to work.

These explorations can be integrated or used separately. Provan highlights that it is important that the organization knows which type of exploration the safety professional is currently conducting. The explorations which take place where the work is conducted are important ways to identify and understand the gaps between Work-as-Imagined and Work-as-Done. Provan concludes that

> through performing everyday work explorations, the safety professional acts as a learner, seeking context and understanding about what's needed to support success on the front line. The safety professional engages with frontline workers, not to make judgments about the safety compliance of their work but rather to update their own and the organisation's mental models about work, risk, and organisational life.
>
> **(Provan, 2021, p. 286)**

Safety professionals also analyze trends, develop organizational processes, and gather external lessons learned. They utilize external expertise and consultants smartly. Safety professionals can also utilize artificial intelligence (AI) and simulations to identify emerging risks and test potential solutions or use digital twins to test solutions and ways of working and to demonstrate lessons learned from real incidents or simulations. Safety professionals can guide the organization in using scenario planning and foresight techniques to proactively address emerging risks.

Increasing diversity and exploring should lead the safety professional also outside the organization. They should look for state-of-the-art knowledge on safety as well as lessons on successes and failures from other organizations.

When: This role is very important for the organization since production pressures and drift (see Chapter 3) so often gradually degrade the priority of safety. In the prioritization role, the safety professional confronts unsafe practices and decisions, but by increasing diversity and exploring the system boundaries and challenging assumptions, the safety professional may contribute to safety before any visible degradation in performance occurs.

8.3 SAFETY DEPARTMENT AND LINE RESPONSIBILITIES

The role and authority of safety professionals depends on their position in the organizational structure as already briefly described in Chapter 6. Whatever the formal position, the safety department needs to have access to senior management and adequate independence from "production pressures" of the line organization.

Safety and oversight functions can be organized in many ways. All the various ways have their pros and cons, and the important thing is to be aware of how organizing affects the possibilities and constraints of the safety department. Many accidents have demonstrated the challenges of organizing safety assurance in a manner that would allow the department access, authority, and an awareness of the issues having safety significance.

In addition to access, the safety department needs resources to adequately address and oversee the safety issues, including follow up of recommendations and corrective actions. As illustrated by the accidents at NASA, BP, and Boeing (see Chapters 2, 4, and 8), succumbing safety functions to business units sacrifices the needed authority and independence of the safety function. However, positioning the safety function in the staff function instead of the line brings other challenges. David Woods, based on the lessons learned from the *Columbia* Space Shuttle accident (see Chapter 2), identified that the challenge is how to build a safety organization that would be "independent enough to question the normal organizational decision-making but involved enough to have a finger on the pulse of the organization" (Woods, 2006). Woods proposed the following four principles for a safety organization:

- *Independent* (of financial pressures and able to challenge management decisions);
- *Involved* (in organizational actions that affect safety, e.g. incident investigations, definition and updating of technical standards, readiness reviews);
- *Informed* (about how the organization is actually operating and what factors are changing it); and
- *Informative* (by providing information about organizational weaknesses and directing interventions).

The challenge comes from the fact that, as Woods notes, these requirements are in conflict. Often the safety function is a hybrid between a matrix and a line organization model, where certain safety professionals work in the line organization but report in a matrix to the safety department. This solves many of the problems of involvement but can be challenging in terms of independence. A further challenge comes from the fact that the safety professional may feel more part of the line organization than the safety department, potentially causing tensions when a more directive approach is needed (see above). The safety department needs to actively build and foster a social identity (see Chapter 2) among all its members, despite some of them interacting mostly with the line organization.

In addition to functional independence, other types of independence need to be considered. In our study of internal independent nuclear safety oversight (Viitanen et al., 2024, 2025), we differentiated between three types of independence:

1. *Functional independence* – freedom from responsibility for the activity being audited or not participating in the work being assessed or overseen;

2. *Financial independence* – freedom from economic interest affecting behavior;
3. *Cultural independence* – freedom from the constraints of sharing the same assumptions, beliefs, and/or values.

We further specified that independence is not a yes/no question but rather a continuum. Similar to the line organization's need to make sacrifice decisions to balance conflicting requirements, safety organizations need to sacrifice independence to contribute to organizational decision making but from time to time sacrifice involvement by distancing themselves from operational decisions.

Some tasks require more independence than others. For example, incident and accident investigations need to be independent, like quality control (QC) activities and various inspections. Risk assessments in turn can be facilitated by the safety department, but these are done in collaboration with the line organization, not independently. Sometimes the safety department is involved in safety analyses of technical modernizations or redesigns, that is, line activities. In those cases, there is a need for an independent verification, either from the safety department, an oversight function, or an external party.

Viitanen et al. (2022), in their study of safety culture specialists' work, noted that direct access to top management was one recurring consideration for organizing the specialists' work. No other optimal ways of organizing the safety culture specialists were identified, with each organizational configuration having its advantages and disadvantages. Instead, leadership issues in general, senior management commitment, and access to senior management were identified as essential for success. We noted similar preconditions in our study of internal independent safety oversight in the nuclear industry (Viitanen et al., 2024, 2025).

Safety is always a line responsibility, but the line has contrasting goals as well as a limited understanding of the safety effects of their own work as a whole, as well as of hazards in general (e.g. how the work of others affects their safety). For this reason, the safety function needs to both advise the line and act as a barrier, or a monitor, in case of missed warnings and degrading conditions. However, it should be absolutely clear to everyone that the safety department is not responsible for safety; the line is. These responsibilities should be clearly defined in the management system.

Dekker (2014) has proposed several issues that need to be addressed when building the safety department as a "concerned outsider who knows the inside". First, the safety department needs to have significant and independent resources. Second, the department needs to have a constructive involvement in management activities, including production-focused decisions and safety trade-offs. He highlights that safety department can never be fully impartial in a sense of just observing and remaining on the sidelines. That would slowly decouple the department from the actual work carried out at the line organization.

Dekker (2014) also argued that safety departments should have no weekly or monthly or even quarterly targets. The safety department does not produce safety, just as the competence development department does not produce competence. They *facilitate* it. Especially counterproductive is setting performance targets for the safety department that actually relate to the line organization – for example, reducing the number of lost-time incidents. This not only shifts responsibility in the wrong direction but creates a vicious cycle where no-one really cares if the indicator represents reality or not (see Chapter 6).

More important than monthly targets is the qualitative intelligence of the department on what is going on at the organization in terms of safety. This means that the safety department needs to maintain a continued grounding in operational reality. Diversity in competence profiles and competence in the actual work carried out at the shop-floor level is important for maintaining this understanding of the workplace reality (cf. ERA, 2020).

Dekker (2014) highlights that in addition to practitioners' knowledge, the department needs academic knowledge on safety management as well as practical skills such as presenting and writing clear and concise reports.

The safety function needs to be positioned at the staff function level instead of the line organization (Dekker, 2014) in order to be as independent of current economic concerns as possible and in order to have a direct access to relevant organizational decision-making levels without having to pass through various levels of middle management. Also, inside the safety department there needs to be a way to bypass the department manager if safety so requires.

I have often discussed with my clients the responsibilities of the safety department and its various independent functions. For example, is it enough to recognize a safety issue and inform the line organization of it? Or how much responsibility should the safety function take for following up on the issue, or assisting in its solving? Can the safety function consider their work effective if the issue has been recognized but not solved? This is what happened at NASA concerning the O-rings of the solid rocket booster joints and the possibility of a foam strike during launch. Both issues were recognized, and for the O-rings, there was even a redesign of the solid rocket boosters planned. However, it was not implemented in time to avert the disaster. But at least the problem was recognized, right?

Regarding the previous example, if the safety department would take responsibility for solving an issue, the next question is whether this can degrade the line responsibility for safety. I believe the answer is yes, it can. Thus, safety departments should mainly act in an advisory and monitoring role, challenging the line to carry out their responsibilities rather than taking over them.

8.4 LAYERS OF SAFETY OVERSIGHT THE ORGANIZATION

Various oversight and control structures always create an additional burden for the organization. The more distant the controller is from the actual work, the heavier the burden in terms of coordination and communication it creates. This is the reason why responsibility for safety has to lie in the line, where the work is actually conducted. The more responsibility and the more safety work the line is able to carry out, the more the system is not only safe, but also efficient. There will always be a need for verification, checks, inspections, audits, and so on, but these should complement and not replace the line's responsibility for assuring and continuously developing safety. Similarly, assuring safety requires that there is a systematic way to challenge the line concerning their decisions and priorities as well as to advise in case of limited understanding of hazards or safety significance of various activities. As discussed in Chapter 6, there are several different ways of monitoring safety in an organizational context.

In response to the Rogers Commission Report (Presidential Commission on the Space Shuttle Challenger Accident, 1986) after the Challenger *accident, NASA established what is now known as the Office of Safety and Mission Assurance at Headquarters to independently monitor safety and ensure communication and accountability agency wide. The Office of Safety and Mission Assurance focused on monitoring unusual events like "out-of-family" anomalies*[1] *and established an agency-wide Safety and Mission Assurance policy.*

The Columbia *Accident Investigation Board concluded that the safety system structure left the Office of Safety and Mission Assurance ill-equipped to hold a strong and central role in integrating safety functions. NASA Headquarters had not effectively integrated safety efforts across its culturally and technically distinct centers. In addition, the practice of "buying" safety services established a relationship in which programs sustain the very livelihoods of the safety experts hired to oversee them. These idiosyncrasies of structure and funding prevented the safety organization from effectively providing independent safety analysis (CAIB, 2003, p. 186).*

In a similar manner, in the US nuclear industry after the Three Mile Island accident, independent technical committees were required by the Nuclear Regulatory Commission (Viitanen et al., 2025). These committees were, however, gradually replaced by other ways of independent verification, the latest of which is called Internal Independent Nuclear Safety Oversight (WANO & IAEA, 2018).

The International Atomic Energy Agency and the World Association of Nuclear Operators have produced guidance on independent internal safety oversight (WANO & IAEA, 2018). Their focus is on oversight of nuclear safety, but many of the lessons are applicable to any safety-critical domain.

The guidance differentiates four layers of oversight. The first layer, in-process line organization oversight, refers to performance monitoring carried out by the workers themselves or their immediate supervisors. This can include self-checking, peer checking, supervision, or various self-assessments carried out before commencing work (such as mini-risk assessments or pre-job briefs). Chapter 9 presents some tools for this.

The second layer, functional oversight, refers to the various organizational functions conducting assessments and monitoring from their respective perspectives. These include quality assurance (e.g. audits), incident investigations and learning from experience activities, and other organizational assessments. These are also elaborated in Chapter 9. Assessments of culture for safety can belong to this category. Cultural assessment is introduced in Chapter 11.

The third layer represents the internal independent oversight, called INSO in the nuclear industry. The main point here is independence, meaning that this function does not have line responsibility and does not participate in line activities. It oversees and challenges the line in safety issues. We developed a model for internal independent nuclear safety oversight (INSO) by utilizing the lessons learned from implementing the WANO approach in the Nordic nuclear industry and globally (Viitanen et al., 2025). Figure 8.2 illustrates an extended version of the oversight model applicable to all types of safety.

Figure 8.2 is naturally a simplification of the various interrelationships between the organizational levels and functions. What it fails to illustrate is the fact that independent oversight should also oversee the dependent oversight activities. However, this should not be their main focus, otherwise they would be merely checking the checkers.

The amount of oversight needed depends on the accepted risk level. This in turn is related to the hazards that the organization needs to control (see Chapter 4), and regulation as well as management objectives (Chapter 6). The influence of coxtext is further investigated in Chapter 10. Based on our study we concluded that independent safety oversight should primarily be seen as an assurance function to oversee and facilitate that the line organization takes care of its operative and non-independent oversight responsibilities (Viitanen et al., 2025). We also noted that:

> The INSO function should … consider that independence is not a black-and-white phenomenon, and that extreme independence is not always desirable. In situations where INSO becomes passively independent (i.e., merely points out issues and does not participate or proactively interact with the rest of the organization), there is a risk that INSO becomes isolated, irrelevant and ultimately dismissed by the line organization. Consequently, the INSO function should continuously balance between independence and involvement.
>
> **(Viitanen et al., 2025, p. 8)**

Total independence of the safety function is not possible or even beneficial for safety. There is always a need to balance the involvement of the safety function with its independence (Viitanen et al., 2025).

The key issue to consider when organizing for safety is the added value of various levels of oversight. Approaching from technical viewpoint and defence-in-depth philosophy, each independent layer of defense contributes to safety, even if the contribution is lower for each successive layer of defense. Still, it is (relatively) easy to calculate the desired safety level and assess how many independent levels of defense are needed.[2] The situation gets more complicated when we are dealing with a complex social system, not a technical system. As proposed by the NAT (Chapter 3), each "independent" layer adds to the complexity of the system. Simultaneously, each extra defense and layer of supervision changes the way personnel perceive their own work. For example, they may feel more secure if they know someone else is checking their performance or, vice versa, this may create extra pressure. Introducing new "layers" into a complex organization always changes the organization, often in unpredictable ways.

FIGURE 8.2 Oversight of safety and its relation to the functions and system levels

An often-heard statement from the line organization is that the safety function comes in to "circle the bullet holes" (cf. Viitanen et al., 2025), that is, to tell the line something they already know very well. This can mean that the safety function is not able to produce added value for the organization due to either the excellent capability of the line organization to perceive and manage risks or to the poor capabilities of the safety department to bring the needed diversity and requisite variety to the line organization views. However, it should be noted that sometimes the bullet holes need to be circled, if the line organization has perceived the problems but has not reacted and corrected them in time. The line organization may have gotten used to and even grown fond of their bullet holes. Safety organization needs to be adaptive in their approach to facilitate the line organization's safety responsibility. A cultural approach with its emphasis on meanings and underlying assumptions helps the safety organization to both reflect on its own premises and question those of the line organization.

In addition to conducting independent safety activities, organizations must ensure sufficient separation between various functions, for example, operations and engineering. Safety organizations should assess whether other disciplines within the organization can operate based on their specific principles or if they are influenced by the same pressures faced by the production department (or its equivalent in the respective domain).

> *McDonald, director of Morton Thiokol's Solid Rocket Motor Project during the* Challenger *accident, describes how his duties included both managing an R&D project where a new solid rocket booster (SRB) was being developed and simultaneously overseeing the operational shuttle program and the older SRB utilized by the program.*

There are two main challenges to organizing internal yet independent safety organizations: First, as has been seen in many safety-critical fields, organizations tend to reduce slack and redundancies. After accidents there is typically pressure to develop some form of increased independent oversight, and various semi-autonomous committees are established with the aim to oversee safety of the line activities. These committees often gradually drift into more paper-based exercises and distance themselves from the daily operational activities, making themselves simultaneously less useful. This leads to them being even less utilized and informed of daily activities in the line. Gradually, these committees fade away, only to be revived in a slightly altered form when something negative happens again. Thus, internal independent oversight should always be established with a clear goal in sight and with a clear role in mind. It is not a silver bullet to corrective actions after an accident, and it is susceptible to the same organizational phenomena as the line organization, including normalization and drift (see Chapter 3).

Second, independence in social systems differs from independence in technical systems. Social systems are aware of other, even independent, actors and change their expectations and behavior accordingly. This tendency contributes either to shifting responsibilities back to the line, or giving them more and more to the safety department, making it de facto a line activity. To succeed, safety oversight always needs the full backing of senior management, which is paradoxical since they should also be able to challenge senior management if their decisions do not adequately take safety into account. Without senior management commitment, internal oversight has little room to breathe. Internal oversight thus works worst in organizations that would need it most, and vice versa.

8.5 SUMMARY

The chapter looked at roles of safety professionals and departments in managing safety in complex organizations. Safety professionals have several roles, including experts, analysts, coordinators, and facilitators. The challenges faced by safety professionals are diverse, often arising from conflicting goals and the necessity to navigate between management expectations and workplace realities. They need to adjust to local circumstances and avoid strict adherence to outdated practices, advocating for a more adaptive approach to safety management. Safety professionals should not only ensure compliance with regulations but also actively contribute to the development of a healthy culture for safety within the organization.

Safety departments should have direct access to senior management to effectively oversee safety issues. The balance of independence and involvement is crucial, as safety professionals must challenge management decisions while remaining engaged with operational realities. Responsibility for safety should primarily lie with the line organization, supported and, if necessary, challenged by oversight functions that complement rather than replace line responsibilities.

REFLECTION QUESTIONS

What safety management principles do the safety professionals in your organization emphasize in their work?

How independent is the safety department from the line organization? Is the level of independence optimal from the organizational safety point of view?

Is the success of your safety organization measured by what the line organization does or does not do? What would be better ways to indicate how well your safety organization performs?

What is the added value of your safety department to organizational safety? How could the value be increased or made more visible?

What kind of culture exists in the safety organization? How well does that culture support the organizational culture for safety?

When there is line pressure for faster and cheaper work, how does your safety organization compensate?

BOEING 737 MAX CRASHES

The Boeing 737 MAX aircraft was grounded worldwide on March 13, 2019, after two crashes, one in Indonesia in October 2018 and the other four months later in Ethiopia in early 2019.[3] The crashes killed a combined total of 346 people. Apart from the human tragedy, it was a huge blow to Boeing's business, since the company had thousands of 737 MAX orders in its books.

On October 29, 2018, Lion Air flight 610 dove into the Java Sea shortly after takeoff from Jakarta, Indonesia, killing 189 people. That aircraft was almost brand new, having arrived at Lion Air three months earlier. Problems began immediately after the plane had left the ground. The nose mysteriously dipped, and the crew compensated by pulling the nose up. The nose lifted but then dipped again. The flight crew repeatedly commanded nose-up trim over the final ten minutes of the flight, not knowing that due to the erroneous angle-of-attack (AoA) data, the Maneuvering Characteristics Augmentation System (MCAS) was continuously activating and commanding an automatic nose-down trim. Unbeknownst to the pilots, an AoA sensor was misaligned. The MCAS relied on only one sensor, which had a fault. The flight crew was not aware of the existence of the MCAS.

The second crash occurred on March 10, 2019, when Ethiopian Airlines flight 302 departed Addis Ababa Bole International Airport. The aircraft crashed 60 kilometers from the airport, six minutes after it left the runway. Aboard were 149 passengers and eight crew members. The aircraft involved was only four months old. In both accidents, the crews were unaware of the existence of the MCAS and faulty information from the AoA sensor, which resulted in non-commanded activation of the MCAS.

The Maneuvering Characteristics Augmentation System

When Boeing designed the 737 MAX, it made the engines larger to increase fuel efficiency. This necessitated positioning them slightly forward and higher up on the plane's wings than the engines in the previous version. Early testing revealed that the relocated engines caused the plane's nose to pitch upward in some situations (e.g., low-speed flight, or flight with a high AoA when the plane is being flown manually). This could cause the plane to stall, that is, to lose its flying capacity. Both software and hardware fixes (such as redesigning the tail) were considered for these undesirable aerodynamic changes, but eventually Boeing settled on a software solution.

The Maneuvering Characteristics Augmentation System (MCAS), a piece of software, was designed to improve the airplane's handling characteristics and to decrease the pitch-up tendencies at elevated AoAs. The objective was to avoid stalling (loss of lift), which can happen if the plane flies at too steep an angle. MCAS activates automatically, without the pilot's input – and, as was realized after the two accident cases, without the pilots' knowledge.

Originally, the MCAS was designed to activate only in high velocities and high AoA situations. In 2016, one year prior to the Federal Aviation Administration's (FAA) certification of the 737 MAX, when Boeing test pilots were finally able to fly the actual plane, they found that it was not handling well when nearing stalls at lower speeds. As a solution, Boeing redesigned the MCAS to enable its activation at lower speeds. Moreover, the new version of the MCAS could move the horizontal stabilizer a maximum of 2.5 degrees (as opposed to 0.6 degrees, as originally designed).

Boeing was worried that the FAA would treat the MCAS as a new function and require simulator training for the pilots – a condition that Boeing had promised its customers would not happen. In 2013, Boeing decided to downplay the significance of MCAS externally, and to present it as just a part of the old system (Robison, 2021, p. 140). Thus, Boeing did not include any information on MCAS in the pilots' training material, and its redesign was never communicated to the FAA (Defazio & Larsen, 2020; Robison, 2021).

During the testing of the 737 MAX, mounting evidence pointed towards challenges with the MCAS system. In 2012, simulator tests indicated that pilots were unable to take control of the aircraft within the required time following the MCAS activation. The results were considered worrisome, but they were not communicated to the regulator.

The official investigation of the two crashes concluded that they resulted from a culmination of faulty technical assumptions, lack of transparency in Boeing's management, and "grossly insufficient" oversight by the regulator (Defazio & Larsen, 2020). The investigation discovered the following root causes of the two accidents:

- Production pressures leading to updating of the Boeing 737 design swiftly and inexpensively;
- Faulty design and pilot performance assumptions;
- A culture of concealment;
- Conflicted representation in the system that deputizes Boeing employees to act on behalf of the government;
- Boeing's influence over FAA's oversight structure.

Oversight of Safety Issues during the Boeing 737 MAX Certification

The Boeing 737 MAX was approved under the FAA's "Organizational Designation Authorization" (ODA) program. The FAA created the ODA program in 2005 and finalized it in 2009 to standardize its oversight of aircraft manufacturers that have been approved to perform certain functions on the Agency's behalf, such as determining compliance with aircraft certification regulations (Office of Inspector General, 2015). The ODA program effectively allows aircraft manufacturers to certify parts of their own designs with limited federal oversight. According to Boeing data, as of March 2017, the FAA delegated all 91 certification plans to Boeing's ODA (Gotcheva & Ylönen, 2021).

The ODA program enhanced the authority that the FAA granted to aircraft manufacturers to perform FAA-mandated certification activities. Before 2009, the FAA selected the authorized representatives (ARs) to the aircraft manufacturers, who were authorized to act on the FAA's behalf. After 2009, the selection of the ARs was done by the manufacturers, in this case Boeing (Robison, 2021).

Under the ODA program, Boeing served as a buffer between the ARs and FAA technical experts to help funnel the information regarding certification issues to the FAA in a more effective and efficient manner. While this change relieved the FAA of an administrative burden, it limited the interactions between ARs and FAA staff, which prevented free communication of issues and concerns. If there had been more fluid and frequent communication between Boeing's ARs and FAA officials in the FAA offices overseeing the certification of Boeing 737 MAX, FAA's knowledge of the ARs' concerns might have dramatically improved the safety of the airplane and enhanced the FAA's certification scrutiny of the 737 MAX program (Defazio & Larsen, 2020).

There were indications that Boeing employees who were supposed to be representing the interests of the FAA under the ODA program were instead representing the interests of Boeing. Further, Boeing designees involved in critical issues regarding the certification of the 737 MAX program failed to keep the FAA adequately informed of key issues, although these same designees did attempt to raise these issues internally at Boeing. The FAA's own employees reported in a survey that they perceived too much external influence on the FAA and that the delegation of certification responsibilities to "external FAA designees" (e.g., Boeing employees) had not been done adequately.

After the Lion Air crash, and before the second crash, the FAA learned about the redesigned MCAS and its vulnerability to a single point of failure in the AoA sensor. The FAA conducted an analysis and estimated that without a software fix to the system, there might be as many as 15 similar accidents during the lifetime of the Boeing 737 MAX. The FAA gave Boeing ten months to come up with a software fix. A month later, the second crash occurred. Regulators in China, the EU, India, Australia, Singapore, and Canada grounded the 737 MAX. Three days later, the FAA followed suit.

Under the ODA program, the FAA did not prioritize oversight of "the highest-risk areas", such as new aircraft designs, and it did not have an adequate system for determining whether the teams that were overseeing certifications were sufficiently staffed. The lack of independent design verification by experts also contributed to the "crash of the regulatory system" (Sgobba, 2019).

The "regulatory capture" of the FAA by Boeing was evident. The head of aviation safety at the FAA was a former lobbyist for Boeing, and before the accident, he actively promoted the transfer of power over to Boeing. Contributing to this regulatory capture was the so-called "Compliance Philosophy" adopted by the FAA in 2015, which aimed at facilitating an open and transparent exchange of information between the FAA and aircraft manufacturers. According to this philosophy, deficiencies and non-compliances identified during inspections would not lead to control action if the licensee was able to demonstrate how the deficiencies would be rectified and would also carry out appropriate analyses. The practice reduced audit requirements by 90% but weakened the FAA's oversight of safety-critical corrective actions, resulting in an update to the program in 2021. It is now limited to deficiencies and deviations of minor safety significance. In this case, typical corrective measures are process improvements and additional training.

What Did We Learn about Safety in Complex Organizations?

Systemic and inter-systemic nature of risks – to understand an accident at an African airline requires consideration of both the history of the American airline industry as well as the competitive and regulatory frameworks influencing the industry. Complex organizations are systems of systems, and to understand their safety, we need to understand the context in which they operate and how that context influences the organization (see Chapter 10).

The influence of strategy on safety – Boeing focused on shareholder price and actively tried to change its culture from an engineering-driven culture to a market-driven culture. This had practical safety consequences in terms of strategic (e.g. 737 MAX redesign) and operational (cost-cutting, outsourcing) decisions. This strategy influenced many operational decisions that ultimately led to the two crashes.

Regulatory capture and the challenges of oversight – The FAA did not have the authority, resources, or competence to adequately oversee the 737 MAX project. Boeing's authorized representatives did not have the authority or the identity to act as independent regulators.

Ironies of automation revisited – The challenges of automation for maintaining "situation awareness" of human operators has been acknowledged since the 1980s (Bainbridge, 1983). The two crashes serve as a reminder of the importance of designing systems with the operator in mind by using adequate Human Factors Engineering expertise in the system design phase. The accidents also remind us about the importance of simulations and system tests when designing and commissioning complex systems.

NOTES

1. An out-of-family event is an operation or performance that is outside the expected performance range for a given parameter or has not previously been experienced.
2. We will not go into all the assumptions included in these calculations in this book.
3. Description of the accidents is based on Viitanen et al. (2024), except where otherwise noted.

9 Tools of Safety Management

This chapter describes various tools and methods of safety management that can be used and that are typically used to manage safety in complex organizations. The chapter is not intended to teach the reader how to use these tools from scratch. There is plenty of existing literature for that. Rather, the aim is to highlight the explicit and implicit roles these tools have in creating, maintaining and changing (for better or worse) culture for safety. The tools are categorized into tools that 1) facilitate leading and planning, 2) support work in line organization, 3) help in monitoring work and safety in the organization, and 4) aid in organizational development. Despite this categorization, many tools contribute to multiple functions. Tools also contribute to one or more of the safety management principles introduced in Chapter 7. The chapter ends by introducing the DISC model for developing and assessing culture for safety.

9.1 LEADING AND PLANNING

9.1.1 Safety Policy and Safety Objectives

Why: The official values, priorities, and objectives of the organization lay the groundwork for all activities in the organization. They are the starting point, but not the end, in managing safety in complex organizations. They are one of the main tools for the *Direct* domain in the adaptive safety management framework introduced in Chapter 7.

What: The policy should be clear and unambiguous in communicating expectations. Consider how the policy will be implemented and enforced already when writing it. Also involve employees in the process of creating the policy and give them the opportunity to comment on the policy later.

Goals can be qualitative or quantitative. Goals and indicators are different matters. The indicators tell about the current state of safety, the goals of what should be achieved (see Chapter 6).

Safety objectives should be set for the entire organization, not only for the safety department. Success or failure to reach them should be attributed to the entire organization instead of the safety department (see Chapter 8). Safety objectives should reinforce the line responsibility for safety. It is especially important that the line managers have objectives that relate to safety issues (Monteiro et al., 2020b).

Safety objectives should be defined as far as possible in positive rather than negative terms. Objectives should also be attainable, and their significance understood by employees. In order to commit to the safety objectives, personnel need to experience them as meaningful and see how their own work contributes to attaining the objectives.

Other issues to consider from the cultural perspective when setting safety objectives for the organization:

- Set achievable yet adequately challenging goals.
- Seek a balance of process and outcome goals.
- Keep in mind that if both numerical and qualitative goals are set, employees naturally focus more on the numerical goals than the qualitative goals.[1] Similarly, if some of the goals include rewards (see 9.2.3), those goals receive more attention than the other goals.
- Avoid simplistic numerical goals combined with "whatever it takes" attitude as this contributes to unethical behavior (cheating, rule breaking etc.) either when aiming at the goal or when reporting results (Bazerman & Tenbrunsel, 2011, p. 106)

DOI: 10.1201/9781003491156-9

- Avoid glorifying the instances of absence (e.g. "x days since last accident") when aiming to promote the presence of safety.
- Discuss the goals with employees and ask for feedback on how they perceive the goals and whether they have in mind potential side effects that the goals can create

When: Safety policy and safety objectives need to be kept up to date at all times. In practice, this means that they need to be revised annually at minimum.

9.1.2 Key Performance Indicators

Why: Organizations need information on both the current safety level in the organization and the progress of various safety initiatives. Key performance indicators provide information to *control* and *direct* domains in the adaptive safety management framework (see Chapter 7).

What: Safety indicators can be divided into three types: outcome, monitor, and drive indicators (Reiman & Pietikäinen, 2012). Outcome indicators are also called lagging indicators. Monitor indicators are typically referred to as leading indicators. Drive indicators are a special type of leading indicators that measure the progress of safety activities rather than safety as such.

The purpose of outcome indicators is to measure the outputs of the system and see trends across time. When trends are shown, the organization can dig deeper to understand why the numbers are changing. The outcome indicators themselves do not reveal that. Strictly speaking, outcome indicators tell the organization very little about safety.

In addition to monitoring changes, attention should be paid to anticipated changes and whether they occur – for example, whether an increase in drive indicators (e.g., more training, more leaders-in-the-field activity) is reflected in decreases in lost-time incidents, in the unavailability of key systems, or in other outcomes of significance.

Monitor indicators are used to monitor how the system is currently performing and what the capability of the system is to perform in future. Monitor indicators can be both quantitative and qualitative. For quantitative indicators, monitoring the trend is important. Qualitative indicators typically produce a more in-depth understanding of specific issues (e.g. culture for safety, quality of the management system).

Drive indicators are objectives of the organization's safety work. Fulfillment of these indicators can be used for rewarding but not for maintaining a view of the current safety level. Similar to annual objectives in organizations, some of the drive indicators should be changed annually, based on the evolving view of safety and the current organizational challenges.

Indicators need to be perceived in context. As they are proxy measures of the underlying concept (safety), they tell a story only when they are seen all together. Each indicator's value may seem insignificant in isolation but gains importance alongside other indicators. Conversely, a significant change in one indicator might be explained by examining others or additional organizational performance sources. From a cultural viewpoint, indicators provide organizational performance clues that necessitate further analysis.

When: Drive indicators are used as objectives but cannot be used as neutral indicators of system performance. Monitor indicators are used as proxy measures of system phenomena.

Indicators are decision-support tools that provide relevant information without dictating specific actions. All relevant indicators as well as other information should be taken into account when making decisions. A common mistake is to see a change in one indicator and react to that with immediate corrective actions without looking at the whole picture. Indicators are tools for leading and planning (drive indicators) and monitoring (monitor and outcome indicators).

TABLE 9.1
Examples of Safety-related Indicators (Reiman & Pietikäinen, 2012; Hopkins & Maslen, 2015)

Indicator type	Examples
Outcome	Lost time incidents, unavailability of the key systems, sick leaves, staff turnover, fatalities, adverse events, violations of technical specifications or safety limits, loss of containment events, number of approved deviations from engineering technical practices
Monitor	Employee job satisfaction, number of repeat events (same type, same place, or same cause), trends in causes and contributing factors of incidents and observations, number of significant safety observations, results from 360-degree leadership evaluations, level of understanding concerning workplace hazards, competence levels, the ratio of preventive versus corrective maintenance on equipment and systems important to safety and/or production, availability of PPEs, mean scores and deviations in safety survey responses, organizational safety assessments, safety culture self-assessments, inspection and audit findings, regulatory notices, amount of overtime worked
Drive (objectives)	Management safety walks, number of leader-in-the-field rounds, amount of safety observations per employee, amount of training hours given, percentage of corrective actions carried out on time, average time to carry out a corrective action on significant findings, coverage of safety observations (how many in the organization have reported at least one), ratio of minor versus significant safety observations, percentage of incident investigations conducted on time, response rates in safety surveys

9.1.3 Rewarding

Why: Safety needs to be one of the issues the organization considers important enough to reward and spend money on. Even though rewarding for safety almost always has side-effects, not rewarding for safety is even more risky, especially if the organization rewards financial goals. Rewarding contributes to the domain of *direct* in the model of adaptive safety management (see Chapter 7).

What: Safety incentive programs are as routine as safety indicators in companies nowadays. The U.S. Government Accountability Office (GAO) conducted an investigation of reward practices in safety after the BP Texas City refinery explosion in 2005 indicated underreporting of incidents due to fear of losing bonuses. GAO (2012) identified two types of safety incentive programs: rate-based programs, which reward workers for achieving low rates of reported injuries or illnesses; and behavior-based programs, which reward workers for certain behaviors such as recommending safety improvements. They conclude that how the incentive programs work depends on the culture of the company (see also Schulman, 2020).

> Safety incentive programs exist in the context of a workplace's safety culture. Some types of programs, particularly those that are tied to low injury and illness rates, may discourage injury or illness reporting. However, the same programs in workplaces with positive safety cultures may have no effect with regard to reporting.
>
> **(GAO, 2012, p. 23)**

There are clearly many challenges in rewarding for safety. The problem with rewarding is that if you do not reward for safety at all, then all rewards will be from financial targets, and that is not good either. *Rewarding for safety is always risky* (Hopkins & Maslen, 2015). *But so is not rewarding.*

Rewarding in safety should focus on leading indicators (drive indicators) and effort instead of lag indicators (outcome indicators) and results. It is also good to remember that financial rewards motivate somewhat, but failing to reach those objectives demotivates even more (see loss aversion in Chapter 2).

It is good to remember that both rewards and sanctions are always socially constructed. Gneezy and Rustichini (2000) showed in their study that when a day-care center tried to curb late pickups of children by requiring the parents to pay a fine every time they picked up their kids after hours, the late pickups increased. Previously, parents saw picking up their kids on time as the right thing to do but after the introduction of the fine, the ethical dimension of the issue was removed, and the question became more of a time versus money where the option of a late pickup was perceived as an extra service from the day-care center.

When: Rewarding for safety should be implemented whenever the organization implements a rewarding system for achieving strategic goals. Reward systems should explicitly recognize both core-task-related and local hazards (see Chapter 4), ensuring that salient but less easily counted risks are not neglected.

9.1.4 Decision-Support Tools

Why: Various decision-support tools are intended to facilitate the integration of safety into operational decisions as well as adequate consideration of various risks in decision making. They also help to avoid groupthink and counteract the negative effects of typical human decision-making heuristics (see Chapter 2). Decision-support tools contribute mostly to *create* but also to the other domains of *control*, *direct*, and *facilitate* (see Chapter 7), depending on the tool and the type of decision.

What: The *Cynefin framework* (Kurtz & Snowden, 2003) is a sensemaking tool for decision making. Cynefin provides a shared language with which the members of the decision-making group can discuss situations, perspectives, and possible actions. The idea is to analyze the situation and the context in which the decision is made. The framework is based on systems theory and complexity theory (see Chapter 1) and it differentiates four decision-making contexts:[2]:

- Chaotic situations, where there is no cause and effect;
- Complex situations, where cause and effect are unpredictable;
- Complicated situations, where cause and effect are knowable;
- Clear/simple situations, where cause and effect are obvious.

In clear situations, decisions can be made by *sensing* the incoming data, *categorizing* the data, and *responding* in accordance with predetermined practice. Having a structured approach is key in clear situations. Complicated situations in turn require *sensing* the incoming data, *analyzing* the data, and then *responding* in accordance with expert advice or interpretation of the analysis. The key is to identify cause–effect relationships and act accordingly.

Complex situations should be approached by first *probing* the situation to discover patterns and make them more visible, then *sensing* the patterns and responding by stabilizing those patterns that are desirable, by destabilizing those that are not, and by seeding the space so that patterns that are wanted are more likely to emerge. Understanding this space requires decision makers to gain multiple perspectives on the nature of the situation. Narrative techniques are particularly powerful in this.

Chaotic situations require *acting* first, *sensing* to find the effect, and then *responding* to the situation. The idea is to act quickly and decisively, to reduce turbulence, and then to sense immediately the reaction to that intervention so that we can respond accordingly.

These domains are not stable features but may change as situations unfold. For example, an emergency situation can begin as a chaotic situation requiring immediate action (e.g. shutting down reactor, evacuating people), but gradually settle to complex (troubleshooting – probing – what is happening), from which the situation can evolve to complicated when the situation is under control but requires active management, and finally to clear when the emergency is over. The main point is to identify what the correct domain currently is, and what that means for sensemaking and decision making.

In simple and complicated situations, the most important boundary for sensemaking is that between what can be used immediately (what is known) and what needs time and energy to find out about (what is knowable). For complex and chaotic situations, distinctions of knowability are less important than distinctions of interaction: what we can pattern (what is complex) and what we need to stabilize in order for patterns to emerge (what is chaotic).

Six thinking hats (De Bono, 1985) is a decision-making tool that employs the idea of different hats that represent different perspectives on the same issue. It promotes parallel and deliberate thinking.

The white hat represents facts and information; it is neutral and objective. The red hat represents emotions and gut feelings, even anger. The black hat is gloomy and negative. It represents critical thinking (caution, risks, identifying problems). The yellow hat represents positive possibilities and benefits. The green hat indicates creativity and new ideas.

The blue hat is concerned with control and the organization of the thinking process. The blue hat guides the thinking process; he is the choreographer who designs the steps, but he is also the critic who watches what is happening (De Bono, 1985, p. 187).

There are several variations of the method. Typically, the entire team uses the same hat (except for the person with the blue hat) for a certain duration, and then everyone changes to a hat of another color. Alternatively, each member is assigned a hat of a specific color for a while and then members change hats.

The thinking hats method may help to avoid some of the biases described in Chapter 2. It gives a social permission to be critical, emotional, creative, positive, and factual. However, it cannot compensate for poor group dynamics or lack of psychological safety.

The nuclear industry has introduced a process called *operational decision making* (ODM) to give guidance to operations managers in handling emerging issues. In dealing with the complexity of nuclear operations, the industry recognized the need to address this issue based on its own analysis of industry events. Given that the ODM process is used for issues that are not perceived as an immediate threat to safe operation, time is available for operations managers to assemble a team with the expertise necessary to address the problem. The ODM process is a formalized procedure for decision making with different levels based on the safety significance of the decision. The level defines who needs to participate in the decision-making process.

ODM refers to the processes and attributes employed by the management to ensure proper response to degraded conditions (e.g. reduction of safety margin, increased primary system and containment leakage that remains below operational or license limits, problems with plant equipment that may cause an unplanned power reduction or plant shutdown, fuel defects, and the aggregate of equipment or material deficiencies).

Operational decision making is critical for safety, and organizations benefit from making this process more systematic. Some issues to consider when setting an operational decision-making process include (Viitanen et al., 2022) that

- Management should be committed to safety and support the ODM process;
- The process should be documented and everyone involved has to be familiar with the documentation;
- The atmosphere in the meetings needs to be open and everyone is expected to share their opinion;
- Decisions that are made should be justified and these justifications need to be written down;
- Disagreements should be noted in the minutes of meetings;
- Decisions need to be widely communicated in the organization;

- Self- and independent assessment of the ODM process, including the decisions, should be conducted regularly to improve the process.

Furthermore, the premises of the decision need to be made as transparent as possible and they should be documented as well as the decision itself. Only then can the decision be revisited and revised or at least used for learning purposes. It is important to document also the options that were not selected but were considered. The "black hat" function (or "devil's advocate") can be used during the ODM process to counter groupthink (McDougall, 2014).

McDougall (2014, p. 36) notes that

> Even though the ODM process still requires experienced participants to make judgements about the suitability and prudence of recommended courses of action, the ODM process provides a framework that encourages discourse and building of a common understanding about the issues at play. Diversity and respect for difference, the gathering of information, and the consideration of consequences, all support a stronger proactive response to emerging issues in a complex environment.

The *pre-mortem exercise*, developed by psychologist Gary Klein (2003) and promoted by Kahneman (2011),[3] is a strategy used to anticipate potential failures in a project or decision-making process before they happen. It involves imagining that a project has already failed and then working backward to identify the reasons for that failure.

In the exercise, participants assume that the project or initiative that is currently being considered has failed, and each participant needs to write down possible reasons for why the project might have failed, focusing on factors that were overlooked, ignored, or not anticipated when the project was implemented.

After collecting everyone's ideas, the team discusses and categorizes the identified reasons for failure, focusing on the most critical and actionable ones. Based on the identified risks, the team creates strategies or plans to address these issues before they can lead to failure.

The main benefit of the pre-mortem exercise is that it encourages proactive thinking about potential pitfalls and helps teams identify risks early, leading to better decision making and improved planning. It differs from the traditional post-mortem (which looks at a failure after it happens) by addressing issues before they occur.

When: Decision-making facilitation tools can be used before any major non-time dependent safety-critical decision is made. If a decision is needed fast, these tools are typically too time-consuming. From a cultural perspective, these tools are useful for surfacing assumptions and also resisting typical human tendencies for groupthink and anchoring (see Chapter 2). However, many of the tools are based on the rational decision-making paradigm (cf. McDougall, 2014) which may limit their usefulness in complex environments.

9.2 SUPPORT TO WORK

9.2.1 Pre-Job Briefs and Mini-Risk Assessments

Why: Pre-job briefs are used to identify hazards associated with a job and to verify that appropriate conditions and countermeasures are in place. They contribute to *control* of organizational performance and to *facilitating* local performance (see Chapter 7) by localizing general principles based on the conditions in the field.

What: Pre-job briefs are structured discussions or meetings conducted before starting a job or task. They are designed to ensure that all workers and involved personnel are aware of the potential risks, hazards, safety procedures, and specific tasks associated with the job.

Pre-job briefs should cover the following topics (DOE, 2009b; Viitanen, 2021):

- Review of the goals and content of the work
  - What are the major steps or subtasks of the job?
  - What are the success criteria for this job?
- Review that adequate tools, equipment, and spare parts exist for the job
- Review relevant documentation and permits
- Review hazards and their controls
  - Ask "what if" questions
  - What is the worst-case scenario in the work? What to do in that case?
  - How do we know that the risks have been averted?
- Review competence and clarify roles and responsibilities
  - Are all the workers qualified for the job, and does everyone know what they are supposed to do?
  - Does everyone know each other and each other's roles and responsibilities?
  - Are contractors/external employees involved and what is their role and responsibilities?
- Review operating experience
  - How has the job been done previously (operating experience)?
  - Are the conditions now the same as previously? What has changed since the last time? How do the changes affect the work itself?

Oedewald et al. (2014) studied the use of Human Performance (HU) Tools in the maintenance activities in the nuclear industry. Maintenance personnel considered pre-job briefings among the most useful and accepted Human Performance Tools,[4] and they are often naturally implemented even before their formal introduction as part of an HU program. Personnel saw the pre-job briefing as a competence development tool to improve understanding of the plant, its systems, and their safety significance; a coordination tool to improve smooth execution of tasks; and a collaboration tool that helps participants become acquainted with workers from other units (Oedewald et al., 2014).

Mini-risk assessments are typically small cards that personnel carry with them to the job site and check before starting the job. Sometimes they are printed (or electronic) forms that need to be checked before starting work. The way these cards work depends very much on the culture. Similar to slogans, these can support good culture for safety, but as such they cannot create one.

Pre-job briefs can also be a simple list of four questions, which every employee needs to be able to answer before starting work (La Duke, 2020, p. 55):

1. Do I know how to do this task?
2. Have I been trained to do this task?
3. Am I able to do this task?
4. Do I understand what I am being asked to do?

When: Pre-job briefings can be given on many levels depending on the complexity and safety significance of the work. For highly significant tasks, the briefing should be held formally, and it should be documented in detail. Mini-risk assessments can be used for any work, but they work best if used in a graded manner (meaning that a more thorough assessment is done for the more significant jobs).

9.2.2 Post-Job Reviews and Debriefs

WHY: Post-job reviews – sometimes also called after-action reviews (AAR) or debriefings – are important tools to facilitate organizational learning by reviewing both successes and failures after a task has been completed. They are an opportunity to praise good performance and identify issues for further improvement. They contribute to *creating* new knowledge and *facilitating* local performance, and to a lesser degree to *directing* behavior and making sure adequate *controls* are in place (see the four adaptive management domains in Chapter 7).

What: Post-job briefs can cover the following topics (DOE, 2009b):

- Did the work go as planned? Why?
- How did we feel about the job? Why?
- Were the instructions adequate for the job? Why?
- Were the practices and competence of the group adequate? Why?
- Were there any surprises? What? Why?
- What needs to be done differently next time?

For post-job reviews to become normal practice, it is important that the personnel understand their significance for organizational learning also beyond their own work group.

A different type of debrief is organized after something negative and potentially stressing happens. There should be a possibility to discuss and get support after a near-miss or an incident. The possibility for a debrief is important for psychosocial wellbeing.

Safety-critical crews and operators involved in an incident are typically taken off duty temporarily. This is due to the potential effects of the incident on their subsequent performance, and it is a wise precaution. It should be clearly communicated to personnel that this action is standard protocol aimed at ensuring safety, rather than a reflection of their individual capabilities.

When: Post-job briefs should be organized by the line organization after safety-critical tasks have been completed. The reviews of major events should be documented and the information from them should be utilized by the safety department. Post-job briefs can be done also with less bureaucracy by keeping the issue of learning from past events on the agenda.

9.2.3 Risk Analysis

Why: Conducting risk analyses is an integral part of safety management (Chapter 6) aiming at identifying hazards, assessing their significance, and planning countermeasures. They are a decision-support tool that can be used to *direct* activities but also to *create* new understanding, *facilitate* local safety behavior and *control* organizational level performance (see Chapter 7).

What: Risk analysis aims to identify hazards and estimate the risk they cause by looking at their potential consequences and likelihood of hazards happening. Risk analyses can unify perspectives from different disciplines by giving them a common language for risk. When done in collaboration with experts from various functions and roles within (or outside) the organization, risk analysis can help to identify trade-offs, gradual degradation, and interdependencies in complex organizations. Risk analyses can help groups to collectively make sense of hazards and prioritize the risks caused by them.

There are multiple more or less formal ways of conducting risk analyses, ranging from SWOT analysis to probabilistic risk assessment of complex technical systems. From a cultural perspective, risk analysis should be seen as an organizational learning opportunity and a support tool for managing daily activities. Both for purposes of learning and the outcome, it is imperative to involve people with diverse views of the work in question (diversity in roles) as well as diverse expertise (diversity in knowledge). Different people perceive risks differently due to, for example, personality, experience, and work role (see Chapter 2).

Risk analysis workshops facilitated by a safety professional provide an effective method for collaboratively evaluating the risks associated with specific tasks or processes. These sessions also enable consideration of preventive measures, actively engaging employees while simultaneously enhancing their awareness and understanding of potential hazards.

In networked environments, other organizations can also be involved in risk analyses. The facilitator of the risk analysis session has an important role in having expertise in the particular analysis technique as well as in creating the necessary psychological safety in the analysis group. Especially important is to have an open mind towards new risks or surprising findings and avoid the impression that someone higher up the hierarchy has already decided that the work is safe, and the risk analysis merely needs to "prove" that.

Different hazards require different risk analysis methods. When choosing a risk analysis method, the different functions, industries, and scope of operations of the organization must always be taken into account. For more information on hazard analysis techniques, see for example Leveson (2023) and Chapter 6.

When: Risk analyses are needed for new tasks and new settings. However, it is important to revisit and revise existing risk analyses as work and environments gradually change. Furthermore, risk analysis should not be done to prove things are safe but rather to aid decision making on how to manage risks.

9.2.4 Formal Guidance: Instructions and Checklists

Why: Prescribing work, the necessary steps to complete it, and the various hazards, as well as the ways to prevent and mitigate the risks, are needed for safety-critical and complex tasks to guarantee a common approach, independent of who is conducting the particular task. They are also needed to provide support and guidance for complex tasks. Rules and procedures formalize personal knowledge and lessons learned about the work. They can be used both to *control* and *facilitate* work performance (see Chapter 7). Safety professionals write instructions mostly to *regulate* organizational performance (see Chapter 8).

What: Formal guidance serves different functions (Provan, 2021; Reason, 2016):

1. Rules set the safety margins – employees do not have to decide the margins themselves under situational pressures and conflicting goals if they are stated in official documents.
2. The complexity of the job requires written guidance on the individual steps to take as well as how the steps are connected with task completion and other tasks. Relying on rules saves cognitive resources and enables the worker to better react to various contingencies.
3. The information necessary to perform tasks correctly is usually too much to be carried reliably in human memory.
4. Routine tasks are carried out routinely (with little conscious attention), and steps may be omitted without formal guidance.
5. Tasks that require communication and collaboration also require a common point of reference – instructions can be used as communication and coordination tools.
6. People change, and a different person may conduct the same task next time. Instruction can be used to relay standardized task information (including lessons learned) from person to person.

Instructions always embed assumptions about power, autonomy, trust, and individual responsibility. It is important to reflect what is the cultural significance of the various rules and procedures in the organization and whether these send the same message as official policies or other espoused values. In addition, the following more concrete issues should be considered when designing and writing instructions or procedures:

- How understandable is the procedure for the persons that need to apply it?
- What is the required background information, language comprehension, and education level?
- How easily accessible is the procedure?
- How can the user be certain that they are using the latest version of the procedure?
- How it has been verified that the procedure can be used in practice with all other procedures?

Checklists are a special type of formal guidance that is typically used right before the task execution. Checklists can act as reminders of issues to consider in the given task and give social permission to verify issues that normally are taken for granted. Their use requires vigilance and leadership in order to avoid a mindless "check-the-boxes" approach. Thus, they can be a good tool to support a vigilant culture, but it requires more than a checklist to create a vigilant culture.

Edgar Schein[5] gave an example in an interview of a mindful use of a checklist in surgery:

> When [the surgeon] arrives at the OR, he may discover that there are strangers there whom he's never worked with. I asked him recently, "What do you do?" He said, "I found that I can use the checklist that way. Instead of using the checklist as a bureaucratic device to skip through, I ask the chief nurse to go through the items slowly, with a lot of eye contact, and I look at each person on each item and invite questions so that during the 3 or 4 minutes that we're doing the checklist, I'm making myself available to questions, I'm maintaining a lot of eye contact, and de facto with body language saying, 'We're going to have to trust each other, and I'm here. I'm going to hear you if you have something to say.'"

This approach requires humility and emotional intelligence from the surgeon which is key in creating psychological safety at a team level (see Chapter 2). The WHO Surgical Safety Checklist and its implementation manual (WHO, 2009) is available online at www.who.int. Gawande (2009) has written a useful book about the use of checklists in medicine.

When developing rules or checklists, it is essential to include the personnel responsible for the work in question in the creation of procedures that will govern that work. They have the best knowledge of how the work is carried out and what are the typical work conditions. Rules and procedures should not prescribe conditions that cannot in reality be fulfilled. This leads to drift and normalization of rule deviance (see Chapters 3 and 6). In addition to the experts of the particular work, process experts or experts who have a wide range of knowledge about the entire system are needed to check the procedures for connections to other tasks, potential contradictions, or omitted perspectives. This is especially important for core task–related hazards (see Chapter 4).

Bozeman (1993), in his study of government "red tape", has made a useful distinction between two types of ineffectual rules: Rules that *are born* bad, and good rules that *have gone* bad. According to his treatise, rules are born bad out of

- Incomprehension (misunderstanding the relation between means and ends);
- Illegitimate functions (i.e., not serving a function that promotes the organizational core task);
- Negative sum compromise (when a rule is designed to serve several purposes, ending up serving none of them well);
- Overcontrol (when managers seek control through formalization);
- Negative sum process (when everyone participates in making the rules because this is what the rules dictate).

Good rules may go bad according to Bozeman (1993) because of issues such as

- Rule drift (when the meaning of the rule or the rule itself inadvertently changes);
- Rule entropy (when rules are passed from person to person and their original meaning gets lost);

- Change in implementation (when individuals start implementing the same rule in a different manner, e.g. more or less stringently);
- Change in the functional object (when the functional object of the rule changes in ways that render the rule obsolete or otherwise useless);
- Change in the rule's efficacy (when changing circumstances erode the rule's usefulness or effectiveness);
- Rule strain (when several rules need to be applied, it can create strain and inefficient use of resources);
- Accretion (when the increase in rules leads to inconsistencies or cross-purposes);
- Misapplication (when an otherwise good rule is used in a wrong manner).

When: Instructions and rules are needed for safety-critical activities. However, when creating safety-related rules and procedures, care must be taken to not contribute to "safety clutter" (Rae et al., 2018), which was discussed in detail in Chapter 8. As discussed, bringing more rules to an already tight system can in fact make the system more complex and increase the gap between Work-as-Done and Work-as-Imagined (the rules and procedures). It should also be noted that rules and their interpretation change in time as the culture develops. Rules, including their interpretation, applicability, and relation to other rules, should thus be revisited from time to time.

9.2.5 Human Factors Engineering

Why: Human Factors Engineering (HFE), also known as ergonomics, is an interdisciplinary field that focuses on understanding how humans interact with systems, tools, and environments in order to design them in ways that enhance safety, performance, and wellbeing. The goal of HFE is to optimize the fit between people and their work environments, taking into account human capabilities, limitations, and behaviors. HFE mostly contributes to *control* and *facilitate* domains in the model of adaptive safety management (see Chapter 7).

What: HFE focuses on understanding and improving the relationship between humans and their work environments, with the ultimate goal of reducing errors, enhancing performance, and ensuring safety. By designing artifacts for humans, HFE contributes to embedding (and potentially changing) assumptions. The design must consider the needs of end users, including maintenance personnel, and the safe use of the product. HFE aims to design systems and interfaces that accommodate human capabilities and limitations, thereby minimizing the likelihood of errors and contributing to effective task accomplishment. HFE seeks to design systems that promote clear communication, information prioritization, and intuitive displays, which help operators maintain situational awareness.

HFE ensures that systems, equipment, and tools are user-friendly and tailored to how humans interact with them. For example, in nuclear power plants, HFE principles have been used to design control rooms where displays and controls are intuitive and comprehensible, reducing the chance of operator errors during both normal operation and emergencies.

When: Human Factors Engineering is needed when designing or redesigning a sociotechnical system. It is also good practice to utilize HFE in case of challenges reported by the operators, or usability challenges noted in incident investigations.

9.2.6 "Leader-in-the-Field"

Why: The main purpose of the "leader-in-the-field" approach is to promote the presence and availability of the supervisor for the workers. For the supervisor, this is an opportunity to listen to the views of personnel and learn about how work is carried out in practice and what are the conditions in the field. The "leader-in-the-field" approach contributes mostly to *facilitate* domain in the model of adaptive safety management (see Chapter 7).

What: The "leader-in-the-field" approach means that the leader participates in the activities of their team and has dialogue with the group, gives positive feedback, and monitors the team performance. While in the field, leaders clarify management expectations and reinforce existing good practices. "Leader-in-the-field" is much more about coaching and being present than making formal observations or identifying deviations. It is about activities every leader should do even without such a formal program.

Good questions for leaders to ask during their field visits are (cf. Hopkins, 2012, p. 134):

- Tell me about your job. What do you do?
- What could go wrong? What are the greatest dangers you face?
- Do you think we have these dangers sufficiently under control?
- Do you think there are any safety issues here that we are not dealing with adequately?
- Are there times when workers around here feel they need to take short cuts?
- How could the management better support good quality work in this organization?

My experience is that often leaders suddenly appearing in the field can cause confusion among the personnel. The practice can also be experienced as irritating by the supervisors themselves if it is seen as an extra burden to an already demanding workday.approach In my view, this has to do with how leaders interact with people as well as with confusing "leader-in-the-field" with safety walks (see below). If the visits become bureaucratic and are done to satisfy performance indicators, they can have detrimental effects on employees' views on management safety commitment. As the visits are easy to count, they can create an illusion of safety if used as a safety performance indicator. Smith (2024, p. 143) argues that "management conversations should be approached with a specific purpose to understand something specific about workplace health and safety management". Personnel need to understand that leaders are in the field for a purpose other than ticking a box.

Sexton et al. (2021) conducted a large survey study in healthcare to determine the association between Positive Leadership WalkRounds (PosWR), an organizational practice in which leaders conduct rounds and ask staff about what is going well, and about wellbeing and organizational safety culture. Before the data collection, leaders were trained in a two-day training session. A typical PosWR visit would occur monthly, last 30–60 minutes, and involve between three and ten healthcare workers, a local leader, and a senior leader. PosWR prompts included the following kinds of questions:

1. From your perspective, could you share three things that are going well in this work setting, and one thing that could be better?
2. Is there anyone you would like to give a shoutout to today, and can you elaborate with specific examples?
3. Are there any celebrations of success, in your opinion, that you could share?
4. Who do you think deserves special recognition for going above and beyond in their role? Please provide specific examples that I can quote, as I will handwrite a letter to this person detailing the glowing comments from their colleagues.

Sexton et al. (2021) found that healthcare workers involved with PosWR reported not only better wellbeing (emotional exhaustion and work–life balance), but also a better safety climate. In their previous study of more traditional leadership walkrounds, Sexton et al. (2006) notes similar results when these walkrounds had been conducted well, with feedback to those involved about the actions taken based on the walkrounds.

9.3 MONITORING

9.3.1 Safety Walks and Task Observation

Why: Safety walks are a way for managers to monitor the conditions in the field and the workplace reality, and to identify unsafe conditions. Task observations are a way for managers to verify how tasks are performed in practice. They differ from "leader-in-the-field" in their emphasis on monitoring performance. They contribute to *control* and *direct* domains in the model of adaptive safety management (see Chapter 7). Safety professionals can use safety walks and task observations to *challenge* the line organization (see Chapter 8).

What: Safety walks and task observations facilitate the leader going to where the work is conducted and where the workers are. During the visit, the leader makes observations about potential hazards, improvement opportunities and positive findings. The leader needs a good understanding of the local work conditions as well as requirements and applicable rules and procedures. A checklist is usually created to guide attention and act as a memory aid.

Typically, observations that are made during the safety walks are reported in a specific system, such as corrective action program, and their implementation is followed there. It is important to make sure personnel whose work has been observed get feedback and that they are aware of the observations that have been recorded.

There is some evidence that safety walks are not very effective ways of verifying compliance. Viitanen et al. (2022) studied safety walks in a nuclear power company. They noted that the safety walks were mostly focused on systems, technical aspects, and "housekeeping" instead of interaction with personnel (ibid., p. 54). They further noted that there were some indications that numeric goals for the safety walks may be overriding the quality. These are quite typical findings for safety walks (Hopkins, 2011; see also Chapter 6 and the effect of safety performance indicators on performance). For this reason, "leader-in-the-field" is a more important method for supervisors whereas safety walks are better carried out by safety professionals (see Chapters 7 and 8).

Both "leader-in-the-field" approach and safety walks provide information on how the work is actually conducted and what the gap is regarding official procedures. They can also shed light on the reasons for the identified gaps. This provides important information on the gap between Work-as-Imagined and Work-as-Done.

Safety walks and task observation are effective for identifying local hazards (see Chapter 4). They work less well in managing other types of hazards. However, if they are combined with some of the questions from "leader-in-the-field" (see above), they can provide managers and safety professionals with information on the conditions in the field (the workplace reality) and shed light on the gap between Work-as-Imagined and Work-as-Done.

When: Safety walks are a regular tool for managers. Consistency is key. If managers show in the field only before the end of the quarter (when indicator data is due), it sends the wrong message to employees. Safety walks and observations during the walks can also be used as one method in assessments of culture for safety. Using observations in analyzing culture is discussed in Chapter 11.

9.3.2 Auditing

Why: Audits aim to verify compliance against predefined requirements and identify areas for further improvement. An organization has to ensure that its products, services, and processes meet established standards and regulations. Audits are also an important method of verifying the effectiveness of the safety management system (Chapter 6). Auditing the management system aims to identify potential deterioration of the safety management system before it actualizes in safety incidents. Auditing contributes to *control* domain in the model of adaptive safety management, and to a lesser degree to *direct* and *create* (see Chapter 7). Safety professionals can use audits to *challenge* the line organization (see Chapter 8).

What: Audits are among the most used methods of verifying the safety of organizational activities. However, audits are only one method of a more general safety assurance (see Chapters 6 and 8), and the other tools of monitoring presented in the chapter are needed to complement audits.

Audits reveal mostly the artifact level of culture (see Chapter 1), but they still affect culture and can be used to steer culture. The information that audits produce is also useful for annual summarizing and, together with other sources of information, can be utilized to make some conclusions about changes in culture.

Culture as such cannot easily be audited, but the structural features of the organization supporting a healthy culture for safety can be verified. Care must be taken to verify not only that the structures exist but also that they are applied in practice. Even if the audit itself does not reach the deeper levels of culture, it is also a way for the organization to demonstrate which issue it values (enough to audit them). Thus, rather than an assessment method, from a cultural point of view, audits are development methods as they can influence where the audited function's attention is devoted, and what kind of issues the function or department considers relevant for safety. They should aim to identify also issues and hazards the organization initially considers as not significant. Auditing requires a good understanding of the technical features of the systems as well as the social system (creating hazards through human action or inaction).

Some issues to consider when planning and conducting audits include the following (Hutchinson et al., 2024a, 2024b; Viitanen et al., 2018):

- Audits verify compliance against predefined standards or requirements. The typical assumption is that compliance with the requirements ensures safety, which can be questioned on many grounds (Chapter 4).
- Auditing can put too much focus on verifiable and standardized issues and downplay the role of employee competence and adaptability. This can steer the culture towards easily verifiable solutions (a "tick-the-box" approach).
- Audits often focus too much on the paper trail and too little attention is paid to the actual work (Work-as-Done versus Work-as-Imagined). External auditors typically audit safety professionals or managers, not the workers. There is a risk of marginalizing the practical knowhow of the field (Almklov et al., 2014).
- Audits are only as good as the auditors' competence and skills in using the audit method.
- Audits can help the organization in identifying problems, but often organizations still struggle with solving them. Here the auditing approach can again tempt the organization to implement simple solutions that the lead auditor can verify and easily close. This corresponds to optimizing numbers instead of solving issues (Chapter 3).
- Passing an audit can create a false sense of security in the organization. However, audits should be treated similar to safety in general; a lack of evidence of danger should not be taken as evidence of safety.

To conclude, audits provide information on what is being done to ensure safety in the organization and what the outcomes have been so far. Audits themselves cannot ensure or verify safety but they can be used to identify hazards and unsafe activities.

When: Audits are one tool to build an understanding of organizational capabilities. They are typically used to verify organizational arrangements for safety and quality, but they should also be seen as methods of facilitating and steering safety improvement. They can point out improvement areas, but they do not work well in verifying safety if they are the only method applied to verification.

Selecting the functions and areas to be audited as well as the specific topics of interest guides the organizational attention. This selection process should be integrated with the overall approach to safety management to ensure that the various support functions, such as quality and safety, have an integrated approach which builds on joint data collection. Audits can also provide information that is utilized and integrated by safety professionals in, for example, summary reports or meta-analyses (see below in this chapter).

9.3.3 Reporting of Observations and Incidents

Why: This method helps organizations to proactively identify potential risks as well as to react to any actualized risks. Reporting contributes to *control* and *create* in the model of adaptive safety management (see Chapter 7).

What: The purpose of a safety observation is to identify and record unsafe acts, conditions, improvement areas, and best practices in the workplace. Observations can include noting PPE use, spotting hazardous conditions, and recognizing safe behaviors. Reports can be used in identification and correction of vulnerabilities that are gradually developing in the organization, in addition to correcting the individual cases reported.

Personnel do not automatically know what to observe and report. Without guidance from safety professionals and a strong message from senior management, the observations will most likely be very simply-to-observe issues such as housekeeping or technical breakdowns. Employees should be given examples of what to report and understand that the goal is to learn and identify potential risks proactively. The objective is not to assign blame or identify specific individuals.

Effective handling of observations (including root cause investigation for serious findings) and feedback is important. For motivation and subsequent reporting, it is also important to consider what kind of corrective actions are designed and implemented. If the corrective actions are seen in the field as making ordinary work more difficult, employees will be much less willing to report in future. As a precondition for reporting, there needs to be adequate trust among the personnel that reports are treated in a just manner.

When considering the issues that are reported by personnel, it is good to consider the complexity of tasks (Grote, 2012) and the type of hazards. Local hazards are typically easier to observe if the employees are vigilant. Reporting is then a good way of engaging the workers to observe and develop work and working conditions (Grote, 2012). However, many core task–related hazards and complex tasks are such that it may be difficult to observe deviations based on one worker's perspective. Thus, in (relatively speaking) simple environments and local hazards, reporting can be used as a corrective mechanism based on individual concrete observations, but in more complex environments and more complex hazards, reporting needs to rely on trending and weak signals that are then further investigated to uncover what the problem may be and how to proceed with corrective action. These observations then seldom deal with the hazards directly but rather with various work conditions that may influence the organization's ability and willingness to manage safety.

Some issues to acknowledge and consider when implementing reporting systems include the following (Viitanen et al., 2018, 2023):

- People can only report issues that they perceive. The importance of understanding risks and perceiving hazards at work is an important but often neglected success factor of reporting.
- Reporting needs to be as easy as possible for the person who makes the report. If possible, there should be a way for the analyst to contact the person who made the report for clarification.
- A culture of trust and fairness needs to be established for voluntary reporting systems to function.

- Issues that are reported need to be addressed in a timely manner. If issues are not addressed, the motivation to report in the future declines.
- Substance knowledge is needed to ensure that the personnel are aware and have the feeling that something is significant enough to report.
- Giving positive feedback and rewarding reporting is important but should be used carefully to avoid overreporting.

When: All organizations should make sure employees can report concerns freely and also anonymously if needed.

9.3.4 Incident Investigations

Why: Incident investigations, when properly conducted, offer a window into the daily realities of work at the organization. When things do not go as expected, there is a need to understand what happened, how it happened, and what needs to be done differently in future. Investigations are tools for organizational learning. Investigations contribute to *create* in the model of adaptive safety management (see Chapter 7) but also to a lesser extent to the other domains (*facilitate, control, direct*), depending on how the results are utilized in subsequent development of the organization.

What: The line organization should be the primary owner of the investigations, as they are also the ones to carry out the corrective actions. Without line organization ownership and commitment, investigations will probably define abstract corrective actions that will not get implemented by the line organization, who can then dismiss the investigation as "rubbish".

Even though the line organization owns the investigation process, the methodological competence comes from the safety professionals. Their role is to act as facilitators and make sure the process is just and that it systematically looks at human, organizational, and technical issues and their interaction.

Key issues are to avoid the hindsight bias (Chapter 2) and excessive blame, and instead aim to understand what happened and how, not who made a mistake. I have noticed that recommendations from incident investigations often still target front-line workers and propose either more training or more rules and procedures. This needs to change. Recommendations should improve the work and work conditions, not punish the workers.

Root cause analysis, in its many applications, aims to uncover the reason why an incident or an accident took place. Care must be taken not to be tempted by the lure of a single root cause, which is very uncommon for complex events (Leveson, 2023). The name *root cause analysis* is a bit misleading because it implies the possibility of tracing the event back to a single cause. Some companies have re-labeled their event investigations as *learning reviews* (Pupulidy & Vesel, 2023). This has been done to highlight their main goal, learning, and to downplay the negative, individual blame-oriented connotations typically associated with investigations.

Investigation methods such as AcciMap can help to highlight the contributing factors "higher up" in the organization (Svedung & Rasmussen, 2002). The Functional Resonance Analysis Method (FRAM) can help to highlight interactions and non-linear effects (Hollnagel, 2012). System-Theoretic Accident Model and Processes (STAMP) looks at the accident from a system-theoretical perspective, building on Rasmussen's (Rasmussen & Svedung, 2000) concept of abstraction hierarchy (Leveson, 2023).

From 5 WHYs to 4 HOWs: One commonly used investigation method is the so called 5 WHYs, where the idea is to go deeper by asking why each event happened. Asking why leads us easily into human errors and other abstract explanations too early in the process. Answers to *why* questions are always based on our beliefs and assumptions. Instead of asking why something happened, it is useful to ask *how* it happened. We will end up with the *why* by asking the *how*, but by asking only the *why*, we easily miss the *how*. An additional benefit of asking how something happened is that we open ourselves to the possibility of multiple contributing factors as well as interactions and coincidences necessary for the incident to take place. Asking *why* easily directs us to a single root cause and may hide the complexity of the event and its interactions.

When conducting investigations, it is important to examine how the situation made sense to people at the time and how to improve the sensemaking and response capabilities of people in future. During the investigation, try to understand how things made sense to the people involved in the event, and avoid judgment with the benefit of hindsight. What was their reality and under what conditions were they carrying out their work? Think also about the wider implications of the incident and what it says about organizational capabilities. Ask questions such as: How well did we anticipate and respond in the event? How could we monitor better in future? How did we (gradually) end up in this situation? What issues did we notice and what issues we did not notice at the time of the event? What mechanisms shape our practices and culture? Did we make the right assumptions, did we take something for granted or normalize it? How did the situation make sense to the personnel involved? How can we increase monitoring and coping capabilities, including developing our culture? Try to seek improvements in the work and organization primarily and only secondarily in the methods of safety assurance.

To adequately learn from events, the following issues should be addressed:

Acknowledge the complexity – There are multiple contributing factors to any event, and the search for the root cause is typically driven by practical need to conclude rather than reality. Do not stop when you find deviations or errors, as these should be only the starting point for a deeper and wider understanding of the system and the context of the event.

Avoid focusing on counterfactuals – These are things that did not happen but, with the benefit of hindsight, should have happened to prevent the incident or accident. Focus on human errors and various other omissions easily leads to overemphasis of these "should have done's" that do not explain what was done in reality and why.

Consider the context – What were the constraints and possibilities for action in the given context? What were the cultural expectations and norms guiding people? What was different in the context in comparison to previous performance? Why did it make sense to the people to act the way they did? What has changed in work or in the environment since before the accident? Separate contextual lessons (in this organization with this group of people operating in this environment) from the general lessons (e.g. drift in complex systems).

Seek multiple perspectives to the same event – In complex systems, no one person understands the whole (see Chapter 1). This also means that each person has a different view of the event, based on their local understanding. Listen to people's stories and acknowledge local rationality of people's actions as well as the fact that they have rationalized their behavior with the benefit of hindsight.

Learn about the work and the organization, not only about the particular event. Go "up and out" instead of "down and in" (Dekker, 2015, p. 51): Look at the system more widely and consider the conditions (structures, resources, environment), assumptions, and interactions – instead of digging into the "root cause". Use events to learn about normal work and its success factors, not only about the current incident or about what can go wrong.

Target the corrective actions on work and organization rather than safety assurance – Improve work and the organization and only secondarily the various defenses and barriers.

When using these principles as guidelines, you should be able to learn more about your organization and not only about the specific event that happened. You will learn about the aspects of the organization and its culture that influence performance.

1. What happened? (the incident and its immediate causes)

2. How did it happen? (timeline of events and their interrelations)
3. Why did it happen? (contributory and latent factors)

In their efforts to avoid blaming, some organizations have stepped too far in the other direction and are afraid of any findings that can be attributed to a certain group of people or an organizational unit. This makes improvement suggestions and corrective actions vague and can hinder development (and paradoxically increase blaming in the organization). There is a fine line between findings that are actionable and findings that are unnecessary blaming. A culture of trust and openness is key when employees and others read the investigations and make sense of the recommendations.

When: Investigations should be conducted after each major incident. However, in addition to major incidents, it is good practice to investigate several minor incidents to uncover weak signals. Furthermore, successes should also be investigated from time to time to elaborate on organizational strengths.

9.3.5 Organizational Assessments and Meta-analyses

Why: Organizational assessment aims at improving organizational performance by gaining information about the current state of the organization and reasons for problems, as well as the functioning of the system in general. The assessment provides information on how the current way of organizing supports safety. It thus contributes to *create* and *challenge* in the model of adaptive safety management but also to a lesser to *direct* by providing an overview of the current challenges from the point of view of the organizational core task.

What: *Organizational assessments* include various types of assessments that look at the organizational factors and safety management at the organizational level (Reiman & Oedewald, 2009). These can include management system evaluations, safety culture assessments, certification assessments, and various inspections focusing on organizational issues. They can rely on a wide variety of methods and be conducted either internally (self-assessments), independently (management system evaluation by QA), or externally (assessments carried out by a third body or inspections by regulator or authorized inspection body).

In external assessments, it is easier to look at organizational and management level issues, including the possibilities and constraints of the safety function as well as potential shared blindspots. However, the more involved the assessor is with the work being assessed, the better the understanding of the realities of the workplace and the job in question.

The safety case (see Chapter 6) can use organizational assessment as one source of evidence that risks are adequately managed. However, organizational assessment is always a snapshot of the current organization, and this view needs to be regularly updated. This update can be done by continuous monitoring, meta-analysis, and periodic re-assessments.

Meta-analysis means an assessment that relies on existing information and documents. Organizations collect a lot of information but typically they are bad at integrating this information and looking the organizational performance on a system level. Meta-analysis builds on existing information and seeks to integrate the information into a view of the current organizational capabilities. Meta-analyses can be done, for example, annually as part of the management review process to update the understanding of the current state of the organization.

In Chapter 11, methods and approaches for cultural assessment and monitoring are presented. The methods can be utilized in other types of organizational assessments also. The chapter elaborates various indicators of degrading culture that continuous monitoring and meta-analyses should consider (see Table 11.4).

When: Organization should conduct assessments on a regular basis, with varying depth and methodology. In-depth assessments should be complemented with periodic focused assessments as well as continuous monitoring of performance.

9.4 IMPROVEMENT

9.4.1 Safety Training

Why: Safety training is needed to ensure that the competencies and working practices are adequate to manage risks. Training is also a way to continuously improve an organization's capabilities to identify and manage both old and new risks. Training contributes to both *control* and *create* domains in the model of adaptive safety management (see Chapter 7), depending on the training contents.

What: Safety training comes in many forms and can range from short talks that are embedded within induction training or pre-job briefings to extensive multi-week seminars and lecture courses on technical, human, or organizational topics. Safety training serves many purposes depending on its content and focus. For instance, training can develop culture-related awareness, behavior, or attitudes (Harvey et al., 2001) or job-specific skills. Training can be targeted at any level of the organization, from shop floor to senior management. Training should always be conducted as close to the job as possible. Simulators and mock-ups are an excellent tool for this, as are various virtual reality (VR) tools that allow us to safely practice hazardous activities. Likewise, various games can be utilized to better involve personnel in the training.

An important function of training is for a specific group to learn to work together, including appreciating each other's expertise and understanding their role in the shared core task. For example, team leadership training is more effective if it includes the whole team and not only the leader (Hynes et al., 2006).

Kirkpatrick's model (1994) is a simple, widely applied model and analysis tool for evaluating the effectiveness of training programs in organizations. The model consists of four subsequent, causally linked levels: 1) reaction, 2) learning, 3) behavior, and 4) results (Kirkpatrick & Kirkpatrick, 2005).

- Level 1 (reaction) is the initial level and indicates whether the participants feel interested in the training session or perceive it as engaging and relevant. This level corresponds with typical "customer satisfaction" surveys.
- Level 2 (learning) refers to whether the content that has been taught has been understood, whether new skills have been developed or acquired, or whether attitudes have changed.
- Level 3 (behavior) refers to whether on-the-job behavior has changed after the training program – whether the participants apply what they learned during the training.
- Level 4 (results) refers to whether positive organizational outcomes have been achieved as a result of the training.

In terms of developing culture, levels 1–2 are dependent more on the contents of the training (and participant expectations), whereas levels 3–4 depend more on culture and contextual issues than on the specific training content.

When planning training initiatives, it is good practice to start with level 4 and consider what results are wished for, and whether training is the right tool to accomplish those results. For example, in the nuclear field, IAEA has developed guidance documentation to facilitate SAT, a Systematic Approach to Training (IAEA, 2021b). SAT consists of five phases:

- Analysis: The identification of training needs, tasks, or competencies selected for training, and KSAs required to perform a particular job.

- Design: Developing the training objectives and questions to assess their achievement, and organizing the training objectives into a training program.
- Development: Preparation of the training materials.
- Implementation: Delivering the training.
- Evaluation: During this phase all aspects of training effectiveness of both the training program and training system are evaluated, followed by feedback leading to training and facility performance improvements.

SAT is quite a heavy process, and I have found that it is easier to apply it to specific tasks and behavior rather than general organization-wide principles or practices. However, even if it is not utilized as a full-scope application, its principles can serve as a valuable framework for developing competence management and training solutions.

It is also good practice to coordinate safety training contents with other technical or non-technical training programs, as there may often be conflicting messages or assumptions. The more unrecognized contradictions and conflicting assumptions there are in the material, the more clearly training teaches the personnel the Work-as-Imagined instead of the Work-as-Done (see Chapter 3).

It is important to differentiate traditional classroom training from on-the-job training, exercises and simulations. Exercises and simulations are extensions of training that enable safe learning from dangerous situations and accidents. However, the simulated situation never fully corresponds to normal work. VR and augmented reality can help in making the simulations as close to reality as possible. On the job training, on the other hand, is an efficient way to induct new people in the ways of the organization, both good and bad. Instead of Work-as-Imagined, on the job training familiarizes workers with Work-as-Done. If there is a large gap between the two, this may create a cognitive dissonance in new personnel. Typically, Work-as-Done is the winner in this competition. This means that on the job training almost always trumps the more theoretical classroom training.

When: Safety training should be part of the continuous development of organizations. It needs to be based on analysis of competence needs and conducted as close to real work settings as possible. Refresher courses on familiar topics are needed in addition to training on new and unfamiliar topics.Training should never be the only method of improvement, as studies show limited effectiveness of purely training-based safety interventions, especially the ones that rely on one-way communication (Dyreborg et al., 2022, pp. 31-32).

9.4.2 Safety Promotion: Slogans, Posters, and Mnemonics

Why: Slogans, posters and mnemonics remind people about safety and hazards as well as about the safety rules, organizational values, and codes of conduct. They contribute to *direct* in the model of adaptive safety management (see Chapters 7 and 8).

What: Slogans can be used to express and amplify cultural features, if they exemplify what the organization stands for (Alvesson & Sveningsson, 2024, p. 45). However, slogans should not be used to create culture but rather to manifest, express, and reproduce the current culture. If the slogan stands contrary to practices and values, it raises negative sentiments among employees and can have the opposite effect on culture than the one intended by management. Slogans can easily become mantras or dogmas, or hollow phrases (Busch, 2016, p. 27, see also Le Coze & Reiman, 2023).

As discussed in Chapters 2 and 3, employees need to understand the risks before they can intentionally avoid them or stop work when observing risks. Likewise, trust and psychological safety are prerequisites for the employees to speak up and stop dangerous work, irrespective of the company slogans or safety rules. For instance, phrases such as "safety first" and "everyone has the right to stop work in case of danger" are meaningless if workers are unaware of the risks or if they are afraid that if they stop work, they will get fired.

In addition to lack of understanding, lack of psychological safety is another reason that slogans such as "everyone has the right to stop the work in case of danger" may fail without adequate consideration of culture. Employees have a high threshold for raising issues if they need to fear consequences – in case of uncertainty, they will probably not stop work but rather try to complete it, in time and as efficiently as they think the management wants them to.

When utilizing slogans, consider what the slogan is supposed to facilitate. Slogans should simplify and make important cultural values, beliefs, and practices easy to remember and apply. A good slogan/mnemonic in the nuclear industry is the STAR principle (mnemonic for Stop, Think, Act, Review) to remind people to self-check the quality of their work and think before doing anything. The U.S. Department of Energy, in its report on human performance tools (DOE, 2009b, pp. 18–19), describes the main phases of the STAR principle as follows:

Stop: Worker pauses, eliminates distractions, and focuses on the task.
Think: Worker anticipates expected outcomes, contingencies to unexpected outcomes, verifies necessary conditions and actions, and reviews documentation and planning.
Act: Worker performs correct action on correct component, following relevant documentation and planning.
Review: Worker verifies that outcomes match expectations or performs contingencies in the case of mismatch.

Similar slogans are used in, for example, the oil and gas and mining industries, such as "take 5", with a similar idea to step back and think about what might go wrong and how to prepare for it.

Visualizing safety is challenging and visualizations are always subject to interpretations (Le Coze & Reiman, 2023). Visualizations are artifacts of their creators and embed their assumptions about safety. It is always good to reflect what the various signs and posters imply in addition to what they visually show.

When: Slogans, mnemonics, and visualizations are usable for reinforcing cultural beliefs and practices, but not for creating them. They can be used to remind people of things that are already understood to be important.

9.4.3 Behavioral Modification

Why: Behavioral modification and behavior-based safety is needed in environments where individual behavior can directly contribute to accidents. This applies especially to local hazards (see Chapter 4) and behaviors that can visibly be seen as hazardous or in situations where human errors can have immediate effects. Behavioral modification mostly contributes to *control* and *facilitate* in the model of adaptive safety management (see Chapter 7).

What: Behavior-based safety (BBS) refers to a collection of approaches that share a focus on employee behaviors and their safety effects. BBS raises strong opinions, and it has its opponents as well as proponents. On the surface, the approach seems antithetical to the cultural approach. And in some respects, it is, but that does not mean that BBS would not in certain circumstances improve safety. And recent applications of BBS have gotten closer with the systems view of safety in their focus on organizational issues, leadership, and organizing (Cooper, 2009). In many ways, the focus of modern BBS is more on practices than behavior, making it more compatible with the cultural approach. Focus on practices requires taking into consideration the culture at the workplace and how employees currently work. I still think the strong focus on behavior limits the usefulness of BBS as a culture development tool, but since many companies utilize it anyway, it is good to consider the pros and cons of the method from the cultural perspective.

BBS approaches safety from the individual perspective and can be used as one method of improving safety if these programs are used in a modern way. It is good to note, however, that behavioral change is not cultural change unless there is deeper change at the level of meanings, beliefs, and values. Looking back at the definitions in Chapter 1, if culture is about practices that are executed even when no-one is watching, a test of behavioral change is when management ends supervising or rewarding for the specific behavior. If the behavior continues, something has changed in the culture – what this is cannot be known without closer inspection.

BBS considers the visible behavior of the individuals as the primary target for development. BBS interventions are typically based on the ABC model of behavior originally proposed by B.F. Skinner: Antecedents, Behavior, Consequences. The model also proposes that the most powerful determinant is the consequence (actually, the expectation of certain consequences based on experience). Antecedents are triggers for behavior, as the famous bell was a trigger for Pavlov's dogs to salivate, indicating an expectation that food was about to be served.

Observations are at the root of the BBS programs. The observations can target workgroups, individual workers, or work outcomes. Feedback from these observations is critical for the success of the behavioral modification. Different ways of giving feedback contribute to improved injury prevention (Cooper, 2009).

BBS programs run the risk of assuming that the unsafe acts of shop-floor personnel are the causes of accidents. Adopting this approach would have negative consequences for the culture of safety in the organization. According to Hopkins (2006), a drawback to programs centered on worker behavior is that they place too much emphasis on "observable" behavior, even making it one of their guiding principles. These programs tend to focus attention on observable behaviors to the exclusion of other less or even non-visible yet equally essential components of risk management. Thus, the tools also underplay behavior in abnormal situations (Hopkins, 2006) or behavior that affects safety indirectly. BBS tools easily ignore the role of understanding in safe behavior.

When: BBS is applicable to local hazards (see Chapter 4). BBS has proven somewhat effective in controlling easily perceivable and conceivable occupational hazards (however, see Dyreborg, 2022, p. 57) but is less effective when the hazards are complex or not easily perceived. BBS is focused on behavior modification, and as such it does not automatically increase personnel's understanding of hazards or requirements of their work. What it does develop is compliance to standards. BBS can be used to counteract the human tendencies to normalize risks and adapt behavior to, for example, schedules and production pressure by teaching personnel certain automatic routines, for example, holding the handrail when walking on stairs. These risks are otherwise easily perceived negligible by employees (see Chapter 2), which does not mean people would not regularly fall down stairs at workplaces.

9.4.4 Seminars and Events

Why: Various seminars and discussion sessions focusing on safety issues are needed to develop cooperation and share information and communicate the importance of safety. They can facilitate dialogue and awareness of risks if carried out in collaborative way, contributing to both *facilitate* and *direct* in the model of adaptive safety management (see Chapter 7).

What: These sessions maintain awareness of work conditions and ongoing tasks, reminding people about hazards and safety and offering an opportunity to listen to employees' opinions, concerns, and ideas. They need to be forums for inquiry rather than advocacy (Acosta, 2024, p. 71). Seminars can sometimes become one-directional events, where leaders communicate directives to employees without fostering dialogue or collaborative learning. This approach does not fully utilize the potential of group meetings since gathering people together provides an opportunity to generate new insights collectively, rather than merely disseminating information.

Personnel seminars, which focus on safety issues, work in the development of cooperation and the sharing of information, for example, concerning new practices. It is important that these are implemented in a participative manner so that employees can engage in dialogue with the management.

When: Seminars and events are useful when there are changes in the organization, or challenges that need to be addressed together. They can also be used periodically to facilitate continuous development.

9.4.5 Culture Ambassadors

Why: These groups can be used to support and steer various culture-related activities, including human performance and human factors initiatives. They can be used to involve employees in safety development as well as to monitor current culture. This means that culture ambassadors can contribute to all four domains of adaptive safety management (see Chapter 7). However, since it is difficult to promote all four domains at the same time, ambassadors should define the focus of their attention based on what is needed at the moment.

What: Some organizations establish, as part of their "safety culture programs", a group of safety culture ambassadors. Safety culture ambassadors are a group of people nominated from various parts of the organization (usually functional departments) whose purpose is to support managers in developing safety culture. Their activities may include, for example, bringing safety culture messages to the meetings of their respective units, ensuring that safety is taken into consideration in decision making, and observing the atmosphere to identify trends that could be averse to safety. The employees act as ambassadors in addition to their other duties, so being an ambassador is (at best) a secondary duty for these people. This has both pros and cons, as it means they are deeply involved in the work in question and know the practical safety challenges. On the other hand, being deeply involved in the actual work means they have limited time for other (extra) duties. Discussions are thus needed on an organizational level to guarantee that these people can actually commit resources to culture development.

The following issues need to be considered when implementing a safety culture ambassadors' group (cf. Viitanen et al., 2018):

- A clear definition of the purpose and expectations with regards to the ambassadors group;
- Definition of tasks and activities of individual ambassadors, for example in relation to communication, training, monitoring, promotion, and influence on safety and safety culture;
- Top and line management commitment, which would support, for example, setting expectations for the ambassadors' group, and help in finding solutions for allocating resources for ambassador responsibilities and balancing between the primary job responsibilities;
- Practices for continuous management and coordination of the ambassadors' group, which may include, for instance, joint meetings;
- Mechanisms for following up the functionality and effectiveness of the ambassadors' group against the organization's development needs, and for the renewal of the purposes and activities if needed;
- Adequate training of the ambassador group in organizational factors and culture.

When: Culture ambassadors are best formed around a project such as a development program or self-assessment. They can also work as part of continuous improvement if their roles and resources are clearly defined. In this case, their specific goals should be defined based on the current requirements (see safety management principles in Chapter 7).

9.5 SUMMARY

Safety management tools are often introduced into organizations with the expectation that they provide neutral, objective ways to control hazards. Checklists, indicators, audits, and risk assessments are typically presented as if they were technical tools detached from social or cultural influence. Yet, from a cultural perspective, tools always embody the assumptions, values, and priorities of the organization that created or adopted them. In practice, they are never value-free but function as cultural artifacts. They both manifest and influence the current culture it.

Safety management tools can be used in various ways. Organizational culture has an effect on how tools are adopted into use. As was emphasized in chapter 7, to change culture, safety professionals and other leaders need to "step outside the culture" to identify which tools are needed and how they should be used to change culture and improve safety. Similarly, it is useful to reflect on the premises of the current safety management tools: What norms, beliefs, and assumptions about safety do the current tools promote?

The main purpose of the chapter was not to teach how to use the tools but rather to illustrate them from a cultural perspective: how the tools can support safety in complex organizations, and how the tools can facilitate the development of culture and how they may have unintended consequences for culture, depending on how they are used.

Many more tools are available in the field of safety management, and thus the chapter did not attempt to cover all or even most of the existing tools. The aim was to give an overview of the typical tools and look at their pros and cons. Each tool has a purpose and an intended use, but our ideas of how a tool should be used and what problems the tool helps to solve are rooted in culture and in our history of using these tools. The cultural approach advocates mindfulness when utilizing the various tools of safety management – mindfulness of their pros and cons as well as their potential side-effects in the organization.

Tools contribute to one or more of the safety management principles introduced in Chapter 7. It is important for safety professionals to be aware of how the tools that they use affect the organization and how the tools can support or hinder the implementation of the safety management principles (Reiman & Viitanen, 2019a).

DESIGN FOR INTEGRATED SAFETY CULTURE

Design for Integrated Safety Culture (DISC) has been originally developed at VTT by me and others (Oedewald et al., 2011; Reiman & Oedewald, 2009). DISC depicts safety culture as something more than the correct attitudes and mindset. The right mindset is necessary, but safety also requires well-designed and functioning structures and processes to ensure good preconditions to carry out the activities with sufficient quality.

Unlike most other safety culture models, the DISC model pinpoints the importance of knowledge and understanding of safety and the core task requirements and hazards inherent in the system. Without a thorough understanding of safety and risks, the organization might focus on irrelevant challenges, make risky decisions, or be blind to new threats, all the while being emotionally committed to safety.

The DISC model includes seven criteria for a good safety culture as well as ten organizational functions that are needed to fulfil the criteria. DISC is an assessment and development framework, not a model of culture. Thus, an understanding of cultural dynamics is needed to properly apply the framework in actual cases.

Organizational functions

Strategic management
Safety leadership
Work process management
Work conditions management
Change management
Competency management
Supervisory support
Proactive safety management
Hazard management
Management of contractors

The DISC model states that certain organizational functions are necessary to develop high safety potential in an organization. These include, for example, hazard management practices (such as risk assessments, redundancy of safety systems, and personal protection equipment); competence management practices (such as training courses on the specific technologies or treatments used, mentoring of newcomers); pro-active safety development practices (such as reporting and analyzing incidents, periodical organizational assessments); and work condition management practices (such as assessing the adequacy of the staffing, and ensuring necessary equipment for work).

When assessing safety culture, the focus is on the seven criteria,[6] whereas when developing culture, the focus shifts to the functions. Organizational functions are the way to influence culture. They are heavily context-dependent, meaning that their manifestations differ between industries and organizations, but they all should contribute toward the seven criteria of a healthy safety culture.

Criteria of a healthy safety culture

Safety is a genuine *value* in the organization that is reflected in the decision making and daily activities.
Safety is *understood* to be a complex and systemic phenomenon.
Hazards and core task requirements are thoroughly *understood.*
The organization is mindful in its *practices.*
Personnel *take responsibility* for the safe functioning of the whole system.
Activities are *organized* in a *manageable* way.
Internal *group dynamics* support organizational adaptation and change.

The following characterization is based on the literature (Oedewald et al., 2011) as well as my experience of using the DISC model in several safety culture assessments during my career.

Safety is a Genuine Value in the Organization that is Reflected in Decision Making and Daily Activities

This criterion means that safety manifests in the organization's practices, documents, and discussions. Safety is visibly a value in the organization that guides activities and motivates. Employees and management consider safety to be of genuine value and something worth striving for in the organization. Safety is taken into account in daily work and decision making at all levels of the organization. Safety is considered in the objectives and performance targets of different functions as well as in resource allocation and work planning.

The organization has set goals related to safety at different organizational levels. Management emphasizes safety, and leaders personally set an example. Leaders at different levels of the organization articulate safety objectives and safety expectations.

Management at all organizational levels understands that safety and productivity are intertwined, and safety takes priority when in doubt. Senior management is committed to safety and allocates significant resources to proactive safety measures such as risk assessments, incident investigations, safety observations, safety walks, and management of change.

Safety is Understood to be a Complex and Systemic Phenomenon

Safety is understood as a complex and systemic phenomenon, meaning that the prevailing safety conception in the organization encompasses the following aspects:

- Safety is a dynamic property of activity that requires constant effort. It does not equal a lack of accident or good incident statistics, and it cannot be decomposed into factors or evaluated factor by factor, independently of other factors.
- Safety is influenced by the interaction of multiple stakeholders (individuals, groups, management, support functions, society). Each person and task have a potential to influence safety. This safety significance is clarified by different organizational practices so that each person can describe the effect of his/her work on safety. The interfaces of the actors are important to safety; thus, the organization aims to promote a good overview of work processes.
- Smooth interaction between humans and technology is a precondition of safety. It is understood that people have a role in every stage of technology implementation.
- There are multiple types of safety in addition to operational safety (e.g., occupational safety, information security, security, environmental safety). Their special requirements (e.g., knowing the different laws, different safeguarding measures) are considered.

Safety is not an isolated phenomenon, but it is linked to how the organization as a whole operates and carries out its core task. Understanding the systemic nature of safety means understanding that the causes of errors and accidents are usually complex, and therefore blaming individuals does not contribute to safety. It is understood that surprises always happen, and not everything can be anticipated or standardized. It is understood that too precise instructions can even lead to paralysis of the system in the face of unexpected situations. Safety requires good foresight, a good understanding of current operations, good responsiveness, and the ability to learn from mistakes and succeed. Safety requires comprehensive development of the organization and technology.

It is also understood that organizations and their operating environment are always changing, and therefore situational awareness must be constantly maintained. Similarly, safety is an achievement that must be constantly worked on. The organization understands the challenge of measuring safety, as well as the potential side-effects of overly simplistic metrics in an organization.

Hazards and Core Task Requirements are Thoroughly Understood

Hazards and core task requirements are thoroughly understood, meaning that the organization has a good outline of things that need to be managed to carry out the core task well and manage hazards that are part of the activity.

The personnel understand the key risks related to operations. They understand both the risks to which they themselves are exposed and the risks that their actions may pose to others, or that arise from the interaction of different actors. The organization understands the major risks involved in its operations and is able to distinguish them from, for example, occupational safety risks.

A good understanding of hazards exists in all the organization's functions and at all the personnel levels. Each actor knows the possible risks embedded in their work in relation to all types of safety. People can perceive how their work connects to the organizational core task.

For organizations that do not fulfil this criterion, they are typically unsure of their strategy as well as the major risks involved. Personnel take risks unintentionally as they lack awareness of what is dangerous or how safety-significant their own work is.

Further, the identified reasons for incidents in these organizations are typically competence gaps or deficiencies in risk perception (e.g. "risky behavior" or "unnecessary risk-taking", which may indicate that risks are not really understood).

The Organization is Mindful in its Practices

The organization is mindful in its practices, meaning that the organization can maintain a mindset that views the knowledge and practices as maybe being imperfect, even though they are developed continually. The tolerance for expressing uncertainties is good, and organizational practices encourage questioning habits when they relate to identifying possible hazards. Risks are constantly monitored with the help of a variety of competencies and methods because the organization is alerted to the possibility of new hazard mechanisms. The organization uses, for example, field personnel, research, and competence development to gain a better overview of the risks.

The organization, operating methods, technology, and personnel are developed continuously and proactively. Safety is developed proactively before incidents occur or before the authorities require it. When defects are identified or dangerous situations occur, their root causes are investigated, and corrective measures are defined. The implementation and effectiveness of the measures are monitored. The organization encourages its personnel to report on both small and large issues, and the personnel report actively. These safety observations and development suggestions are utilized in the development of operations both by correcting individual deviations and by examining broader trends.

Since it is never possible to predict everything and safety is a complex phenomenon, a questioning attitude towards prevailing assumptions and operating methods must prevail throughout the organization. The organization must encourage people to be critical, raise concerns and observations about safety, and provide feedback on how these issues are addressed. Individuals at every level of the organization tend to avoid excessive complacency, question assumptions, and heed counterarguments. Discussion and expression of opinions are encouraged. Employees feel that they have the courage to express their opinions and concerns. The organization also tries to question their perception of risks and how well they think they are managed. The organization understands that risks change, which is why it is important to actively challenge current risk perceptions and try to identify new, or changed, risks.

The organization understands that the "reality" described in the plans and instructions always differs from how the work is actually carried out. It may also be the case that the actual way in which the work is carried out is at least as safe as that described in the guidelines. Such differences must be analyzed without assuming that every deviation from guidelines is a deliberate violation. More important than trying to completely eliminate circumvention and "interpretation" of instructions is to try to understand those aspects of the organization's activities that encourage and sometimes force one to go against the guidelines.

For organizations that do not fulfil this criterion, there are challenges in acting in an alert way, and mindless routines exist. Uncertainties are not perceived, and risky conditions are not challenged. The organization does not put effort into gaining a better understanding of risks.

Responsibility is Taken for the Safe Functioning of the Whole System

This criteria means that the organization possesses an idea that every member has the possibility and responsibility to act for the safety of the whole system. Even though the official task descriptions are clear, there are practices and mindsets that encourage juridical accountabilities to be exceeded when safety may be affected for better or for worse. Employees at all levels of the organization are responsible and feel obligated to take care of safety and be aware of safety requirements.

The formal roles and responsibilities related to safety are clear both in normal situations and when responding to emergencies.

For organizations that do not fulfil this criterion, there are unclarities or inconsistencies concerning responsibilities. Official descriptions concerning responsibilities and expectations for employees are unclear or inconsistent, or they are not known or for some other reason do not guide the actual practices in a systematic way toward a broad responsibility of the plant's functioning.

Activities are Organized in a Manageable Way

Manageable activities means that there are sufficient resources for carrying out the jobs with good quality and according to the plans. Staffing levels, competencies, and work process knowledge are sufficient. In addition to the working conditions, timetables and tools are adequate for the work. Even exceptional work situations can be managed without total chaos.

The organization anticipates the future resource and competence needs as well as the required working practices. It is prepared in advance for forthcoming changes.

If activities are not manageable, in many activities, the actual course of action does not correspond to the plans or other descriptions due to the lack of personnel or material resources, or due to poorly designed work processes. The employees report mild or moderate difficulties in terms of being in control of their tasks. The organization may also not have defined the needed competencies, resources, or work processes, or it has significantly failed in its estimates. The organization acts in a fire-fighting mode because of insufficient resources and management, and individuals autonomously decide their working practices. The spread of resources across the organization may also be uneven, with some parts having enough resources and some too few.

Internal Group Dynamics Support Organizational Adaptation and Change

The atmosphere of the organization should be such that it allows for open discussion about possible dangers and the raising and processing of various concerns and development ideas, as well as a general discussion of common rules of the game. Leadership and interaction are important for positive group dynamics. Trust and psychological safety are recognized as important phenomena, and personnel strive to maintain them.

The importance of both formal and informal networks is also important for this criterion. These networks need to be formal for making decisions, but they can also be informal for asking advice, discussing technical issues, sharing hunches and lessons learned, and getting to know and understand each other. It is important that there is an opportunity for positive interaction in the organization.

Internal dynamics concerns issues such as normalization and drift (see Chapter 3), and how these manifest in the organization. Drift and normalization can concern any level of the system, from practices to interaction patterns to basic assumptions. Here the question is not so much about their content, but rather how they gradually develop, and whether the organization is able to steer its culture toward practices, interaction, and assumptions that support safety.

NOTES

1. This tendency is called the quantitative fallacy and sometimes the McNamara fallacy, named for Robert McNamara, the United States Secretary of Defence from 1961 to 1968. The fallacy involves making decisions solely on quantitative data and ignoring all qualitative information. Daniel Yankelovich (1971) noted that in practice this means measuring what can be easily measured and presuming that what cannot easily be measured is not important and does not actually even exist.
2. A fifth context is *confusion/disorder*, where it is uncertain which of the other four domains apply.
3. One of the few things these two scholars agree on. See also their joint article Kahneman and Klein (2009) – subtitled a "failure to disagree"! See also Chapter 2 for Kahneman's theory of System I and System II and Klein's ideas of intuitive decision making.
4. HU tools mainly focus on anticipating, preventing, or catching human errors. Several HU tools have been defined and implemented in the nuclear industry. The US Department of Energy's Human Performance Improvement Handbook (DOE, 2009a, 2009b) defines a total of 27 HU tools for shop-floor personnel (individuals and teams). In addition, the DoE handbook defines 14 management tools such as self-assessments, operating experience, and change management. The most used HU tools include pre-job briefing, post-job review, self-checking, peer checking, independent verification, three-way communication, phonetic alphabet, procedure use and adherence, and questioning attitude (Viitanen, 2021).
5. https://catalyst.nejm.org/doi/full/10.1056/CAT.18.0185
6. There were six criteria in the original publications, but I noticed in the assessments that I conducted that there is a need for a summarizing criterion. This seventh criterion anticipates the direction that the organization is developing in and brings an additional dynamics into the assessment.

10 Safety and Culture in Context

This chapter looks at the external factors shaping organizations. It aims to show how safety is constructed in interaction of the system and its environment and that when the environment changes, safety changes also. It ends with a consideration of national cultural influences on safety.

No organization is an island. Organizations are systems of systems, existing in and adapting to the wider influences of society. Organizational cultures are formed and reshaped through forces residing beyond the direct control of the organization, such as macro-economic, institutional, regulatory, technological, regional, or ethnic/national cultural influences (White, 2017, p. 133).

10.1 CHANGING LANDSCAPE

Today's world is quite different from the one in which most of the safety theories described in the previous chapters were developed. Le Coze (2021c) identifies five trends shaping a new landscape of high-risk systems. These trends are both products and producers of globalization, and they are present in high-risk systems today: digitalization, externalization (outsourcing), financialization, standardization, and self-regulation. Le Coze (2021c, pp. ix–x) notes that since Perrow published his Normal Accident theory,

> globalisation has restructured companies' environment and the role of states, finance is at the heart of capitalism and markets, Internet constitutes the core infrastructure of the digital age and the notion of anthropocene indicates that humanity now acts as the equivalent of a geological force, transforming our environment in an unprecedented scale, challenging the distinction between nature and culture.

Le Coze (2023) argues that Perrow's notions of coupling and complexity (see Chapter 3) need to be scaled up to address the global interconnectedness and systemic risks of high-risk technologies, such as those seen in the COVID-19 pandemic and climate change. High-risk technologies now pose global risks and systemic threats, affecting flows of people, goods, and data, and leading to issues like ecological degradation and pandemics. Furthermore, global flows reveal power asymmetries and unbalances among nation-states, impacting their ability to manage and regulate these flows effectively. COVID-19 exemplifies the global interconnectedness and the need for shock-absorbing and shock-diffusing properties in nodes (states) and networks to manage such crises effectively (Le Coze, 2023).

Le Coze proposes Post Normal Accident Theory (see summary of Normal Accident Theory in Chapter 3), which puts more focus on networks, strategy, and the eco-socio-technological nature of disasters (Le Coze, 2021c).

Bieder et al. (2024) explored how significant *global changes* impact industrial safety. They provide an overview of the significant changes that have occurred since the beginning of the century and analyze their potential impacts on safety management. They focus on four changes: climate change, digital transition, geopolitical tensions, and demographic changes. They note that there is a need to move beyond traditional approaches and develop new strategies to manage safety in a complex and uncertain world.

Andrew Sharman described the current landscape of safety in the following way in 2019, before the COVID-19 pandemic and the war in Ukraine:

> In addition to potential financial and reputational worries, in recent years, we have seen epidemics (such as the SARS virus), pandemics (like swine and avian 'flu), natural disasters (such as earthquakes, volcanic eruptions and tsunamis), man-made disasters (from oil spills to factory collapses), product recalls

DOI: 10.1201/9781003491156-10

> (from everything from cars to children's toys to baby milk) and workplace accidents all of which have the potential to bring organisations to a shuddering halt.
>
> **(Sharman, 2019, p. 5)**

Sharman (2019, pp. 11–17) also notes the many negative consequences of globalization for safety. He notes how over the last few decades, globalization has seen many instances of multinational organizations moving their manufacturing plants to developing countries to capitalise on cheaper labor costs and avoid the more stringent health and safety legislation of developed countries. This externalization (Le Coze, 2021c), or outsourcing, not only changes risks but also transfers them from one organization to another.

> *The collapse of an eight-story commercial building named Rana Plaza in the Dhaka District of Bangladesh on 24 April 2013 led to over 1,000 deaths, mostly clothing factory workers working for major multinational fashion brands.*[1] *The building contained five clothing factories: most of the people in the building at the time were garment workers.*
>
> *Since 2008 all new buildings in Dhaka have been required to apply for a final "occupation certificate" following a visit by inspectors to check they have been constructed in accordance with their original plans and used for their stated purpose. Rana Plaza was neither built according to its plans nor used for stated commercial purposes. It was also never given the final safety clearance.*
>
> *Instead of the six floors the owner was permitted to build, there were eight, and while he was given permission for a commercial office block, he rented its floors to industrial companies. These additional floors, the machinery they housed, and the heavy power generators on the roof were factors in Rana Plaza's collapse.*
>
> *The day before the collapse, large cracks appeared in the building, and an engineer who was called to inspect the building determined it was unsafe. Rana and the factory owners, however, ordered workers to return the next morning as they were under pressure to deliver products to their customers in time. When the generators were switched on that day, the building collapsed.*
>
> *Five months prior to Rana, another factory building (the Tazreen Fashion factory) collapsed, resulting in the deaths of over 110 individuals. However, the magnitude of the Rana Plaza collapse garnered significant global attention to the perilous working conditions faced by numerous workers within the garment industry. Furthermore, the disaster prompted scrutiny regarding the obligations of American and European companies and governments that utilize labor in Bangladesh and other low-wage markets globally.*
>
> *In many developing countries, occupational safety and health laws as well as labor laws in general are underdeveloped, and the workforce is rarely unionized. Lack of government oversight leads to reliance on weak self-regulation. For multinational organizations, this offers an opportunity to transfer risks and reduce costs as long as consumers are not too bothered by, or aware of, the suffering of the workforce that manufactures their products.*

The challenges of *outsourcing* and use of contractors have already been discussed in Chapter 6, but a few key points are worth repeating here: Outsourcing allows specialization, which has both pros and cons for safety. Current safety models do not very well consider the networked nature of modern businesses, although at the same time many accident investigations demonstrate the communication, coordination, and learning challenges created by organizational boundaries (see also Chapter 3). The contracting process and the contract itself are key in assuring the safety of outsourced services. Furthermore, some degree of monitoring of the contractor's performance by the client is always needed. All in all, the better the company itself understands the nature of the task (including its interconnections, necessary competence, and resources, as well as risks) that it is in the process of outsourcing, the better the outcome of the contracting process should be for all parties if sufficient trust and goodwill exist between them in the beginning (meaning no-one cheats).

Dekker (2021, p. 21) gives an example from rail about the added complexity created by the privatization of railways from the point of view of the rail surveyor (privatized, formerly part of the state rail company):

> the anxieties around having private contractors stomping around live train traffic (on a railway owned by another company, with trains running on it that belong to yet another company) can be readily imagined. This is why documentation requirements are fairly intricate. Not only that: the contractor has to supply its own full-time safety officer while their surveyors are working anywhere within 50 meters of a train track, but the rail track–owning company sends its own safety official as well. These two are sometimes complemented by a third safety official, from the train operator (if there's only one exploiting that piece of track) who assists in cases of more complex train schedules. The risk assessment forms used by the contractor (which are different from those used by the rail track owner and the train operator) are not developed in-house, but supplied by yet another consulting contractor.

Dekker goes on to list other contractors that are involved in assuring the safety of the rail surveyor. Previously these issues could be managed at scale by the state railway organization. Now each contractor must do all these things. Dekker highlights that this extra bureaucracy occurs after securing a contract. The complexity of the bidding process and its associated bureaucracy is a separate issue entirely.

Systemic risks refer to potential threats that endanger the functionality of systems of critical importance for society and their scope in time and space (Renn et al., 2022).[2] Their impact may extend beyond the system of origin to affect other systems and functions. They define complexity, uncertainty, ambiguity, and the propensity for ripple effects beyond the system of origin as characteristics of systemic risks. Renn et al. emphasize the role of global institutions that systematically investigate the environment for weak signals that are potentially related to major threats, such as WHO or the Center for Disease Control were in the case of the COVID-19 pandemic. They conclude that to be effective governance of systemic risks requires interdisciplinary and cross-sectoral cooperation, a close monitoring system, and the engagement of scientists, regulators, and stakeholders.

Climate change is a systemic risk that impacts all individuals, although to varying extents based on geographical location and various other determinants. Climate change can create new hazards and change old ones. For example, power plants need to take into account both rising temperatures and rising sea levels. Increase in extreme weather events such as storms and heat waves challenges the safety of sociotechnical systems as well as the critical infrastructure in a new way.

The effects of *geopolitical tensions* on safety are difficult to predict. The rising instability may affect how willing companies are to invest in safety when global priorities shift. A related tendency is the rise of authoritarian regimes around the world. Corruption and organizational wrongdoing (Chapter 3) are connected to authoritarian regimes, with several negative safety consequences.

The changing global landscape creates new external hazards (see Chapter 4) and changes the old ones. Global changes are also reflected in other types of hazards in organizations, and, in many cases, blur the distinction between safety and security. They also contribute to, and coincide with, changes in work.

10.2 CHANGING WORK

Work has changed considerably since the adoption of the concept of organizational culture in the 1980s. Traditional manufacturing jobs have decreased especially in the Western countries, while knowledge and service sectors have grown. Simultaneously, the physical demands of work in many industrial sectors have decreased while the psychological demands have increased. These psychological demands include requirements for networking, teamwork, perceiving emotions of others, and controlling one's own emotions (Väänänen, 2024).

Digital transition means, for example, the adoption of digital technologies and artificial intelligence (AI) in all aspects of work. These profoundly change many existing jobs, create new jobs, and eliminate others. AI and digital technology also change the internal dynamics of organizations and will most likely affect how they learn and develop competencies in future. There are several open questions: As AI does more and more "thinking" in organizations, will it affect how well the employees understand the basic phenomena and risks associated with the core task? How does AI change the way organizations learn, innovate, and develop? Will AI increase bureaucracy, or flatten hierarchies? Will it bridge or widen the gap between thinkers and doers? Probably there will be no clear-cut answers to these questions, but nevertheless AI needs to be managed also from a safety perspective.

Today's knowledge work is characterized by a "high level of individual freedom to make decisions and plan one's work, but also by a high level of temporal interdependency and social ties" (Väänänen 2024, p. 124). This "illusion of autonomy" (ibid.) or the "*autonomy paradox*" (Mazmanian et al., 2013) among knowledge workers has several consequences for both work wellbeing and quality (Andersen et al., 2022). Combined with the ubiquitous technology and the ability to work anywhere and anytime, the increased requirements for self-structuring and flexibility have had consequences for the mental health of knowledge workers (Väänänen, 2024, p. 125). There are increased opportunities for self-development, but for employees, these opportunities can produce anxiety and exhaustion due to further increase in options and decision-making requirements. At the same time, missing these opportunities may be experienced as depressing (Väänänen, 2024, p. 131).

Temporary work and project-based organizing have increased in almost all areas. Sandberg and Löwstedt (2025) note that project-based organizing can contribute to overwork. They note, based on their study of construction site managers, that they had internalized the pressure to overwork and tended to experience it as self-chosen. However, they did not have the ability to change their behavior even if they wished to do so. They described their work as highly autonomous and considered this autonomy as a significant motivation. Paradoxically, the interviewees overworked either to keep their current jobs (if they were satisfied with their role) or to advance to a better position (if dissatisfied with their current role). The autonomy paradox shows in the interplay between autonomy and internalized organizational control. Sandberg and Löwstedt (2025) conclude that overwork cannot be treated solely as an individual nor a structural issue. One example of a structural issue maintaining overwork was the high responsibility of site managers for the end results of the building process. Sandberg and Löwstedt (2025) concluded that project-based organizing contributes to the autonomy paradox and overwork.

Individual flexibility and resilience are increasingly required to compensate for organizational and societal inflexibility. This effect is not limited to white-collar workers but shows in work in general. Employees are expected to adopt new approaches, use social skills, show emotional receptiveness, and have the ability to handle mental pressure (Väänänen, 2024, p. 135). According to Väänänen (2024, p. 135), in modern work:

> Work and self-values reach down into the identity of the employees. Failure to realize oneself is not just experienced as a professional failure, but also as a personal failure.

At the same time, so-called blue-collar workers in many industries have faced increased workload due to staff reductions, increased demands to incorporate new technologies and tools to existing work processes, and less than optimal working conditions.

Hybrid or remote work has increased since the COVID-19 pandemic. The pandemic made the separation between white-collar and blue-collar workers apparent and in many ways widened the gap. Hybrid work is typically not possible for blue-collar workers, and the decreased presence of office workers at the office or in the field risks further alienating these professions. The integration of automation and artificial intelligence in blue-collar industries has the potential to reduce manual workloads but can simultaneously contribute to increased knowledge requirements.

Vartiainen and Vanharanta (2024) reviewed the discussion concerning hybrid work (HW) during and after the pandemic in Europe. They looked at studies concerning the relation of HW to wellbeing and performance. Concerning wellbeing, the studies they reviewed showed a mostly positive relation on average, but most also noted potential adverse effects of hybrid work.

According to their review, long working hours can cause musculoskeletal disorders, and blurred work–home boundaries may increase fatigue and mental demands, risks to mental health, professional isolation, and feelings of loneliness. Positive wellbeing outcomes on the individual level were shown as job satisfaction (N =6/14, meaning in six out of 14 reviewed studies); reduced stress, strain, and exhaustion (N =5/14); and increased positive emotions, affective attitudes, work related flow, and happiness (N = 5/14). On the organizational level, the positive wellbeing outcomes were seen in organizational commitment, work engagement, and support (3/14). Positive impacts were noticed in organizational performance and productivity (N = 5/14) and supervisor-reported (N = 3/14), self-reported (N = 2/14), and objective (N = 2/14) performance. Also, other studies have gotten mixed results about the effects of hybrid work on wellbeing (Choudhury et al., 2022).

Vartiainen and Vanharanta (2024) also note the possibility of a "hybrid work divide", creating a group of privileged professions that enjoy autonomy and flexibility as well as the challenges that hybrid work puts on the supervisors. Hybrid work, like any work, needs to be designed and managed along with other issues in the organization. It thus presents a leadership challenge (see Chapter 7). Table 10.1 presents potential safety consequences of increased remote and hybrid work.

Working away from the "field" has had safety consequences. Hopkins (Hopkins, 2000, p. 51) noted already some 25 years ago that one of the contributing factors to an explosion in the Esso Longford plant outside Melbourne was the fact that during a reorganization, engineers were moved from the site into headquarters several kilometers away. The accident happened during the plant startup. Before the reorganization, the plant engineers were available for operators' questions and concerns during startups and had an in-depth understanding of the plant. This was an important unofficial competence that was lost in the reorganization. Similarly, hybrid work will most likely change many official and unofficial practices in organizations. This change should be managed and assessed as any other organizational change (see Chapter 6).

Overwork and stress have increased in most occupations (Sandberg & Löwstedt, 2025). According to the World Health Organization (WHO), depression is the single largest contributor

TABLE 10.1
Potential Safety Consequences of Increased Remote and Hybrid Work

Positive	Negative
Flexibility in working styles and times increases the quality of work when employees can more easily choose how and when to work.	Induction and mentoring of new personnel may be more difficult and rely more on self-learning and tele-meetings. This may lead to competence issues later.
There is the possibility to stay away from toxic office environment politics and focus on work contents.	There may be a loss of weak social signals in meetings and informal discussions.
There is the possibility to organize meetings with more participants than previously – the capacity of meeting rooms is no longer a limiting factor. For example, staff meetings or morning meetings for all personnel facilitate information flow.	White-collar workers may lose touch with the daily reality of blue-collar workers. This may increase the gap between Work-as-Imagined and Work-as-Done, when remote workers are making plans, designs, and procedures for those at site to implement.
The possibility to focus on work without interruptions and stress created by the open offices many organizations utilize today improves quality, especially in tasks requiring concentration.	The loss of chance encounters and sharing of information via unofficial channels may affect information flow.
For people living far from the workplace, hybrid work can be an important way to maintain a better work–life balance.	Hybrid work can create additional coordination burden (and stress) for supervisors leading employees who work remotely. Supervisors may also find it more difficult to monitor the wellbeing of their subordinates or identify the early signs of fatigue or other problems.

to global disability. Especially worrisome is the prevalence of mental health problems in young people.

Mental health and its problems are challenging from a cultural perspective for multiple reasons. First, organizations have an effect on the mental health of their personnel. Second, the mental health of employees influences safety in an organization in a complex interactive way, where a decrease in physical safety can contribute to further decline in mental health. Finally, mental health in society at large has an effect on the environment in which the organization operates.

Finnish professor Ari Väänänen has studied the rise of *mental vulnerability* and mental health–related problems at work (Väänänen, 2024). He argues that we need to question our assumptions about mental health. He criticizes the dominant paradigm of mental health and work stress as relying on simplistic exposure–outcome thinking whereby individuals are expected to react biologically and psychologically to the external stressors they perceive. This type of reductionist approach misses the cultural and historical nature of mental health.

Väänänen (2024, p. 156) noted that changes in mental health of various occupational groups have been related to the wider evolution of the occupation, which may or may not correspond to the evolution of the market and other occupations. Organizations do not operate in vacuums; they are closely connected with the wider systems through their core task, technology, and professionals (cf. Reiman & Oedewald, 2007; White, 2017).

A train accident took place in Greece in March 2023. A passenger train carrying some 350 people crashed with a freight train shortly after leaving Larissa on its way to Thessaloniki. The incident happened as the passenger train emerged from a tunnel in the municipality of Tempi and collided with the freight train. The first four carriages of the passenger train were derailed, and the first two caught fire and were almost completely destroyed. More than 50 people were killed.

The accident led to wide demonstrations against the Greek government. The government was blamed for neglecting the safety of the railway system and for putting blame on the accident on the rail operators and "human error". The Greek railways were privatized step-wise during the financial crises in 2009–2018.

In April 2025, a guerrilla group carried out a bomb attack in Athens in front of the office of the Hellenic Train company. Fortunately, no-one died in the explosion. In its statement, the guerrilla group said that the attack was a response to the 2023 train disaster and it criticized the underfunding and safety deficiencies of the railways.[3]

All in all, changes in work and the landscape in general coincide with decreasing employee commitment to organizations. According to Gallup's (2025) State of the Global Workplace report, around one fifth of employees feel engaged at work. In the United States, employee engagement fell to its lowest level in a decade in 2024, with only 31% of employees engaged. Europe in turn was found to have the highest percentage of unengaged employees (73%). These figures paint a gloomy picture of the current state of the workforce and the organizations relying on the workforce to succeed or fail. These changes also challenge the management of safety in complex organizations; if safety is seen as an organizational capability (Chapter 4), employee wellbeing and mental health should be an integral part of it. Like external hazards, psychosocial hazards are only partially attributable to the organization itself. However, organizations remain responsible for addressing these hazards and are accountable for any shortcomings in promoting psychosocial safety and wellbeing within the workplace.

10.3 CHANGING EXPECTATIONS

It has been noted that societies tolerate risks associated with different technologies in a variety of manners (Amalberti, 2013). Nuclear industry and aviation are expected to be ultrasafe, whereas some fields are allowed to have a lower safety record (e.g. civil transportation, healthcare, mining). Also, fields that develop rapidly can cope with accidents by emphasizing that those took place in an earlier, more immature phase than the current phase. However, there are signs that society does not accept the risks associated with new technologies in a similar manner to some decades ago. On the other hand, the higher the safety level of a given industry, the higher the societal expectations are. This can paradoxically lead to a situation where accidents in generally safe industrial domains are more costly to the particular company than those in more risky domains (Amalberti, 2013). Thus, from a societal viewpoint, safety can always be better, but accidents matter most when they are rare (when they happen in mostly safe industries or organizations).

History of Boeing and the 737 Family

Founded by William Boeing in 1916, Boeing evolved from a military and government supplier to a leader in the civil aviation market, known for its engineering focus. Boeing's history is marked by competition with Douglas Aircraft Company (later McDonnell Douglas) and Airbus Industries. The Boeing 737 first flew in 1967 and became one of Boeing's most sold airplanes. The original models included the 737-100 and the longer 737-200, followed by the successful 737-300. By the mid-1990s, the 737 was the world's best-selling jetliner. In the early 1990s, Boeing started updating the 737 into the 737NG series, which featured significant improvements while maintaining compatibility with earlier models (Robison, 2021; Serling, 1992).

A turning point in Boeing's culture was 1997, when Boeing and McDonnell Douglas merged to form a single company under the Boeing name. Harry Stonecipher, the CEO of McDonnell Douglas, became Boeing's president and COO. Stonecipher had joined McDonnell Douglas in 1994 as the first external leader of the historically family-run business. Prior to this, he worked at General Electric, where he mastered cost-cutting strategies and enhancing shareholder value (Robison, 2021). In 2001, Boeing moved its headquarters from Seattle to Chicago, into a "new, leaner corporate center focused on shareholder value" instead of "how-do-you-design-an-airplane stuff", as the CEO of Boeing who implemented the merger and the relocation, Philip Condit, succinctly put it (Robison, 2021, p. 79).

Boeing's revamped business model emphasized extensive outsourcing and had a strong focus on efficiency and costs. For example, the simulator training department was established as its own company, selling its services to Boeing as well as to airlines. Harry Stonecipher boasted in an interview:

> *When people say I changed the culture of Boeing, that was the intent, so it's run like a business rather than a great engineering firm. It is a great engineering firm, but people invest in a company because they want to make money.*
>
> **(Callahan, 2004)**

The Boeing 787 Dreamliner was the first new civilian aircraft developed using a streamlined business model, with its development approved by the board in April 2004. This project faced significant cost overruns and numerous outsourcing issues, hinting at the challenges that would later arise with the Boeing 737 MAX (Robison, 2021).[4] Initially scheduled for release in 2008, the Dreamliner finally entered commercial service in late 2011. The delays and budget overruns also impacted Boeing's plans for the successor to the 737NG. Boeing considered both creating a completely new airplane and re-engineering the 737NG with more efficient engines and other features. A team was set up in 2006 to evaluate these options, but no decision had been made by 2010. In December 2010, Airbus announced their A320neo family – a re-engined, more efficient version of the A320 – prompting major American airlines to consider ordering from Airbus if Boeing did not offer a comparable option.

In 2011, Boeing decided to redesign the 737 instead of creating an entirely new airplane. This approach had several benefits: it would take less time (up to 10 years for a new design), cost $2.5 billion compared to $20 billion, and avoid lengthy certification processes since the airframe was already certified. The 737 MAX was to maintain similarity with previous models, allowing pilots and ground staff to use their existing training with some additional instruction. Boeing assured customers that only a brief tablet-based course would be needed instead of full simulator training. All this led to severe pressures inside Boeing to make sure nothing would point to a need for extra training, including the new MCAS system (DeFazio & Larsen, 2020; Robison, 2021).

Management is becoming increasingly aware of how much accidents cost, for example, in terms of production losses and claims for compensation. In addition, safe and reliable operations are an important competitive advantage, a key part of the brand of an increasing number of companies. However, there are also trends in a different direction, for example, the rise of authoritarian regimes all over the world and corresponding downplaying of worker rights and the value of safety in general. The financial value of safety is highly contingent on societal values, policies, and regulation. Hence, it is not feasible to definitively state whether safety is financially advantageous without specifying the context and the timeframe. Nonetheless, from an ethical standpoint, it is indisputable that organizational policies that disregard the value of human life are unethical.

The scientific evidence of the relation between the profitability of a company and its safety is also very inconsistent, partly due to the above-mentioned issues. Several studies find a positive relationship between business profitability and safety, others find a negative relationship, and others have found no relationship at all (Madsen, 2013). In an aviation context, this relation has been studied by Madsen (ibid.), who proposed that the relationship between profitability and safety should differ for firms performing above their financial targets and for firms performing below these targets. His study of the U.S. commercial airline industry between 1990 and 2007 showed that

1) For firms performing below their profitability aspiration, a positive relationship exists between profitability relative to the aspiration and accident rate (increased profitability leads to more accidents).
2) For firms performing above their profitability aspiration, a negative relationship exists between profitability relative to the aspiration and accident rate (increased profitability reduces the accident rate).

Thus, there was no direct relation between the airline firm's profitability and its future accidents; rather it depended on the relation between the firm's aspired profitability and actual profitability. However, the study indicated that airlines with greater financial slack experience fewer accidents than do those with less slack, and organizations under bankruptcy protection are also at increased accident risk. Thus, the effect of profitability on risk taking in the domain of safety risks is the exact inverse of its effect on risk taking in many other domains, such as investment risks. Madsen (2013, p. 770) notes that

> organizational investments in safety are only stochastically related to future accident rates, reducing the likelihood that accidents will occur but never eliminating the possibility of an accident altogether.

Consequently, organizational decision makers who choose to reduce safety investments trade off the increased risk of a costly accident with the certainty of a small financial benefit. Madsen (2013, p. 785) concludes by arguing that his study shows that

> safety fluctuates with profitability relative to aspirations, such that accidents and incidents are most likely to be experienced by organizations performing near their profitability targets.

In the end, safety depends on societal values and perceptions: how much risk the society is willing to tolerate, and how the society values safety against other goals.

10.4 CHANGING REGULATION

Many authors have noted the increasing standardization in many spheres of life (Dekker, 2021; Graeber, 2018; Le Coze, 2021c). Standardization has allowed global trade to increase and is also a factor allowing the increased reliance on outsourcing (see above in this chapter). At the same time, global trade has put pressure on deregulation or at least standardized regulations across countries.

There has been an increase in reliance on industry self-regulation, with the hope that this would lead to less bureaucracy and more efficiency.

So far, the increased reliance on *self-regulation* has not led to a decrease in rules. On the contrary, there is evidence that it has led to an increase in the number of rules and regulations (Dekker, 2021, see also Graeber, 2018). This development is not restricted to safety domain; Graeber (2018, p. 17) notes that regulations and paperwork have increased with every market reform. However, in the safety field there is an especially strong tendency to add rules but never remove them, thus contributing to a layered bureaucracy with an ever-expanding system of safety rules that exist on paper and are adapted, interpreted, and worked around in practice. Often, deregulation has been presented as an answer to the overburden on rules, but evidence shows that it actually brings more rules in addition to new complexities (Graeber, 2018; Dekker, 2021). Dekker notes that moving from prescriptive regulation to performance-based regulation has had unintended consequences:

> Performance-based regulation can quickly lead to more rules, rather than fewer. Free markets, after all, have increased the likelihood of organizations being sued for not meeting their obligations. The vaguer the principle regulations are under which [the] obligations [that companies need to fulfil] sit, the more seductive it becomes for an organization to overspecify the way it demonstrates meeting those obligations.
>
> **(Dekker, 2021, pp. 60–61)**

Overspecification of practices leads to an increased gap between Work-as-Imagined and Work-as-Done and contributes to organizational drift (see Chapter 6).

Dekker proposes that the current reality of having more compliance requirements and more bureaucracy than before stems from three main factors. Macroeconomic factors, such as privatization and deregulation, have led to unfree, overregulated workers and a growth of internal compliance bureaucracy. This shrinking government regulatory footprint shows especially in the increasing rules in the area of health and safety. Meso-economic factors are related to the structures and interorganizational or industry arrangements, and include deprofessionalization, the growth of compliance workforces, and surveillance capitalism. These factors refer to the reduced autonomy of professionals that have been replaced with external surveillance and control by various project managers, HR, auditors, and so on. Micro-economic factors are related to the decisions of individuals and businesses. The factors include auditism, financialization and short-termism, and the monetization of compliance.

Lessons from the privatization of water utilities, railroads, and healthcare show many counterproductive effects, such as increased costs, decreased reliability, and decreased quality of service (Dekker, 2021). At the same time, these fields have introduced more compliance requirements and more bureaucracy, further hampering effectiveness.

Increased bureaucracy and a compliance focus on safety can be frustrating when combined with pressure to perform and be more efficient. As discussed above, the psychosocial demands placed on individuals have been steadily growing during recent decades and stress at work has increased. All this highlights the importance of widening the perspective of safety from physical injuries to mental states and holistic wellbeing.

On 14 June 2017, a fire broke out in the kitchen of Flat 16 Grenfell Tower, a high-rise residential building in North Kensington, West London. Due to a recent re-cladding of the entire building in insulation panels with low fire resistance, the fire was able to quickly jump over to the other flats via the burning external wall. The fire resulted in 72 deaths and more than 70 injuries.

Naderpajouh et al. (2022) *note that the outsourcing of building control and other systemic inspection problems contributed significantly to the loss of lives. The modifications to Grenfell Tower that covered the refurbished building in flammable cladding were approved by a building inspector working for the local authority. During the investigation, the inspector admitted that his work fell below the standards of a reasonably competent inspector, yet he pointed out that at the time he was responsible for reviewing up to 130 projects. Site inspections were carried out on behalf of the local authority by a contracted inspector who said his remit was not to check whether the work met regulations or matched architects' drawings but rather to check that the work was neat and tidy.*

The inquiry into UK building regulation post-Grenfell (Hackitt, 2017, 2018) found that, while the introduction of this arrangement had driven down costs, inspectors have a conflict of interest and experience a "difficult trade-off", failing to gain work if they do not approve more risky designs. After a partial privatization of building control services, these Building Control Bodies (BCB) compete for business, and no longer act as impartial checkers of building safety, but rather "become far too embedded in supporting the building design process rather than being an impartial rigorous verifier of buildings safety" (Hackitt, 2017, p. 54). The inquiry also reported "repeated concerns expressed about the commercial pressures associated with rigorous enforcement of fire safety requirements" (Hackitt, 2017, p. 55). Future business considerations were found to affect the willingness of the BCBs to sign off completed work.

One specific issue that has recently received attention in safety science is the question of regulating safety culture (Bieder & Bourrier, 2013). Some industries, such as the nuclear industry, have international standards for safety culture which many countries have adopted into their regulations. Whether safety culture is explicitly regulated or not, it is good to note that whenever human and organizational performance-related issues are regulated and certain practices such as learning from incidents, reporting, and so on are required from companies, the culture of the organizations is *de facto* a subject of regulation. This is not a negative thing as long as the regulations are clear and allow organizations to bear the ultimate responsibility for safety in a manner that best fits their strategy and culture.

10.5 INDUSTRY CULTURES AND SUBCULTURES

Organizational cultures vary significantly between industries due to several factors, including the nature of work, industry norms, and their operating environment. Cultures form around the organizational core task, including technology and the core competencies necessary to carry out the core task (Reiman & Oedewald, 2007; White, 2017). White (2021, p. 88) notes that "organizations develop cultures by doing meaningful and habitual things", and in this way, "culture follows task".

Industries differ in how important a role ensuring safety has been given in them. Aviation, nuclear power, chemicals, shipping, and rail transport are sectors that have long been considered safety-critical sectors. In these industries, a lot of attention is paid to ensuring and developing safety – and safety incidents or accidents receive a lot of media coverage.

Different industries highlight different hazards and therefore emphasize different "safeties". Safety development can refer to, for example, the improvement of occupational safety in the construction industry or the development of nuclear safety in the nuclear industry. The importance given to occupational safety, environmental safety, and process safety is due not only to the special characteristics of the work object and the technology used, but also to the traditions of the industry.

One can also ask whether organizations of one industry represent just subcultures of a wider industry culture? This question is especially relevant for highly international and highly regulated industries such as the nuclear industry or aviation. However, even in these industries, there are widely differing organizational arrangements and ways of working, and one can thus consider that these are cultures of their own, not merely subcultures. One example is the contrast between "safety cultures" at the Fukushima and Onagawa nuclear power plants in Japan (see Chapter 6).

The variance in industry cultures highlights how cultures form at the intersection of task accomplishment in a given environment by a certain group of people, who need to learn to work together and adapt to the environment. Organizations that have a similar core task and operate in mostly similar environments most likely develop cultures that resemble each other (cf. Chatman & Jehn,

1994) without ever being identical. Again, as examples from Fukushima and Onagawa and numerous other non-safety-critical organizations show, there is ample room for organization-specific differences based on, for example, leadership, organizing, national cultural influences, historical influences, and chance. Furthermore, if an organization has several different tasks or functions (such as operations, maintenance, engineering, R&D), it most likely has subcultures built around the specifics of the task.

Hutchins (1995) coined the term *cognitive artifacts* to denote tools or objects that aid in cognitive processes, such as thinking, remembering, and problem-solving. These artifacts can include books, notes, procedures, maps, and mathematical notations (Hutchins studied navigation in maritime). Hutchins showed how external tools (physical objects as well as abstract systems) support and extend human cognitive abilities, facilitating tasks and decision-making processes in ways that would be challenging or impossible without them. Modern cognitive scientists and cultural researchers speak about embodied cognition (White, 2017, 2021) with the same idea. As technologies differ between industries, it is natural that they develop different cultures. White (2021) demonstrates how culture is technologically grounded. He points out that how people perceive technology is key and that "this is a function of how central the technology is to a firm's means of production, how the technology is put to use, and what meaning people attach to the solutions it provides".

All in all, differences in tasks and technology most likely result in differences in culture. This is natural and necessary for safe and effective work. It does not mean that organizations with similar tasks and technology would share similar cultures. The ways organizations manage external adaptation and internal integration (see Chapter 1) differ significantly, resulting in large variance in cultures even within a specific industry. The cultural differences at Fukushima and Onagawa nuclear power plants described in Chapter 6 are a good illustration of this.

10.6 CONTEXT AND IDEAL CULTURE

Various typologies of culture (Competing Values Framework, Multi-Focus Model, etc.) are not generic ideal models as their effectiveness depends on many factors. This means we should be careful in declaring some types of organizational cultures as universally more effective than others. However, at the same time, safety sets some unique requirements for culture (see Chapter 5 for safety culture models).

The following things need to be considered when deciding on the ideal culture:

- Strategy and the role of safety in it (including the risk appetite of the organization);
- The organization core task, including basic production technology and core processes;
- Type of hazards (local, core task–related, external, psychosocial);
- The size and scope of organization's activities and size and scope of the network and subcontracted activities;
- Current structures for managing safety;
- Current culture, including the history of the organization.

Strategy defines the desired safety level of the organization. This level is partly affected by societal values and expectations. For example, the nuclear industry has to reach nuclear safety levels unimaginable for the transportation or construction industries. It is important to note that the type of hazards differs between the industrial sectors. This is further discussed below.

As discussed in Chapter 6, Le Coze (2021c) highlights the importance of considering strategy in safety management. Strategy must take the operating environment into account. This also means considering the effect of the changes illustrated at the beginning of this chapter.

I have emphasized the importance of considering the *organizational core task* in organizational assessments, and when defining the desired culture. The organizational core task denotes the shared objective or purpose of organizational activity.

The physical object of the work activity (e.g. particular power plant, manufacturing plant, offshore platform, human being), the objective of the work, and the society and environment (e.g. deregulated electricity market, harsh winter weather) set constraints and requirements for the fulfilment of the organizational core task.

Different industrial domains have different outside influences; for example, laws set different constraints on the organization and economic pressures vary.

The content of the work, the nature of the hazards involved in their daily activities, the basic education of the personnel, and the role for overall safety in the company differ. The core task of the organization sets demands (constraints and requirements) for the activity and should be kept in mind when making assessments of the organizational solutions or performance. Culture builds on, and should support, the organizational core task.

We identified three generic organizational core task requirements and three working practice requirements in our research in nuclear power plant maintenance (Oedewald & Reiman, 2003):[5]

Generic Core Task Requirements

- *Anticipating* – predicting the state of the system and the effects of actions, as well as planning the needed actions and resources in advance;
- *Reacting* – responding to unexpected events and deviations;
- *Monitoring and reflecting* – monitoring the effectiveness and results of one's actions as well as of the overall state of the sociotechnical system.

Instrumental Demands

- *Adaptability* – balancing between anticipating and reacting, for example, when non-anticipated events occur (originally called *flexibility*);
- *Systematic response* – balancing between reacting and monitoring, for example, verifying what has been done and documenting the results of the actions after reacting has been needed (originally called *methodicalness*);
- *Learning* – balancing between monitoring and anticipating, for example, changing expectations based on what is learned about the system and monitoring new or changed issues.

The requirements should apply to any complex organization, but their specific manifestations differ. These requirements can be used as normative criteria alongside more safety-specific criteria introduced in Chapter 5, when assessing how well the current organization and the current culture supports safety. For example, how does the organization currently monitor its environment, its internal activities, and the hazards associated with the core task? How is the organization reacting to various contingencies, and how good is the organization at anticipating what may happen? And when something happens, does the organization respond methodically and flexibly, and does it learn after the event?

The core task is also connected to the life cycle of technology in question; the hazards and their uncertainties differ depending on whether the system uses proven technology or innovative new technology. In proven technologies, one further factor concerns the effect of ageing on the reliability of technology, a source of uncertainty in the technical system.

A major influencing factor is the uncertainties in the organizational core task. For example, the maturity of the main technology influences both the existing culture as well as the so-called ideal culture. Misperceptions of maturity can contribute to a dysfunctional culture, as happened at NASA during the 1980s (see Chapter 2).

In chapter 4 we differentiated between core task related, local, psychosocial, and external hazards. The organizational core task is an inherent source of hazards for the organization. For some organizations, these hazards may lead to major accidents with consequences for people and the environment. For others, the core task–related hazards may mostly threaten production with the potential for individual human harm (e.g. manufacturing equipment with injury potential). These hazards manifest in complex manner and require balancing a centralized approach with decentralized one. The strategy for assuring safety should aim at fostering sensemaking at all organizational levels while at the same time anticipating potential accident scenarios and creating guidelines for safe work (see Chapters 7 and 8). Promoting system goals is needed so everyone understands how their work affects the system safety. As the system matures and ages, and units develop local practices, diversity and new ideas need to be brought into the system together with ways of monitoring and assessing the practices and culture.

Local hazards can often be managed by assuming the system is complicated, not complex – or, in Perrow's terms, closer to linear than complex interactions. This also means that accident models focusing on error prevention and hazard control can get the organization quite far in its safety journey. The same does not apply to core task–related hazards. A system that manages those hazards is always complex, and a culture needs to have the requisite variety to handle the complexity. When managing local hazards, the emphasis is on supporting local actions and coordinating work so that it can be conducted safely (see Chapters 7 and 8).

Grote (2012) notes that for process safety issues, "cultural norms that emphasize and support good task fulfillment will in themselves be beneficial for safety", whereas for personal safety (local hazards), "protection against personal risks creates a task separate from the primary work task" (Grote, 2012, p. 1990).

Psychosocial hazards can be connected to multiple issues at work and outside work, making their management a very complex activity. The responsibility for addressing these issues often resides within the human resources (HR) function rather than the safety department. Culture and leadership have a major effect on psychosocial hazards as well as the other hazards, highlighting the importance of collaboration between safety and human resources. Monitoring the organization, including culture and employee wellbeing, may help in identifying weak signals of psychosocial hazards. Networking with healthcare providers is also helpful for safety professionals, who typically lack expertise in this particular area.

External hazards typically require a highly centralized way of managing, focusing on prioritization and anticipation. External hazards and especially systemic risks require networking with many organizations for the purpose of monitoring as well as response. Connecting, diversifying, and searching for knowledge and solutions are central, as well as training and simulation exercises concerning preparedness of the system for high consequence–low probability hazards.

Creating trust and interacting are needed for all types of hazards, but the more invisible and complex the hazard, the more important a psychologically safe environment is for the personnel to speak up. For local hazards, fostering sensemaking in order to facilitate perceiving and assessing the significance of the hazards can be at least as helpful for speaking up as creating trust is. An organization's specific hazards influence how the safety management principles introduced in Chapter 7 should be balanced.

The scope of activities and growth. Growth of the organization and development of working practices and competencies needs to be aligned when managing safety. Too strong growth without adequately embedding the ways of working and competence in procedures and official practices can lead to chaos. On the other hand, practices that have been developed for a company of 30 will not necessarily work in a company with 300 employees. Thus, formal practices need to be revisited from time to time. And when these practices are changed, adequate risk assessment should be carried out and the reasons for the change communicated to the personnel. It is still good to prepare for the fact that some people prefer to work in small companies and may leave when the company grows, despite how well the change is managed. Growth, even when the growth is positive and due to an increase in business, is always an organizational and cultural challenge as well as an opportunity for leaders to build culture.

The current culture and history of the organization influence how safety should be managed and led. Complex systems are history-dependent and irreversible as we established at the end of Chapter 1. Interventions, successes, and failures of the past have an impact on how today's activities are perceived. Furthermore, as the organization matures, it tends to specialize in activities it has been successful in. This benefits safety as long as the activities are sound, and the context remains the same. However, cultures always include blind spots and dysfunctional features. Continuous monitoring and periodic assessments (see Chapter 11) are needed to create awareness of the strengths and weaknesses of a culture from a safety perspective.

Current culture influences safety work in the organization in two ways: First, it influences what are the most effective and accepted ways of developing safety – how employees prefer safety to be developed. Second, the current culture influences what safety professionals and leaders should do in terms of enhancing safety – what should be developed. The strengths and weaknesses of the current culture influence how the organization deals with safety issues, and the safety professionals need to contribute to maintaining the strengths and improving the weaknesses.

An important strategic decision concerns the way organization manages uncertainties (Grote, 2012). Uncertainty may be minimized by high levels of standardization, central planning, automation of work processes, and high levels of specialization, with few degrees of freedom for employees. This approach aims to maximize stability. Another approach is that all actors in the organization may be empowered to cope with uncertainty locally by providing them with options for action rather than fixed plans and standards, with ample opportunity for cooperation and learning, and with decision-support tools rather than algorithmic automation. This approach aims to maximize flexibility. Grote (2012) notes that the predominant approach to managing uncertainty in any organization needs to be considered and reflected upon when designing, implementing, and evaluating safety management systems. Requirements for flexibility and stability differ between industries, but may also change across time, and vary inside the industry among the various functions (Grote, 2012). Grote writes that

> stability is required due to tightly coupled processes, need for traceability of decisions and processes, low fault tolerance, or low qualification levels of personnel. Flexibility, on the other hand, is needed for handling high levels of uncertainty caused by frequent external changes and/or variances and disturbances in work processes, for fulfilling high demands on innovation or also for avoiding overroutinization and complacency in basically very stable processes.
>
> **(Grote, 2012)**

It is important to remember that many requirements for managing safety in complex systems stem from complexity. As discussed in Chapter 4, there will always be surprises, and the system is always changing. Supporting local actions, monitoring system boundaries, exploring the system to learn more, and promoting the shared core task and system level goals are always needed in complex systems (see adaptive safety management principles in Chapter 7).

Safety management has traditionally relied on rules, procedures, and barriers to prevent harm by preventing the hazard sources from contacting people. This is done by affecting both the behavior of the people (safety as absence of errors and adherence to procedures) as well as design solutions (safety as hazard control), as illustrated in Chapter 4. This type of safety work requires a rule-focused culture. However, as illustrated in this book, these two approaches alone are not enough to guarantee safety. The third approach, promoted by the Resilience Engineering paradigm (see the end of Chapter 11), and variously called Safety II, Safety Differently, or Human and Organizational Performance (HOP), focuses on increasing the capability of the organization to succeed. It further emphasizes safety as the presence of certain organizational capabilities, which are connected to everyday work and not only to separate barriers or defenses.

After defining the above-mentioned issues, the organization should revisit its strategy for assuring safety and consider what the safety they are managing is. How wide a definition of safety should be adopted? Is safety more a local issue at the various work sites to be managed in a decentralized manner, or is it directly connected to the production technology and core processes, or is it both? What is the role of human in safety – should safety include employee wellbeing and psychosocial issues? What external threats are there to safety? How are these issues connected with each other? Forming a joint understanding of these issues is important for creating a healthy culture for safety suited to the particular context.

An important thing to keep in mind is that as the context changes, so does the ideal culture. What was optimal yesterday may not be so in the future. Once the organization has created a culture for safety, it immediately needs to start recreating it.

10.7 SUMMARY

The landscape of organizations is changing due to globalization, digitalization, and other macro trends. There is an increasing reliance on self-regulation in industries, which has not resulted in reduced bureaucracy but rather an increase in compliance requirements.

Systemic risks, defined as threats to critical societal systems, are characterized by complexity and the potential for ripple effects beyond their origin. Climate change is highlighted as a significant systemic risk that alters existing hazards and introduces new ones, impacting safety across various sectors. Geopolitical tensions also create unpredictable effects on organizational safety investments.

The "autonomy paradox" refers to increased individual freedom in knowledge work coexisting with heightened psychological demands, leading to anxiety and exhaustion among workers. Hybrid work arrangements have gained popularity post-COVID-19, but they also present challenges such as professional isolation and blurred work–home boundaries.

Organizational cultures differ significantly across industries, influenced by the nature of work and industry norms. Organizations must continually adapt their safety strategies to align with changing societal expectations, regulatory environments, and the evolving nature of work. A proactive approach to managing safety, considering both physical and psychosocial factors, is essential for fostering a culture for safety in complex organizations.

Safety is constructed through interactions between organizations and their environments. The organizational core task and the hazards that need to be managed should always form the backbone of the organization's approach to safety. The core task and the hazards influence what a healthy culture for safety looks like. There are no general ideal culture models that can be copied from organization to organization.

REFLECTION QUESTIONS

How have global changes affected your organization? Have the risks of these changes been analyzed?

How much does your organization rely on outsourcing or networks? How has outsourcing and network organization changed the risk profile of your organization? Have the risks been shifted to contractors, or have new risks been created?

What is your organization's current policy concerning hybrid work? How have the risks been analyzed? What pros and cons have you perceived in the hybrid work model?

Does your organization have more or less rules than ten years ago? Are there areas that have not been standardized that should be?

How does your organization's strategy take safety into account? How does strategy influence safety in your organization?

What is the core task of your organization? How clear are the hazards that are associated with the core task?

How well does the culture of your organization support monitoring, anticipating, and reacting to various contingencies at work? Does it support learning?

NATIONAL CULTURE AND SAFETY

Schein (2010, p. 5) argues that due to, for example, globalization, "understanding culture at any level now requires some understanding of all of the levels". He also proposes that as the world becomes more interdependent and global, "it will be differences in the deep assumptions about authority that will be most problematic" (Schein, 2010, p. 104).

Geert Hofstede has suggested that national cultures differ in six basic value dimensions. These dimensions are:

- Power distance;
- Uncertainty avoidance;
- Achievement orientation (formerly *masculinity*);
- Individualism/collectivism;
- Indulgence;
- Long-term orientation.

Hofstede et al. (2010) defines *power distance* as the extent to which those with less power in social institutions and organizations accept and believe that power is not distributed equally. In high power distance cultures, organizations centralize power as much as possible in a few hands, and subordinates are unlikely to approach and contradict their bosses directly (ibid., p. 61). Power distance has also been found to negatively affect speaking up on errors and safety concerns (Peadon et al., 2020).

Individualism describes societies in which the ties between individuals are loose – everyone is expected to look after him- or herself and his or her immediate family (Hofstede et al., 2010, p. 92). In highly individualistic cultures, individuals are free to speak their minds and openly disagree on issues (Hofstede et al., 2010).

Uncertainty avoidance refers to the extent to which members of a cultural group are prepared to deal with unanticipated events and the extent to which people feel threatened by unknown or ambiguous situations (Hofstede et al., 2010). Uncertainty-avoiding cultures shun ambiguous situations. People in such cultures look for structure in their organizations, making events clearly interpretable and predictable. Uncertainty-avoidant cultures rely more on rules and procedures. However, uncertainty-avoidant cultures do not mind familiar risks, and people in them are in fact willing to take risks to reduce ambiguities, for example, preferring action to waiting (Hofstede et al., 2010, p. 198). On the other hand, cultures with weaker uncertainty avoidance are more willing to accept also unfamiliar risks in, for example, engaging in activities for which there are no rules.

Long-term orientation describes how societies promote future-oriented virtues, such as perseverance and thrift. Short-term-oriented societies focus more on the near future, including short-term success and gratification (Hofstede et al., 2010).

Indulgence is the disposition to freely gratify fundamental human desires associated with life enjoyment and fun. High indulgence is associated with greater unrestrained self-expression (Hofstede et al., 2010).

Achievement orientation refers to the desirability of assertiveness, toughness, and material rewards against the desirability of modesty. In low-achievement-orientation cultures, cooperation and quality of life are valued.

Power distance is an especially important dimension from the safety point of view. High power distance affects psychological safety negatively (*The Culture Factor*, 2025). People are less likely to speak up if their bosses (or their bosses' bosses) are present in meetings, or if they know their statements will be recorded in minutes of meetings, or otherwise communicated upwards in the organization. There is also some empirical proof that high power distance is detrimental to safety (Reader et al., 2015). Further, in high-power-distance cultures, subordinates avoid pointing out their leaders' errors (Lei, 2018).

Uncertainty avoidance is another interesting safety-related dimension. High uncertainty avoidance may endanger safety because it narrows down attention and behavior opportunities. Schein has noted that "as environments become more turbulent and occupations become more technically complex, the ability of leaders to tolerate uncertainty will become more necessary for survival and learning" (Schein, 2010, p. 120). This may be related to the ability to adapt to changing circumstances – to people being more flexible towards change.

People in high-uncertainty-avoidance countries tend to have a negative attitude towards errors, including less tolerance of errors – also, ritualistic, face-saving, and aggressive communication strategies are used in embarrassing situations such as mistakes (Lei, 2018). The positive side of high uncertainty avoidance is that people are less willing to take risks; in high-uncertainty-avoidance cultures, people depend on structure and rules (both positive and negative consequences, depending on the rules in question).

Collectivism can also have a safety effect. There the potential mechanisms include hiding of problems from "outsiders" due to loyalty to the ingroup, or formation of norms counterproductive to the overall group. The positive side of collectivism is the emphasis on shared values and consensus in, for example, decision making. The ingroup bias (see Chapter 2) is also stronger in more collectivist groups (Sapolsky, 2017, p. 275).

Table 10.2 below shows examples of scores on five dimensions from selected countries.[6]

Pratama and Caponecchia (2025) in their literature review study of the influence of national culture on aviation safety notes that power distance, individualism, and uncertainty avoidance were the three cultural dimensions most found to influence aviation safety. These dimensions were particularly associated with issues in team communication and decision-making. Furthermore, both a high degree of power distance and a high degree of collectivism were most consistently found to be significantly correlated with a higher rate of aviation incidents. Findings on uncertainty-avoidance culture had more mixed results,

TABLE 10.2
Selected scores on Hofstede's five dimensions of national culture (data from theculturefactor.com, retrieved April 2025)

	Power distance	Individualism	Achievement/masculinity	Uncertainty Avoidance	Indulgence
United States	40	60	62	46	68
Israel	13	56	47	81	
Brazil	69	36	49	76	59
Australia	38	73	61	51	71
Belgium	65	81	54	94	57
New Zealand	22	69	58	49	75
Japan	54	62	95	92	42
Singapore	74	43	48	8	46
Germany	35	79	66	65	40
France	68	74	43	86	48
China	80	43	66	30	24
India	77	24	56	40	26
Finland	33	75	26	59	57

with some studies suggesting a positive link between high uncertainty avoidance and accidents but other studies suggesting the opposite. These results are in line with Noort et al.'s (2021) studies of cockpit voice recordings in aviation accidents, introduced in Chapter 2. They (Ibid.) demonstrated that power distance reduces safety voice through less safety listening. In the accident cases they reviewed, they noted that safety-critical staff nearly always speak up across hazardous situations. However, these initial acts of safety voice are frequently poorly listened to, and this reduced the amount of safety voice prior to accident. Power distance predicted this extent of safety voice. Power distance also predicted more safety voice for junior flight crew in low-power-distance countries, whereas senior flight crew voiced more in high-power-distance countries. They (ibid.) highlight the critical role of safety listening in adequately responding to safety voice.

Enomoto and Geisler (2017) showed in their cross-country comparison that power distance was positively associated with the number of aviation accidents. Interestingly, individualism was negatively associated. These associations remained after controlling for the level of economic activity (gross domestic product – GDP) and severe weather conditions in these countries.

National culture influences also *leadership expectations*. Javalagi et al. (2024) found that traits of extraversion and agreeableness (see Chapter 2) were both found to exhibit much stronger leadership advantages in collectivist cultures when compared to individualist cultures. They also note the following:

> *First, in collectivist societies (cultures that value interdependence with one's group), the five factor model traits – and leader Extraversion and Agreeableness in particular – are stronger predictors of leader effectiveness, consistent with the theorized need for enhanced social coordination in such cultures. Second ... leader Big Five trait effects are mediated by leader behavior ... Third, trait Honesty-Humility robustly predicts leader effectiveness beyond the Big Five traits, expanding the trait approach.*
>
> **(Javalagi et al., 2024)**

Employees have different expectations of their leaders across national cultures (Chhokar et al., 2007). Lyubykh et al. (2022) note in their meta-analysis that power distance moderates leadership effectiveness. In high-power-distance countries, employees might be less receptive to leadership behaviors oriented at change, more tolerant of destructive leadership behaviors, and less tolerant of passive leadership behaviors.

All in all, national culture has several potential safety effects, which are illustrated in Table 10.3 below.

Comparisons between countries are made more difficult by the fact that national culture affects response style on the most used instrument: the questionnaire. Johnson et al. (2005) investigated the effects of power distance, individualism, uncertainty avoidance, and masculinity (achievement orientation) on two commonly recognized response biases: extreme response style (ERS, choosing the extreme end of a multiple-choice question) and acquiescent responding (AR, tendency to agree with the questions regardless of their content). They had a sample of 18,000 survey questionnaires completed by employees in 19 nations on five continents. Power distance and masculinity were found to be positively and independently associated with extreme response style. Individualism, uncertainty avoidance, power distance, and masculinity were each found to be negatively associated with acquiescent response behavior. Overall, extreme response styles (ERS) are prevalent in countries with high power distance, but they tend to decline in individualistic cultures (Bonjeer & Vonkova, 2024; Johnson et al., 2005; Peterson et al., 2014).

TABLE 10.3
Some Potential Effects of National Culture Attributes on Safety

Dimension	Effects of low score	Effects of high score
Power distance	Personnel may be more willing to report mistakes or stop unsafe work. Employees are more easily involved in developing safety. A potential negative consequence of low power distance is lack of clarity concerning authority in, for example, emergency situations. Employees are also more willing to question management decisions, which may in some cases lead to risk taking and excessive autonomy/initiative.	There is some empirical evidence that high power distance is detrimental to safety because it may reduce the openness of communication on safety (Reader et al., 2015). In high-power-distance workplaces, people may be more hesitant to challenge unsafe decisions made by their supervisors. They may also be less willing to report their own mistakes due to fear of negative consequences.
Uncertainty avoidance	Low-uncertainty-avoidance cultures are open to new issues and flexible towards change. This can have either positive or negative safety effects, depending on the change. Low-uncertainty-avoidance cultures are comfortable with ambiguous situations and with lack of detailed rules. This improves adaptation but may also lead to risk taking.	People are less willing to take risks; people depend on structure and rules (both positive and negative consequences, depending on rules in question). Employees may show more change resistance to new issues, which can have either positive or negative safety effects, depending on the change. New situations and ambiguous situations with no clear guidelines are experienced as stressful, which may impact performance in, for example, emergencies.
Individualism	In collectivism (ingroup or institutional), potential mechanisms include hiding of problems from "outsiders" due to loyalty to the ingroup, or formation of norms counterproductive to the overall group. A collectivist culture may maintain group harmony by, for example, not challenging unsafe group activity (Reader et al., 2015). Collectivist cultures may share stronger collective responsibility for safety and look after one another.	People in high individualist cultures may be more willing to speak up on concerns and to challenge unsafe work. People in high individualist cultures may voice concerns mostly on issues that relate to them individually, but not to the collective. Personnel may be more willing to develop their own unique working styles, including ways of using – or not using – procedures.

Despite national cultural influences on safety, organizations can and should create a culture of their own (Hopkins, 2018). This may happen by building on the characteristics of the national culture that facilitate safety and hinder, or guard against, dimensions that are contrary to safety. National cultural scores as illustrated in this chapter are mean scores and the variance inside any given country is large (see the comparison between the Fukushima and Onagawa nuclear power plants in chapter 6). Recruitment, promotion, and competence development are viable tools to shape the organizational culture for safety in any national culture.

NOTES

1. Description of the accident is based on (Bergman & Nelson, 2013; Labowitz & Baumann-Pauly, 2014; Savage, 2021).
2. In principle, any risk that threatens the functionality of a sociotechnical system can be considered systemic. However, the concept of Systemic Risks has been used to denote global risks affecting multiple organizations as well as society at large.
3. Helsingin Sanomat 14.4.2025 "Possibly a new guerrilla group behind the explosions in Greece".
4. The lithium battery fires in several Dreamliners in 2013 uncovered similar deficiencies at Boeing and the FAA as the 737 MAX accidents. Among the NTSB's findings were inadequate FAA oversight, a failure by both FAA and Boeing's authorized representatives (ARs) to identify critical deficiencies, and flawed safety assumptions by Boeing that it made in the airplane's System Safety Assessment. (Defazio & Larsen, 2020).
5. These core task requirements were originally created for nuclear power plant maintenance. but have since been utilized as generic requirements for complex safety-critical organizations.
6. Data from theculturefactor.com.

11 Assessing and Monitoring Culture

This chapter deals with the question of how to achieve and maintain an awareness of the current culture for safety. It presents a way of conducting assessments of culture. The assessment framework proposed in this chapter builds on assessments I have conducted as a scientist, a practitioner in the nuclear industry, and a consultant. I also rely heavily on the practices of self-assessment and independent assessment of safety culture in the nuclear industry (IAEA, 2016b). The chapter concludes by presenting the basic principles of Resilience Engineering and illustrating how these can be used to advance the understanding of safety in complex organizations.

The aim of the assessment is to gain awareness of the strengths and weaknesses of the current culture in terms of achieving safety. In addition, assessment aims to identify opportunities for improvement.

Monitoring culture is important to maintain cultural awareness between more in-depth cultural assessments. Both require specific competence in organizational factors and culture in addition to understanding the possibilities and limitations of the various methods of gathering information and making conclusions.

11.1 MAINTAINING CULTURAL AWARENESS

There are several reasons for actively assessing and monitoring culture of an organization. One of the main reasons is the tendency of an organization to gradually drift into a condition where it has trouble identifying its vulnerabilities and mechanisms or practices that create or maintain these vulnerabilities (cf. Snook, 2000; Rasmussen, 1997).

Organizations create simplifications, categories, and narratives in order to maintain an image of a predictable work environment. These help to form beliefs and expectations about future events as well as risks. In narratives, as well as in sensemaking in general, the plausibility of the story prevails over its accuracy, and thus they can hide risks as well as reveal them. These narratives need to be challenged from time to time.

Organizations establish norms regarding acceptable and unacceptable behaviors as part of their culture. Assumptions about safety and risks influence what is considered normal and what is treated as abnormal. It is essential to periodically review and update these definitions of normal and abnormal to ensure they remain accurate and relevant.

Basic assumptions typically get stronger and more unconscious as time goes by, and at the same time the espoused values, norms, rationalizations, and self-image of the organization may become disconnected from both artifacts and assumptions, and overly positive (Schein, 2010, p. 289).

Simultaneously, the context and its requirements continuously change, technology ages, new hazards are created, and the old ones change.

Since the existing culture at the organization influences how risks are perceived, how safety is defined, and what means are considered for achieving safety, it is important to maintain an awareness of culture at the organization including the organization's perceptions of itself and its environment.

There are different ways of maintaining an awareness of the current state of the culture for safety (Table 11.1). These range from continuous monitoring to periodic assessments. The various ways of monitoring and assessing culture for safety should be clearly described in the safety management system (see Chapter 6), including frequency, purpose, the associated methods and analysis.

DOI: 10.1201/9781003491156-11

TABLE 11.1
Types of Organizational Culture Monitoring Activities

Type	Purpose	Examples
Continuous monitoring	Identify early signs of degradation and threats to culture	Corrective action programs and observations
	Ensure that culture and safety-related topics are adequately considered in the organization	Safety walks, site visits, and other observations
	Highlight the effect of various issues on culture, including the management system	Key performance indicators
		Events and investigations
		Employee surveys and questionnaires
Annual summarizing	Summarizes the findings from the organization's safety monitoring activities	Annual summary reports or analysis reports
	Compiles information from several monitoring methods, which reduces bias caused by the method	Reporting findings concerning culture for safety in the annual management review
Periodic assessments	Get an in-depth understanding of the current culture for safety	Self-assessment
	Identify cultural strengths and opportunities	Independent assessment
	Form a baseline for continuous monitoring and annual summarizing	External assessments
	Discover and label organizational phenomena	
Focused assessments	Look at the effect of culture on specific issues	Functional self-assessments
	Identify underlying cultural issues that may have a wider effect in the organization	Incident investigations
	Focus on a specific function, department, or dimension of culture and thus have a more in-depth analysis of that issue	Topical investigations based on weak signals or trends

Continuous monitoring, as well as assessment, is based on information obtained from several methods and on the interpretation of these data sources from a cultural point of view. While assessment aims to create an image of the current culture, the purpose of continuous monitoring is to detect a change from the present. Annual summarizing builds on the fact that organizations typically collect large amounts of information that can be compiled and analyzed from an organizational safety point of view without any additional data collection.

Periodic assessments are the actual assessments of culture for safety that create the baseline for all subsequent monitoring (and summarizing). Table 11.1 differentiates between self-assessments, independent assessments, and external assessments. The difference between independent assessments and external assessments is that an independent assessment can be done, for example, by an internal independent oversight or quality function, whereas external assessments are carried out by a third body.

External and/or independent assessments, when well done, provide useful information for the organization, and these should be carried out from time to time, even without any visible safety problems. An external expert can help highlight issues that have become so commonplace in the organization that they are no longer paid attention to. It can sometimes also be easier for an external expert to bring up issues that are so sensitive that it is difficult to address them within the organization, but which would be important to discuss in order to improve safety. On the other hand, an external evaluator is not familiar with the organization, and it may be difficult for the evaluator to assess the safety significance of the findings. Terms and "slang" are also partly foreign to outsiders, and it takes time to understand them. Even the understanding of the concept of culture probably differs.[1]

Self-assessments of a culture differ from self-assessments of the management system (as required for example by ISO standards). The key difference is that in cultural self-assessment, the first aim is to form an overall picture of the cultural dynamics in the organization, including issues such as what employees consider important and how they perceive safety. The focus is also different: instead of employee perceptions and shared assumptions, the self-assessment of the management system focuses more on the evaluation of work processes and instructions and verification of compliance. Self-assessment of culture requires a good understanding of culture as a phenomenon, safety theories and models, and methods of data collection used in cultural assessment (see below).

11.2 ASSESSMENT FRAMEWORK

The aim of an assessment of the organization's culture is to increase understanding of the structure and dynamics of the culture and identify how those create and maintain the organizational strengths and vulnerabilities. The aim is not to get an exact measure of culture. In fact, that is not even possible. What the assessment can give, however, are indications of the areas or issues that are strengths from the point of view of their contribution to safety, as well as those areas or issues that are judged to be weaknesses. It should also enhance the organization's understanding of the reasons behind its practices and routines. The assessment is always based on expert judgment, and as such it includes uncertainty inherent in human judgment.

Cultural assessment is always qualitative. This means that the assessor must use themselves as an instrument of analysis; the feelings and thoughts that surface during the evaluation are all sources of information to a capable evaluator. One of the hazards of culture assessment lies in this inherently qualitative and subjective nature of the evaluation: an evaluator who is not competent in psychosocial issues can interpret some of their internal reactions and intuitions incorrectly.[2]

An assessment is always based on an underlying theory, whether the theory is implicit in the assessor's mind or made explicit in the assessment framework. The theory influences the methods that are chosen, the focus and the criteria used in the evaluation, and, subsequently, the data that will be gathered and the results achieved. Everybody has a working theory on how organizations function and what issues are important in understanding them, as well as how to gain that understanding. Making this theory explicit enables one to reflect on the premises of the evaluation and compare the findings to those premises. Typical safety and accident theories are presented in this book in Chapters 1–5. It is important that the assessment team has enough competence in safety and accident models.

When planning an assessment, one of the first issues to decide is whether the assessment will be conducted internally or whether an external party is contracted to conduct it. The ability and willingness of an insider to conduct a valid and objective assessment can be questioned on several grounds. The insider usually has overt or covert vested interest in the evaluation. They are accountable to some superiors and must take their priorities into account. Depending on the outcome of the assessment, they may gain or lose power. They are further influenced by the social network they identify with and consider their peers. This social identity (Chapter 2) may influence the assessment process. Furthermore, having worked in the organization, an internal assessor already has preconceptions of the problematic issues (and problematic individuals), which might or might not be correct but can in any case lead to confirmation bias (Chapter 2).

Personnel in the organization also have opinions about the assessor based on the common history, and that may affect what they want to disclose and what topics they want to discuss. The social identity of the insider affects whom they consider part of their group, and, for example, whom they consider experts and people with relevant information to give to the assessment. The insider is also, as the name implies, an insider in the organization; thus, they are prone to the same structural and cultural phenomena that influence and bias the ability of the personnel to perceive their vulnerabilities accurately. They are not independent. A company with a healthy culture may in fact make a more critical self-assessment than a company with a complacent culture.

Trust is an important prerequisite for successful organizational culture assessment. Trust develops in social relationships, and a certain amount of trust is needed for the assessment to succeed. It is harder for an outsider to gain the trust of the personnel, and they might feel doubtful of the motives and competence of the external evaluator. On the other hand, organizations are composed of numerous subcultures. A certain level of mistrust can exist between these subcultures, and one should not presume ad hoc that an insider assessor is trusted more than an outsider. Furthermore, social relationships include phenomena such as power conflicts and groupthink (see Chapter 2). Power conflicts and hidden agendas affect the way individuals share information and cooperate on a wide range of issues. Some employees may perceive the assessment as their chance to advocate their own interests and personal goals. The assessor should not have any personal agendas or hidden motives in the assessment, but they should remain sensitive to the fact that some at the organization might have their own motives for sharing or not sharing certain information or sharing a biased view on a matter. If trust exists, organizations are more willing to share their weaknesses. Typically, people are hesitant to disclose information that puts them or their group in a bad light, especially if they have doubts about how the information will be used.

External assessments have different challenges from self-assessments. Outsiders may not understand culture correctly, and misinterpretations and focusing on irrelevant things are possible. Outsiders also have interests and prejudices. A company is rarely at its most open in front of outsiders, and embellishment may occur. Politics and impression management may also be more difficult to perceive for an outsider. On the other hand, being ignorant of the power struggles or "sacred cows" in the organization can be an advantage from the point of view of conducting an independent assessment.

It is important to acknowledge the political dimension in cultural assessments. The bias that it brings to assessments can be reduced by making the goals of the assessment clear, looking for conflicting interests, seeking second opinions, and asking for explanations and reasons behind statements. Politics and power plays in organizations are one topic of interest for the assessor of culture, thus in addition to bringing potential bias, they always reveal something about the organizational reality. I have noted that often people in organizations have found it easier to talk about problems to an outsider, who they trust will not use the discussions to promote their own agenda in the organization. If there is a lot of mistrust or a lack of psychological safety is suspected, an external assessment is probably a better option than self-assessment.

Whatever the chosen assessment approach, when we assess culture for safety, we need normative criteria for judging the safety significance of the different findings. The typical safety culture models and a combined normative framework that can be utilized in culture assessment were presented in Chapter 5. The DISC method, presented at the end of Chapter 9, is another possible normative framework.

The purpose of the assessment is not to compare how an organization scores on different dimensions compared to other similar organizations. The aim is to increase understanding of how the organization works, including issues such as what it considers important and what it does not, how it reacts to different events, what it has learned, and how it defines its goals and the means to achieve them. The aim of the assessment is to find issues that occur almost unconsciously in daily activities and to raise awareness of them in the organization (IAEA, 2016b). The aim is also to promote discussion and understanding of culture in the organization, and to identify such ways to develop the organization and its culture to which the organization itself can commit.

One of the frequent questions posed to me by the senior management of the companies whose culture I have assessed is what the firm's score on a scale of (input any two numbers here) is. Typically, they also want to know what the mean score in their industry is, or the mean score among their competitors. Safety specialists in turn often are familiar with some safety culture maturity model (see Chapter

5) and wish to know their level of maturity (on a scale of 1–3 or 1–5, depending on the model). These discussions are good to have as they indicate the pervasiveness of key performance indicators in all businesses (see Chapter 6). The temptation for the evaluator to defer back to numbers is high, as this seems to be what the company wants. However, the right approach is to treat this as an artifact of the given (industry) culture, indicating their way of viewing safety and understanding culture. They may want numbers, but they need understanding.

Another typical discussion concerns the items and individual responses from the questionnaire. This is another sign of top management's obsession with quantitative data, but it is also a sign of a real need to monitor the trend in culture development. Typically, too much attention is devoted to mean scores on individual items. Corrective actions are requested to target the exact issue measured by the questionnaire item. This is one of the main reasons that senior management prefers questionnaire results over, for example, interviews. They consider numerical responses as more scientific, as if somehow employee perceptions transform into science when they tick the multiple-choice box, whereas they remain subjective opinions if they speak their mind to the interviewer.

11.3 PLANNING THE ASSESSMENT

The assessment aims to identify both the strengths and weaknesses of the culture in relation to safety. When planning a cultural assessment, the following list of issues should be considered (Table 11.2).

A distinction can be made between the descriptive phase (what the current culture is like) and the evaluative phase (what the strengths and weaknesses of the current culture are from a safety perspective). The evaluative phase requires a normative model (what a good safety culture is). Such models include, for example, the European Rail Agency's safety culture model (ERA, 2020), the IAEA model for the nuclear industry (IAEA, 2016b), and VTT's DISC model (Oedewald et al., 2011). Because an organization's culture is also determined by what the culture ignores or considers

TABLE 11.2
Issues to Consider when Planning an Assessment of Culture (cf. Oedewald et al., 2011)

Reason and purpose of the assessment	What is the reason for the assessment?
	What are the specific goals of the assessment?
	What instigated the assessment? Is there a specific problem to be addressed?
	Is there adequate senior management commitment for the assessment?
Design of the assessment	How should the assessment be done in terms of scope and depth?
	What are the criteria that are used to evaluate the organization, and how are the criteria derived and operationalized?
	What methods are used in data collection?
	What are the required resources in the assessment?
	Does the assessment team have the required competence in using the methods, understanding the context, and integrating findings?
	How is the assessment organized – who leads, who participates?
	Are issues related to anonymity and confidentiality discussed?
Execution and follow-up	How is the progress of the assessment communicated to personnel?
	How are the personnel affected by the assessment process?
	How is data collected and analyzed?
	When should the assessment be conducted?
	How are the results of the assessment to be reported to the organization?
	Is there enough commitment in the target organization?
	Is the assessment team competent enough to complete the evaluation?
	Who defines the corrective actions and who follows up on their implementation?

trivial or insignificant (Weick, 1998), normative models are needed to highlight potential neglected areas.[3] For this reason, interactive and open methods are important; they give members of the organization the opportunity to highlight things that they themselves consider relevant (and downplay issues they do not consider significant). Non-interactive methods such as surveys can only highlight issues that the evaluator has considered important enough to ask. Responses to the questionnaire may indicate that the employees do not care about a specific issue, but this interpretation is difficult to make without further exploration of the issue.

The descriptive part is important because it is a bottom-up description of the current culture. The descriptive part is also crucial for the assessment since many features of culture can have both positive as well as negative influences on safety. Especially the positive influences are easily sidelined if we approach culture with a checklist or ready-made criteria. It is also important to avoid relying too much on Work-as-Imagined; the descriptive part of the assessment is essential in order to reach the Work-as-Done.

The assessment requires a wide range of expertise from the assessment team. Understanding cultural phenomena requires behavioral or social science skills (e.g. psychology, social psychology, sociology, anthropology). Understanding the different aspects of managing organizations is also important (including leadership and management systems). Understanding the safety implications of phenomena requires mastering both the social and technical context. From the point of view of the overall safety of the organization, the evaluator should be aware of the hazards and danger mechanisms that are central to the company's operations and familiar with the typical accident models.

If the assessment is carried out as a self-assessment, the evaluation team must be well trained before starting the assessment. Not all team members need to be in-depth experts in all methods and theories, but the team needs to have sufficient competence.

An important decision at the planning stage is the selection of the methods that are used in the evaluation. In cultural assessment, multiple sources of information and multiple types of data are needed. Thus, several different information-gathering methods should be used. Interactive methods are always needed, where the personnel have a chance to raise any issue into discussion that they feel warrants attention.

Culture is the way a group has learned to react, think, feel, and act (Schein, 2017). Therefore, when mapping culture, it is necessary to highlight the learning experiences on which culture is based and to identify the structures and dynamics that sustain or slowly change (Figure 1.2) the current culture. The key to assessing culture is to go deeper than the surface level of culture (artifacts, Figure 1.1); otherwise the assessment will not differ from the evaluations of the management system or behavioral and management system audits. The use of participatory and qualitative methods ensures that values, beliefs, and assumptions shared by the staff are reflected in the evaluation. In addition to participatory methods, the survey, for example, enables the collection of extensive data from the entire organization.

The selection of assessment methods is always a question of choice. No one method has been shown to be superior to the other methods, but they all provide different kinds of information. When planning the assessment, the number of methods depends on the goals of the assessment as well as practical considerations such as time and resources. Table 11.3 lists the main methods and their advantages and disadvantages in cultural assessment. Surveys, interviews, and focus groups focus mainly on the middle level of culture, whereas documents and observations represent different angles of the level of artifacts (Figure 1.1 in Chapter 1). To reach the level of shared assumptions, all data needs to be analyzed and findings from the different methods compared.

When I conduct culture assessments, I try to use at least three methods. I always do interviews/focus groups and document analysis. Then, depending on the specific goals of the assessment and other considerations such as resources, schedule, and access, I will select focus groups or the survey as a third method, sometimes both. If the organization has good records and safety indicator data, I aim

TABLE 11.3
Different data gathering methods in culture assessment

Method	Strengths	Weaknesses
Survey	Everyone can participate as a survey is easy to implement throughout the organization. It highlights the common issues within the organization and identifies where opinions diverge. It offers the possibility to locate subcultures (departments, professional and age groups). Factor analysis can uncover underlying variables indicating how employees perceive connections.	Asking personnel to rate preselected dimensions of culture in a questionnaire does not necessarily uncover the issues that the personnel neglect to consider or that they consider insignificant for safety. The way questions are formulated can influence the responses, leading to potential misinterpretations that are often hard to detect. If something is not asked, it will not be answered either. A survey does not answer how or why questions. Numerical data is easy to misinterpret.
Interviews	The staff can tell about things they consider meaningful and how meaningful they consider the themes chosen by the evaluator. Open-ended questions allow the interviewees to speak up on various issues not anticipated by the interviewer but significant to safety.	The dynamics and chemistry of the interviewer and the interviewee affect the results. As interviews are heavy to implement, only part of the personnel is often interviewed, in which case the result does not necessarily describe the entire organization. Selection of people to interview is critical – it should be as random as possible but still cover all essential functions and organizational levels.
Focus groups	A focus group is a more cost-effective way of gathering information from a large number of people than an individual interview, and still allows the opinions of the personnel to be heard.	Group dynamics can affect results – for example, if there is an opinion leader in the group that others agree with, even if they could disagree when interviewed alone. On the other hand, if these are identified the information can be utilized in the assessment.
Document analysis	This tells about concrete events or measurable things, not opinions. It works well as an orientation method, for example, before interviews, to better understand the context and guide the follow-up questions.	The documents do not answer all relevant questions about cultural phenomena, informal practices, and so on. Documents are rationalizations of what the organization wishes to be and what it wishes to show.
Observation	This offers the ability to observe "real" behavior in the "real" context, and the opportunity to see the prerequisites and boundary conditions for working in the field.	Attention is focused on easily noticeable things, such as cleanliness and protective equipment. The challenge is to infer the motives, perceptions, and assumptions behind visible action. The presence of an observer can influence behavior (impression management or even deliberate misrepresentation).

to analyze that as well. Observation of work is often difficult due to time constraints and the fact that I typically assess entire organizations with multiple types of work ongoing. However, I always make notes about my meetings with the representatives of the organization in question and note other observations as well, for example, how people treat each other, how well the safety professionals know their organization, and what kind of power dynamics there are between different groups, such as the safety professionals, senior management, and workers.

11.4 COLLECTING DATA

The following is a review of the typical methods for data collection (see Table 11.3). The methods are described from the perspective of cultural assessment, but most issues apply to any organizational safety assessment.

SURVEYS

Surveys collect information about the perceptions, values, beliefs, and attitudes of the personnel – the middle level of culture, as illustrated in Figure 1.1. Surveys allow a wide range of employees to express their views. Since surveys quantify observational information, they make it possible to contrast responses between different groups and levels of the organization, and thus to assess cultural unity.

The use of the survey method requires competence in psychometrics. Survey responses are often viewed through the averages (and standard deviations) of individual items in the questionnaire. However, this can misdirect the organization to focus on individual issues.[4] In the analysis of survey data, besides focusing on average scores, it is important to consider the variability in responses and the factors contributing to this variability, such as age, department, job function, and position. Additionally, correlations between the questions should be investigated. Factor analysis is recommended to investigate the internal structure of the responses and uncover underlying variables.

Corrective measures should be targeted at the phenomenon as a whole, and not at its individual manifestations (questions measuring the phenomenon). This is similar to key performance indicators. An organization may want to increase employee involvement in safety and set an indicator "number of employee safety observations" to monitor it. As outlined in Chapter 6, with quantitative targets, it is common for the original objective to be overlooked, resulting in efforts being solely focused on improving the indicator score.

As already mentioned, organizations often wish to compare their scores with their peers or competitors in the same industry. Care must be taken to ensure that comparison does not become an end in itself. The comparison should serve as an aid in interpreting one's own results, not as a competition in which the best result wins. If comparing survey scores becomes a competition between organizations, it may lead to respondents glossing over their estimates in the hope of rewards or appreciation. It is also easy to focus attention only on numbers and rankings, rather than seeking a deeper understanding of the background to these figures and genuinely developing culture. Danger is that comparisons with other organizations can, in some situations, serve as an argument for doing nothing about the identified weaknesses. For example, findings related to high work-related stress are nowadays typical in many surveys and studies. In that sense, the high level of work stress in an organization can be in line with other organizations. This should not reduce the importance of considering work stress and wellbeing at work. Comparative survey data should primarily serve the internal assessment and learning of the organization. The use of scores as a tool for external evaluation and remuneration should be viewed critically. Survey scores are always an artifact (see Chapter 1) of the given culture, with multiple factors influencing them.[5]

The survey produces information only on what is explicitly asked. This makes the choice of questions and their precise formulation a key. Since surveys produce numerical (quantitative) information, it is often thought that survey information is in some way more objective or scientific or truer than, for example, information obtained through an interview. This is not the case. Quantitative data is of a different type than qualitative material, but it is not more reliable or objective in the study of culture. In addition, quantitative data is also always interpreted – what averages, dispersions, and differences mean in that context.

Many mistakes or overly crude generalizations can be made in the interpretation of survey data. For example, it is typical that management evaluates activities more positively than employees. If the difference is small, then almost nothing can be deduced from it for the organization in question. What's interesting, however, is how the organization views and explains the findings.

The *halo effect* means that a person who is dissatisfied with one thing is also dissatisfied with almost everything else. This is also reflected in the other way around, that is, a generally satisfied respondent answers all individual questions more positively. Surveys thus polarize opinions. The amount of polarization can be estimated, for example, by looking at how much all variables correlate with each other.

Social desirability bias means that the respondent seeks to give a good image of themselves by responding in a manner they think they should. Survey responses are also strongly influenced by experiences from previous surveys; if negative or critical answers have led to negative consequences, staff may have learned to respond in a certain way – this cultural feature does not easily come to light without other methods of obtaining information, that is, *why* questions.

The survey results reveal the respondents' perceptions of the things being asked. These perceptions always have objective consequences, but that does not mean that the perceptions are "true" in respect to representing the actual situation. If the staff is of the opinion that they do not trust the management, this will affect their behavior, and from this it can be concluded that there is no trust. If the staff thinks that the organization is good at learning from events, this does not necessarily mean that the organization is actually good at learning. In both cases, further clarification is needed, that is, *why* the perceptions of the staff are as they are.

Since survey responses are employee's perceptions of how things are, organizations with a deficient understanding of risks combined with strong cohesive culture can score high on them, indicating a "good safety culture" if one only looks at the mean scores.[6] In many cases, safety interventions can show us a drop in survey scores between the first measurement and a follow-up survey. This is typically due to the fact that the employees' safety perceptions and attitudes have improved, and they are able to more critically reflect on their work and the work environment. Again, it is important to interpret the reasons behind the scores, and not just take the declined scores at face value and declare the safety improvement effort a failure.

Survey data provides an opportunity for visualization of significant group findings, which can help focus discussions on issues, concerns, and trends. Quantitative data also makes it possible to track trends as a function of time. When the survey is repeated regularly, questions in which there is a statistically significant change between measurements can be examined in more detail. In order to find the cause, further clarification is typically needed through interviews or focus groups, that is, *why* the change has occurred. As noted above, deteriorating scores in a survey can indicate employees' improved safety perceptions instead of an actual decline in the overall "safety culture". It is recommended to inspect the underlying structure of the responses with factor analysis and compare that between measures; changes in the underlying structure may indicate changes in culture.

Based on the survey scores, it is possible to take a position on, for example, the prevalence or severity of the development issues highlighted in, for example, interviews or focus groups. For example, if communication is criticized in interviews, the survey can be used to take a position on whether the opinion is the prevailing opinion and, if so, how strong it is. Together with trend tracking, this helps to target corrective actions to the most significant issues.

It is useful to include at least one open-ended question in the surveys, where the respondent can write their opinion on the theme related to the topic in their own words. These answers may provide important information that can explain some quantitative findings (e.g., why a question has a low average, a large dispersion, or a lot of missing answers). In this way, it is also possible to identify questions that the staff has felt are poorly formulated or ambiguous.

The survey does not produce ready-made results on what culture is like, but rather it provides one source of information that, when combined with other data, can be used in the cultural assessment. For the survey to be effective, the results need to be processed and analyzed, and the findings should be discussed with members of the organization or other experts. Also, the analysis of open-ended responses to surveys often provides good additional information about culture.

Often organizations want to include "purely safety related questions" in their safety culture surveys as they may already have an employee wellbeing or motivation survey run by the HR department – or even a "culture survey" separate from a "safety culture survey". This is not a good practice, and it indicates a narrow view on safety culture (instead of culture for safety) in the organization, or alternatively a very common turf battle between the support functions such as HR and the safety department. The problem with only including questions with a direct safety implication is that people know what the management wants to hear. In such questions there is a stronger temptation for impression management or self-deception in projecting a better image of oneself than reality would allow. Finally, "problems of safety" are seldom just problems of safety, but rather problems of management, organizing, working practices, and leadership in general.

Interviews

While surveys seek to measure the culture for safety or its various aspects quantitatively, interviews and other qualitative approaches strive to understand how culture has evolved and the factors that sustain or alter it. The primary objective is to delve into the "why" but also elaborate on the "how". Interviews provide insights into the backgrounds, causes, and explanations of different phenomena, as well as emotions tied to various subjects, circumstances, or individuals. They also serve to elaborate on and enhance results obtained from other methods such as surveys and document analysis. Thus, interviews are pivotal, and perhaps the most crucial method in evaluating organizational culture.

The most important tool in interviews is the interviewer him/herself. The conduct of interviews requires from the interviewer both an understanding of the essence of the culture and knowledge of the interview method. Interviewees are sensitive to the interviewer's behavior, and with introductory questions or an attitude (positive or negative) towards the answers or the interviewees, the interviewer without realizing it can have a big impact on the results. On the other hand, it is necessary for the interviewer to understand the essence of cultural knowledge: beliefs, perceptions, attitudes, and values are true to their bearers, and even if they do not correspond to the so-called reality, they have objective consequences when influencing the actions of their plaintiffs.

The degree of organization (structuring) of interviews may vary. The interview can proceed according to a strict preliminary plan, even based on ready-made questions and answer options, in which case it differs little from a survey method. Such a structured interview is based on specific questions that aim to get as unambiguous an answer as possible (usually yes/no).

In semi-structured interviews, a list of questions is used, in addition to which the interviewer asks more detailed questions based on the themes that arise during the interview. The aim is to formulate the interview questions in such a way that they allow for broad answers around each theme. In unstructured interviews, the interviewer minimizes their interference in the dialogue by asking very open-ended questions and allowing the interviewee to lead the interview. Each unstructured interview is thus unique, while semi-structured interviews combine the themes around which the conversation takes place. For cultural assessment, semi-structured interviews are typically best suited.

Interviews are an interactive method. The interviewer should encourage the respondent to speak freely on the topic at hand. The interview must not be an interrogation or an audit in which the respondent feels that they may be right or wrong in answering. The interviewees must have the opportunity to bring up issues that interest them and that they consider important. The interviewer should monitor how the interviewee reacts to various questions. Non-verbal cues in responses (e.g., the interviewee's enthusiasm, caution, frustration, or indifference) potentially say something about the organization and its culture.

The understanding and trust between the interviewer and the interviewee greatly influence the process. The interview may be experienced as a test or interrogation, in which case the interviewee tries to give an answer that he or she imagines to be correct in the interviewer's opinion. On the other hand, interviews can offer interviewees the opportunity to vent long-term frustrations, which can escalate opinions and lead to, for example, exaggeration or dramatization of things.[7]

The identification of shared perceptions, contradictions, and tensions in large data requires competence in the analysis of qualitative data. Recurrent cultural expressions need to be identified. These produce information about the beliefs, perceptions, values, and operating models that prevail in the organization. Of particular interest are findings that shed light on power relations and dynamics in the organization, thereby determining what things are considered meaningful and paid attention to and what is ignored or silenced (see also Edmondson, 2018; Schein, 2017).

In interviews, the following questions can be used:

- How does a typical day at work look?
- What are the biggest risks in your work?
- How does your work affect safety?
- What is the current level of safety in the organization? What could be done to improve safety?
- With whom do you interact regularly in your work? How do you interact with your supervisor, colleagues, and subordinates?
- How do supervisors and management support the raising and handling of safety issues?
- How can you report or raise a safety observation in your work? Have you done so?
- How are safety observations, suggestions, or concerns analyzed in your company? Do you think the right things are being addressed?
- What is the vision of your company's management when it comes to safety? What is your role in/contribution to fulfilling this vision?
- What are the strengths of your organization's culture? What things are working well?
- What is the one thing you would change in your organization if you had the power to change anything? What would you make sure does not change?

The selection of the people to be interviewed is important. Ideally, representatives from all organizational functions and levels of the organization are interviewed (Oedewald et al., 2011). Further, the interviewees should represent different working experiences and educational backgrounds. A balance needs to be struck between adequate coverage and resources. If focus groups are conducted in addition to individual interviews, these can be used to ensure all organizational functions are covered with interactive methods.

It is important to let the interviewees' voices be heard and not immediately correct them if they say something that is "wrong" or "inaccurate". This is challenging especially for internal assessments where the interviewer probably has a position on many of the issues and knows a lot about what the organization has and what it lacks. The interviewer should remain as neutral as possible on any of the more contentious issues and listen how the interviewee perceives them.

Document Analysis

Documentary analysis reveals how the organization appears in written material and records. In the analysis of documents, a distinction can be made between 1) documents belonging to the management system; 2) records such as minutes of meetings, orders, and communiqués; and 3) other documented information, such as indicators. Management system documents say more about how an organization would like to operate in an ideal state (Work-as-Imagined), while records and indicators say about how the organization has actually worked (Work-as-Done, or at least Work-as-Recorded).

Good documents to be analyzed include organization charts and management system manuals, official policies, management reviews, annual reports, incident and non-conformance reports, other investigation reports (including regulatory inspections), HR questionnaires, trend data, and key performance indicators. Also, minutes of meetings from various management and operational teams can be a good source of information. Many companies collect a lot of information; the problem is in finding the relevant data. For example, observation systems and corrective action databases can offer information that is already categorized based on, for example, type of observation, significance, selected corrective actions, and contributing factors.

The aim of the document analysis is not to look for deviations or other shortcomings in individual documents – it is therefore not an audit or an official review. The purpose is to gather information about how the organization thinks and how it plans to act, as well as how it has worked (records and indicators). Various safety indicators imply what the organization wants to measure and manage; thus, what the organization has selected as worth measuring is as important as the actual scores on these indicators (see also Chapters 6 and 9). Document analysis can provide an understanding of how the organization prioritizes safety based on both formal management system documents as well as records of actual events. Analysis also provides information on how the various operating models, policies, and processes are described. The records reveal the other side, that is, how the operating models and processes have worked in practice.

When analyzing documents, the following issues can be considered:

- What is the general quality of documentation – for example, are the guidelines clear and up to date?
- What does the authority and responsibility structure of an organization look like, including roles, responsibilities, and authorities, as well as key decision-making practices?
- How are decisions documented and justified?
- What role does safety play in the official goals and operating philosophies of the organization?
- What kind of activities and behaviors does the organization expect from employees? And what will not be tolerated?
- What practices are there for the organization to learn and develop?
- How are events investigated, and how are corrective measures defined and implemented?
- What does organizational openness and communication look like? What about trust between management and staff? Do these issues come up in the records?
- How does the organization solve problems, conflicts, and technical challenges?
- What kind of choices (trade-offs) are made, and what kind of risks are accepted?
- What kind of contradictions and ambiguities can be found in the material?
- How is the gap between Work-as-Imagined and Work-as-Done acknowledged in the documentation? How does it show in the records?
- Does the organization explicitly state that some issues are not important to them? How about risks? Are some risks assessed as insignificant?
- What safety-related aspects are overlooked in the material?
- What things are not talked about/written about?

Focus Group Method

The aim of a focus group is to explore selected themes in an open-ended, interactive fashion. The focus group method is a form of group interview in which there are several participants (in addition to the moderator/facilitator); the discussions emphasize a certain rather strictly defined topic, and the focus is on interacting within the group and in the joint construction of meaning. Typically, two experts are needed to apply the focus group method, one of whom leads the discussion while the other takes notes. The length of one session varies from a couple of hours to just over three hours. In addition to the two facilitators, the optimal number of participants ranges between five and twelve.

Focus groups are more than just interviews with multiple participants at the same, as the group dynamics becomes one of the additional sources of information: who speaks first, whether the group agree on issues, whether they are familiar with each other's work, whether they are interested in each other's work, how they address each other, and how they behave in general. Facilitators should pay particular attention to where participants seek to focus discussions and which topics they are correspondingly quick to skip. Facilitators must listen to both the content of the group discussion and consider emotions, irony, confrontations, and tensions (IAEA, 2016b).

The aim of the focus group work is to explore the chosen theme(s) in an open and interactive way. The theme can be something that has emerged from the other methods, or a general topic of interest. With focus groups, it is possible to deepen the understanding of why a topic is important to the work community in question, or why something is not important. With the help of the discussion, it is possible to delve into possible causes and justifications, as well as to collect feedback or form development ideas (IAEA, 2016b).

The challenges of focus-group work are largely related to the fact that the use of the method requires skilled facilitators who know how to manage group dynamics, especially if the topic being worked on is sensitive or arouses strong opinions. Power relations may distort the conversation, although such interactions still provide information about the internal dynamics of the organization. The group or its individual members may also have personal goals or a need to vent anger or frustration. Expressions may be influenced by groupthink or peer pressure (see Chapter 2).

Good themes to discuss include the following:

- What makes this work challenging? When is work at its most challenging?
- What makes this work enjoyable? What are the good aspects of the workplace culture or the job itself that are worth maintaining?
- What resources do you need for your work? Are they always available? Why?
- How is the work organized? How do the processes and instructions support work?
- How do you act when the conditions, including resources, are suboptimal? What needs to be compromised?
- How well do you think the risks are under control? What could be done better?
- What are the strengths of the organization and its culture? What works well?

Observation of Work

Observations of behavior and the performance of work tasks can also be used in the assessment of culture. Observations look at the physical manifestation of culture (artifacts, see Figure 1.1). The observations often concern specific situations, such as meetings and the performance of selected tasks. The second option is the so-called shadowing, which tracks a specific team or employee. The purpose is to see how culture manifests itself in natural work situations. Cultural observations differ from task observations, which are more normative and often aim at identifying deviations (see Chapter 9).

In systematic quantitative observation, the behavior of employees is scored by means of a pre-established checklist or scoring system. The systematic scoring of behavior therefore requires that we first define as precisely as possible what kind of behavior is considered safe (Cooper, 2000). However, it is very challenging to define safe behavior and the criteria for scoring it universally, especially when the criteria must be verifiable by direct observation.

Sometimes the observation criteria are made as simple as possible, and the observer counts, for example, how many times an employee without a helmet is detected or how many walk in the stairwell without holding on to the stair railing. In this way, a rather narrow picture of the overall safety of the organization's activities is obtained (Hopkins, 2019). The fact that someone wears or does not wear a helmet cannot be used to determine, for example, whether the employee in question understands the safety significance and risks of his or her activities. Many safety-critical jobs and activities are also difficult to observe, so observation may be biased by the more visible jobs, such as maintenance, or the time between jobs, for example, when people are transferring to the work site and preparing the site for the work.

Observations can vary between passive participation (in which the observer is a passive bystander who records his observations, sometimes even without the object of observation knowing that he or she has been observed); moderate participation (in which the interaction between the observer and those being observed is limited, but the observers know that they are under observation); and active participation (in which the observer actively participates in the activity at hand).

When observing the workplace, pay attention to the following (cf. Sharman, 2019):

- How do the workers interact among themselves and with the management? How are conflicts resolved?
- How do the leaders interact among themselves and with the workers? How are conflicts resolved?
- How does the behavior of the employees at different organizational levels look: engaged, relaxed, isolated, stressed, withdrawn?
- How does the physical layout of the workplace and housekeeping support work?

While observing, try also to pay attention to what is not there, and what is not happening even if it should. The safety culture models introduced in Chapter 5 can help in identifying missing or neglected issues.

Observation can also be combined with an interview method. Good questions to ask include the following (cf. de Boer, 2021, p. 54; Hopkins, 2012):

- What makes this task challenging? When is work at its most challenging?
- What does your work depend on? What needs to be in order?
- What makes this work enjoyable? What are the good aspects of the workplace culture worth maintaining?
- What resources do you need for your work? Are they always available? Why?
- How do you act when the conditions, including resources, are suboptimal? What needs to be compromised?
- What lessons would others have to learn about how you do your work?
- What pointless things do you have to do in your work? Why are they irrelevant?
- What's the worst thing that can happen at your job or as a result?
- How well do you think the risks are under control? What could the organization do better?

11.5 ANALYSIS AND ASSESSMENT

A good overview of how to analyze data in two phases (descriptive and normative) has been produced by the International Atomic Energy Agency (IAEA, 2016b). Although aimed at being used in the nuclear industry, the guidance is applicable to any safety-critical domain.

Integrating the results produced by different methods is a multi-step process (IAEA, 2016b; Oedewald et al., 2011) consisting of six steps:

(a) Identification of common themes from each method;
(b) Identification of differences and conflicting findings within and between methods;
(c) Iterative analyses based on similarities, differences, and contradictions;
(d) Descriptive analysis of culture;
(e) Evaluation against a normative framework;
(f) Presenting and communicating the results.

Identification of common themes (a) is done initially for each method independently (IAEA, 2016b). The idea is to extract issues and themes that emerge from the data. For example, if multiple interviewees raise the issue of blaming or fear of getting punished for speaking up, that can be a common cultural theme in the organization. Themes should be labeled and data points categorized under the emerging themes (Excel is a good help in this). Once the data is analyzed for each method, differences and conflicting findings within and between methods (b) need to be addressed. Differences within a method can be, for example, due to organizational functions, levels, age groups, or another differentiating variable. Differences between methods can be due to the particularities of each method or due to the selection of informants. For example, interviews with workers and focus groups with senior management will probably produce partly different findings concerning culture.

In the next step (c), data produced by each method can be revisited with an increased understanding, or more elaborate questions, based on findings from another method. This can mean, for example, a contradiction between survey scores where employees indicate a perception that management is highly committed to safety rules and interviews where employees criticize the management's safety attitudes. Both results can be true, but they may measure different things. Issues to inspect in this case would be why the interviewees feel management's safety attitudes are bad, if the interviewees discuss how the management approaches safety rules, and whether the survey can give more information on what the safety rules are that the management endorses (or how the question on safety rules correlates with other questions). Some contradictions may be eliminated once data is further analyzed, whereas some contradictions will remain as genuine aspects of the culture. The result is a revised understanding of the culture – what common themes emerge based on the integration of information from all used methods, and what contradictions and differences there are in the culture. These findings are then used to describe the culture (d). In the description, the identified main themes are described as well as the differences and contradictions that were identified.

At the final stage (e), the strengths and weaknesses of the culture for safety are identified with the help of an expert judgment by the project group. A normative model (see Chapter 5) is helpful in this stage to highlight the needed and potentially neglected issues. The results can be presented according to the selected normative model, with each of the criteria assessed, provided that there is enough information to cover all the cultural dimensions. Alternatively, the descriptive themes can be mapped to the relevant cultural dimensions, and the themes are assessed based on this mapping.

The choice depends on the normative model. If the model focuses on organizational level functions (see Table 5.1), it is easier to assess how the cultural findings fit the organizational dimensions. If the normative model focuses on psychological issues and more on the level of beliefs, and basic assumptions, it makes sense to assess how these dimensions reflect in the themes identified based on the descriptive analysis. Unfortunately, many normative models of safety culture include both organizational functions and psychological states without adequately differentiating between them or explaining their connections (see Chapter 5).

When identifying themes, pay attention to how the identified themes show up in the practices and structures of the organization (see Chapter 1). Consider also how the themes are related to each other: do they support each other, are they in contradiction, or are some clearly distinct from some of the others? Note that the themes are not the same as underlying assumptions and logics (Chapter 1) but they can be further analyzed to identify assumptions. These assumptions can concern one of themes or several themes simultaneously, and their extraction may require deep analysis of the data from different methods iteratively. Interactive methods are key in this, but other methods can offer hypotheses or validate selected themes (e.g., the underlying factor structure of the survey answers may hint at how the employees perceive the world).

One good approach is to present the results in a SWOT format (Strengths, Weaknesses, Opportunities, Threats). SWOT is well known by senior managers and results are easier to communicate with a familiar framework. However, it is good to note that very seldom any of the normative dimensions presented in Chapter 5 can be fully concluded as either a strength or a weakness. Typically each dimension has both strength and weaknesses to some degree.

The assessment of culture for safety does not necessarily have to directly define the needed development measures. These must be agreed together with the management and integrated into the development of the entire organization and the continuous improvement of the safety management system. It is also a good idea to consider development measures together with the personnel, for example, by applying focus groups or personnel seminars.

The main idea of the assessment framework is that the best added value of culture concept comes from this combination of seeking to understand the current culture (descriptive phase) and, only after understanding it, making normative judgments about how the given culture affects safety (normative phase).

In one organization where I conducted a follow-up safety culture assessment, senior management requested a numerical value for each content area of culture. When evaluating the leadership component of the company's safety culture model, we rated it as average with a score of 3 on a scale of 1–5. Notably, when comparing these results to the previous assessment, which also assigned a score of 3, significant changes in reasoning were observed. Previously, the mediocre rating was attributed to leadership practices characterized by leading through fear and belittling employees. In the follow-up assessment, improvements in this area were noted. However, the current challenge had shifted to a lack of direction in leadership. Providing just a numerical result would have obscured significant changes in the company's leadership style, which encompass both positive and negative aspects, as is often the case with changes. More crucial than the numeric value is the explanation for the assigned score and the dynamics of how that number was determined.

The results of the assessment, including the agreed development measures, must be communicated openly to the personnel. However, it is important that development measures are accounted for, resourced, and scheduled as part of the continuous improvement of the organization. Culture development should not be a separate project independent from organizational development.

Assessment and development of organizational culture are intertwined functions in many ways. Assessment can already serve as a development intervention. It can help staff structure their work from a new perspective and thus change it. Assessment is also usually seen as one of the key stages in organizational development. The results obtained during the evaluation process guide what issues are developed in the organization. Similarly, the choices made during and before the evaluation, for example in terms of the assessment focus, what methods are used to collect data, and how conclusions are drawn up, affect both the results produced by the evaluation and the choice of development methods. On the other hand, development can support assessment by producing more understanding of the organization's operations when trying to change operations. This information either supports the conclusions of the assessment or proves them insufficient or wrong. At best, assessment and development support each other and are intertwined into a natural whole.

11.6 CONTINUOUS MONITORING OF CULTURE

The continuous monitoring of organizational culture relies heavily on the methods used to monitor the functionality and effectiveness of the safety management system. The difference lies in how the information generated by these methods is interpreted and used. In the monitoring of organizational culture, instead of individual deviations, disruptions, or errors, we are interested in factors, trends, and root causes at the group or organizational level. The monitoring of culture goes deeper than monitoring the functionality of the management system, into the organization's practices and into daily work. In addition to utilizing the methods of the safety management system, continuous monitoring utilizes traditional HR data and various systematically collected observations of practical work.

When monitoring culture, it is important to have an underlying model of what culture is as well as pragmatic indications of what are the potential warning signs of an increase in risk level. Similar to cultural assessments, a checklist approach needs to be avoided, but a descriptive analysis is not done either – monitoring deals with changes from the baseline established in the previous in-depth assessment. For this reason, the assessment and monitoring approaches are complementary, and both are needed.

Table 11.4 describes the possible methods and sources of information for continuous monitoring of the organizational culture for safety. Most of these methods have already been described in Chapters 6 and 9 from the point of view of how they contribute to management of safety. Table 11.4 summarizes the considerations from the perspective of monitoring of culture.

It is a good idea to summarize the results of continuous monitoring, for example, on an annual basis and report them to the management. This can be done in a management review or in another similar meeting. In the annual summary, it is important to look at the common findings and trends, rather than individual deviations or corrective actions. For example, observations can be classified according to the model of culture used in the organization. However, in this case, it is worth taking into account the differences in the significance of the observations. Some findings are more significant than others. Successfully implementing continuous monitoring requires a good comprehension of cultural issues and the characteristics of cultural data.

TABLE 11.4
Examples of Methods for Continuous Monitoring of Culture For Safety

Method/data source	Description and considerations
Audits	From the point of view of culture, in addition to verification of requirements, attention is focused on how the organization responds to questions and what the answers communicate about values, assumptions, and openness.
Safety indicators	Safety indicators give an indication of the state of the organization's safety, but they must always be interpreted with the help of other information.
Observation system: causes and trends of observations	These allow the dentification and correction of vulnerabilities that are gradually developing, and engagement of personnel to observe their work environment. Classification of observations for trending purposes is useful, for example, according to the elements of culture.
Incident investigations	These are used in identifying and addressing vulnerabilities that have already manifested or that are gradually evolving in the organization. They can help in identifying weaknesses in culture, if conducted from a systems perspective.
Observations of safety culture ambassadors	These monitor the state of the culture with knowledgeable employees who have decided to act as ambassadors for culture issues. It is advisable to collect observations systematically during the year. Safety culture ambassadors may also be involved in broader culture assessments.
Joint chats, coffee, monthly meetings, etc.	Meetings and discussions focusing on safety issues serve to develop cooperation and share information and communicate the importance of safety. These are learning opportunities for both management and employees.
Safety walks and task observations	These involve observation of working conditions and work. Observers need to be careful not to focus too much on things that are easy to verify (e.g. protective equipment). They can provide information on how the work is actually conducted and what the gap is regarding official procedures.
Employee surveys	Periodic safety or wellbeing at work surveys typically provide information about the atmosphere, motivation, and wellbeing of employees. Monitoring trends in these can inform about changes in culture more broadly.
Sick leave, turnover, and other HR indicators	Information on the wellbeing and comfort of the personnel, when interpreted correctly, provides information on culture.
360-degree management surveys	These provide information on leadership issues, which are key to the development of culture. It is also a good idea to include questions directly related to safety and culture in the surveys.
Exit interviews	Interviews conducted for people leaving the company can provide information on strengths and weaknesses related to the operation and management of the organization.

If potential degradation of culture in terms of safety is identified, the organization can either devise corrective actions if the findings are clear or conduct an in-depth or focused assessment (see Table 11.1) in order to better understand the problem.

11.7 WARNING SIGNS OF DEGRADING CULTURE

Continuous monitoring should identify potential changes in culture and enable the organization to react to possible deterioration in safety or culture. There are several signs that can indicate a degrading, or degraded, culture for safety. These warning signs are mostly surface-level manifestations of culture, that can alert the organization (or external assessor/regulator) to inspect the deeper levels of culture. Typical warning signs are summarized in Table 11.5. These warning signs should be treated as indicators: they are rarely management targets in themselves. Rather, they guide leaders and safety professionals to investigate underlying logics and cultural dynamics.

11.8 SUMMARY

Cultural assessment aims to identify strengths and weaknesses of the culture in relation to safety. The purpose of an assessment is to understand the structure and dynamics of the culture and to identify strengths and vulnerabilities. Cultural assessments are qualitative and based on expert judgment. Internal assessments can be influenced by vested interests, social identity, and preconceptions, leading to potential biases. Trust is crucial for successful assessments, and external evaluators may face challenges such as unfamiliarity with the organization and misinterpretations. External assessments can be advantageous in highlighting issues that have become commonplace or sensitive within the organization.

Multiple sources and types of data are needed for reliable cultural assessment. Methods include surveys, interviews, focus groups, document analysis, and observation. The evaluation process involves identifying common themes, differences, and contradictions, followed by descriptive and normative analysis. Results can be presented according to the selected normative model, emphasizing areas of dissent, unity, and shared views. Numerical values should always be accompanied by explanations and reasoning. Organizational evaluation and development are intertwined. Evaluation serves as a development intervention, guiding the choice of development methods and supporting continuous improvement.

REFLECTION QUESTIONS

When was the last time your organization conducted a cultural assessment? What methods were utilized?

Has your organization conducted a cultural self-assessment as well as an external assessment? Were the results similar?

Is there adequate trust in your organization to conduct an internal assessment of culture for safety?

What characteristics of the signs of degrading culture for safety do you recognize in your organization?

What characteristics of degrading culture do you recognize in your organization that have been salient for a long time, but for which no corrective actions have been carried out?

How does your organization currently monitor its culture and the potential safety effects of culture?

TABLE 11.5
Danger Signals of Degrading Culture for Safety (revised from Reiman & Rollenhagen, 2018)

	Senior management expectations
Low priority of safety	Safety is of little importance in decision-making. Short-term profits or cost issues dominate management's attention.
Unclear expectations toward culture	The line between acceptable and unacceptable behavior is blurred or varies between individuals. Staff and managers can behave unpredictably and even in a hostile manner towards others, without such behavior being considered exceptional.
	Deficient understanding of safety and risks
"Not my issue" mindset	Safety is seen solely as the responsibility of the safety organization or as something that is only done to please the authorities. The safety function has no authority to influence matters.
Superficial and linear view of safety	Safety is seen from a simplified and linear perspective, for example, as reliability of components, or lack of human errors. Issues are corrected superficially, without addressing the root causes. Safety is measured with one indicator.
"Bad apples" mindset	When something negative happens, the organization tries to find someone to blame. Corrective measures consist of measures aimed at individuals, such as warnings, relocation, or dismissal. Additionally, training and more detailed instructions are proposed.
"One safety" mindset	Safety is associated only with occupational (or any other type of) safety, and other types of safety are neglected.
	Lack of mindfulness and questioning attitude
Ritualistic behavior	The risk assessment is only used to demonstrate that the current risk management measures are adequate, not to identify new risks and design new controls. Rules may be enforced even though they make no practical sense.
Control focus	The tolerance of uncertainty is low among the personnel. There is a strong focus on finding permanent solutions to unclear issues. Discussions without immediate corrective actions are discouraged. Quantitative measures are emphasized over qualitative.
"One size fits all" mindset	The same standard solution is applied to all problems. New events are not considered a sign of new problems. The organization learns to repeat things but does not develop new capabilities.
"Issue solved" mindset	The organization believes that it has already developed all the safety management measures it needs and that it will successfully apply them. Safety is seen as a static state that has already been achieved.
	Inadequate management of changes
Normalizing	Many warning signals, alarms, or deviations are ignored because they are repeated so often. They have become normalized and are seen as part of everyday work.
Chaotic development	Organizational development is not systematic. The organization is changed without examining the effects of the changes on safety. Change projects are not critically evaluated.
Drift	Different department- and team-specific operating methods have formed in the organization that have not been documented anywhere. Official guidelines and practical activities do not correspond.
	Lack of open and safe climate
"Do not rock the boat" mindset	The organization is not open to differing views on how work should be done or what risks it entails. Employees are not encouraged to speak or suggest improvements. People don't like to talk about mistakes.
Taboos	There are a lot of taboos in the organization that cannot be discussed.

(*Continued*)

TABLE 11.5 (CONTINUED)
Danger Signals of Degrading Culture for Safety (revised from Reiman & Rollenhagen, 2018)

"Good news" culture	People only tell good news to their superiors. This can be because they believe that is what the management wants to hear or because of previous negative experiences from telling bad news.
Fear of negative consequences	Punishment is feared if one reports a safety problem, human error, near-miss, or dangerous situation. This may lead to fewer reports of near-misses compared to reports of actual incidents and to actual incidents not being reported.
	Passive approach to safety
Reactive mindset	Problems are fixed only when something negative has already happened. Development is not based on early warning signs or observations.
Recurring issues	Safety observations remain open for long periods of time without adequate corrective actions, or they are transferred from one person and unit to another without anyone taking responsibility for the measures.
Safety apathy	The organization does not encourage new ideas or suggestions from employees. The person presenting the problem is also expected to propose corrective action or not present the problem in the first place. The organization is condescending of its negative conditions and does not seek to improve.
	Suboptimizing and organizational politics
Local optimizing	Departments and teams develop their own practices without considering their impact on others. The aim is to complete different indicators without considering whether the measured issue is achieved (e.g., near-misses are reported if they are requested, not if they occur).
Politics	Safety issues are used as an excuse to promote other benefits such as better salary, working hours, or facilities.
	Deficiencies in organizing for safety
Paper safe	The safety management system exists only on paper and does not guide the actual operations. Procedures and instructions are not followed in real life.
Firefighting mentality	The work is spent dealing with issues that are considered acute, leaving no time for development, for example. Learning is adversely affected when temporary solutions and quick fixes, initially intended as provisional measures, become permanent.
Fragmented approach	The organization lacks systematic planning and management practices for safety development.
	Excessive internal focus
Isolation	The organization isolates itself from the environment, does not share information with others, and does not try to learn from others.
Defensiveness	The organization does not accept criticism from outside experts. External evaluations are labeled "unduly negative" and dismissed.
Vicious circles	The internal dynamics of an organization create a vicious circle in which small problems or discussions quickly escalate into big problems and conflicts.
Groupthink	The internal dynamics of the organization create a complacent picture of the organization's own competence. Groupthink leads to an underestimation of external cooperation and an overestimation of one's own competence.

RESILIENCE ENGINEERING AND SAFETY

Resilience as a concept has been used in many different fields of application, from its original meaning of the property of material to psychological, organizational, societal, ecological, and global level issues. In the early 2000s, resilience was adopted in the safety field by the formation of Resilience Engineering (RE) by Hollnagel et al. (2006). In essence, their mission parallels that of the HRO school (see Chapter 5): to find out why complex systems do not (often) fail. Whereas HRO remained mostly descriptive, RE also asked the follow-up question: How can we design (or engineer) such systems?

Erik Hollnagel has defined resilience as "the intrinsic ability of a system to adjust its functioning prior to, during, or following events (changes, disturbances, and opportunities), and thereby sustain required operations under both expected and unexpected conditions". Pettersen and Schulman (2019) introduced the concept of *precursor resilience*, which they differentiate from *restoration resilience* and *recovery resilience*. According to them (Ibid., p. 461), precursor resilience is about monitoring and keeping operations within a bandwidth of conditions and acting quickly to restore these conditions rapidly as a way of managing risk. Restoration resilience refers to rapid actions conducted to resume operations after a disruption, and recovery resilience, in turn, to putting damaged systems back together to establish a "new normal" at least as reliable and robust as before, if not improved.

Hillmann (2021) conducted a literature review of organizational resilience, and she identified five distinct disciplinary perspectives to resilience in organizational settings: 1) ecology, 2) safety and reliability, 3) engineering, 4) positive psychology and organizational development, and 5) strategy. She places Normal Accident Theory and High Reliability Organizations theory to the second category and Resilience Engineering into the engineering category.

According to the review of Resilience Engineering (RE) by Le Coze (2019b), the roots of RE lie in cognitive systems engineering, human factors, cognitive ergonomics, and system engineering. RE also relies heavily on systems theories, especially cybernetics and complexity theories (see Chapter 1). According to Le Coze, RE relies on Rasmussen's (1997) ideas of "adaptive entities producing self-organized patterns through trade-offs in a space of resources and constraints (within an envelope)". Variability, adaptation, trade-offs, and expertise produced in daily activities are some of the key concepts used in RE (Le Coze, 2019b).

RE emphasizes the ability of organizations to operate flexibly in varying circumstances – the ability to react to various disruptions while maintaining the ability to carry out the basic mission. Every level of an organization must be able to respond to both regular and irregular threats, to monitor these threats and, if necessary, to change its perception of risk, anticipate threats and distractions, and learn truthfully from what has happened. Resilience is therefore about an organization's ability to face and adapt to change. RE is a multidisciplinary paradigm that aims to promote the reliability of sociotechnical systems by contributing to their design and management.

According to the basic tenets of RE, failures and successes are closely related phenomena and not incompatible opposites. This adaptability and flexibility of human work is the reason for its efficiency as well as the reason for why failures occur (Hollnagel, 2006, p. 13).

Cornerstones of Resilience

Hollnagel (2009b) defines four cornerstones of resilience that organizations can use to navigate complex and unpredictable environments. These cornerstones help systems adapt, learn, and improve in the face of challenges:

Anticipation focuses on the ability of a system to foresee potential future threats, risks, or opportunities. It is about knowing what to expect and recognizing signs that might indicate emerging issues or changes in the environment, which could impact system performance. This cornerstone emphasizes planning and preparation. This cornerstone is close to the idea of precursor resilience (Pettersen & Schulman, 2019) and is similar to what HRO theory proposes (see Chapter 5).

Monitoring involves continuously tracking system performance and the surrounding environment to detect any deviations, potential risks, or changes in context. Monitoring is about knowing what to look for. It allows an organization to stay aware of its current state and helps identify whether any corrective actions are needed. Monitoring often relies on data, indicators, and trends. The organization needs to know what to watch. This cornerstone also falls under the concept of precursor resilience (Pettersen & Schulman, 2019) and shares similarity with HRO theory's "sensitivity to operations" (see Chapter 5) but with a wider focus on the system and its boundaries.

Responding refers to the capability of a system to take action and be able to respond to regular and irregular changes, disturbances, and opportunities. It's about dealing with disruptions and managing unforeseen events effectively. The focus here is on flexibility and agility, ensuring that the system can adapt and respond to challenges in a timely manner. The organization has to know what to do in different situations. Depending on the nature of contingency, this cornerstone refers to either regaining functionality after an anticipated event or improvising in case of an unexpected event. The resilience advocated by the HRO theory resembles this cornerstone.

Learning from both successes and failures is necessary to improve future performance by increasing the repertoire of responses. This cornerstone involves gathering knowledge from past experiences, analyzing what went wrong (or right), and applying those lessons to prevent future issues or enhance resilience. Continuous learning helps systems evolve and become more robust over time.

These cornerstones emphasize that people in complex systems are constantly adapting to changing circumstances while learning about both the circumstances and the system performance. They can be seen as system-level abilities for managing external disturbances and internal performance variability by attenuating unwanted variability and amplifying useful variability to align everyday adjustments with system goals and constraints.

Critique of Resilience

In RE, attention is drawn to the ingenuity and adaptability that professionals in high-risk industries display in the continuing dynamic process that is needed to maintain daily operations. Safety is developed together with the operational capacity of the entire organization. However, Resilience Engineering does not make a very clear distinction between different types of safety (Hopkins, 2019), a challenge discussed in detail Chapter 4 of this book. RE has been applied extensively in healthcare, a context where it seems especially suitable considering the complexity of the healthcare system, its multiple goals and tensions, and the nature of patient safety as both a local and system level property (see Chapter 4).

The Resilience Engineering school has also been criticized for neglecting the teachings of earlier HRO theory and reinventing the wheel (Hopkins, 2014, 2019; Le Coze, 2022)[8]. As was illustrated in Chapter 5, HRO theory emphasizes compliance with instructions in normal work combined with the ability to react to unexpected and uninstructed situations. The cornerstones of resilience are also quite close to the HRO hallmarks. RE puts more emphasis on positive organizational capabilities and learning from normal work but otherwise it mainly complements the hallmarks. RE has also devoted little attention to issues such as culture or power (Le Coze, 2019b), whereas culture is a central concept in HRO theory and power has been extensively debated in connection with Normal Accidents Theory (see Chapter 3).

Finally, there is also some confusion about the relationship between resilience and safety. Some scholars, such as Hollnagel, seem to consider safety (in the broad sense) to be a similar feature to system resilience. Hollnagel has named this Safety-II (Hollnagel, 2014). On the other hand, systems can be resilient but not safe (Morel et al., 2009) or safe but not resilient, at least if we utilize the view of safety as an absence of accidents.

In my view, being resilient means being aware of the various functional boundaries (safety, quality, resources, profits) of organizational activity, being able to recognize how the organization is currently performing, monitoring whether one's model of the environment, safety and hazards is adequate, and steering the organizational processes in the necessary manner. These processes include recovery from incidents or even accidents. The ability and willingness of the organization to carry out its core task is thus at the core of resilience. From the perspective of resilience, the key to preventing accidents is in supporting personnel in adapting to the daily trade-offs and in accentuating the positive capacity in the organization, and further, in understanding how the organization functions and realizing the gap between formal and informal organization, making the functional boundaries visible, monitoring the changes in the boundaries, and finally recognizing sources of variability and potential combinations that create and change hazards.

Designing for Resilience and Safety

Resilience is not an attribute but rather a capability that is developed. Some scholars (Woods, 2018a) highlight that resilience is a verb rather than a substantive.

Vogus and Sutcliffe (Vogus & Sutcliffe, 2007) describe resilience as being vitally prepared for adversity by having a collective reserve of capability that allows a team to make adjustments when challenged, rebound, and emerge stronger. Their description of resilience includes both the ability to course correct and the ability to rebound when challenged, often referred to as the latent and active aspects of resilience.

Woods (2019) has proposed several *essentials of resilience*, including adaptive capacity, graceful extensibility, trade-offs, initiative, managing the expression of initiative, and reciprocity. Adaptive capacity refers to the system's potential for adjusting patterns of activities to handle future changes in the kinds of events, opportunities and disruptions experienced. If the system has adaptive capacity, it is poised to adapt, it has some readiness or potential to change how it currently works – its models, plans, processes, and behaviors (Woods, 2019).

Another key concept is *brittleness*. Woods emphasizes that failures are due to brittleness. By brittleness he means how rapidly a system's performance declines when it nears and reaches its boundary. The opposite of brittleness is graceful extensibility – the ability of a system to extend its capacity to adapt when surprise events challenge its boundaries (Woods, 2018b).

Resilience brings an important concept to safety by emphasizing first that *resilience is something that the system does*, and second, by refocusing attention to *how the system normally functions* and what are the *trade-offs, sacrificing decisions, and tensions* that people in the system need to deal with daily in order to succeed.

We proposed the core task requirements of anticipating, reacting, and monitoring in 2003 (Oedewald & Reiman, 2003), with learning being an instrumental requirement needed to accomplish the three critical requirements. These can be considered generic requirements for complex organizations, but how critical they are for the activity depends on many factors, including the core technology, the operating environment, and the type of hazards (see Chapter 10). This book builds on the idea that culture should support accomplishing the core task. The core task requires a certain degree of anticipation, monitoring, and reacting, and the culture needs to support the learning and improvement of these capabilities.

Resilience Engineering, in my view, does not adequately separate what the system needs to do in order to survive (the core task), what the system is habitually prone to do (culture), what is threatening the system (hazards), and finally, what is the selected risk appetite and way of managing the system (strategy).

NOTES

1. Even more so for the concept of "safety culture", see Chapter 5 in this book.
2. Applied social and behavioral sciences are not deterministic sciences. Behavioral scientific theories seldom provide a means for making causal attributions or predicting behavior. Thus, cultural assessments can never provide a quantitative and precise measure of the culture or a causal prediction of a future chain of events (i.e. activities leading to accidents or safety).
3. However, normative models are useful for identifying neglected issues if used after the open-ended descriptive phase has been conducted. Utilizing the models as checklists without conducting the descriptive part can result in the assessor issuing a non-conformity for some specific area, but without the realization of why the issue had been neglected.
4. The basic idea of the survey methodology in behavioral and social sciences is to map specific phenomena with several questions that deal with different aspects of the same phenomenon. From these questions, so-called summated scales are formed, which consist of average scores from multiple individual questions. Factor analysis is typically used to identify the survey items that correlate (and thus measure the same underlying variable). The average value of the sum variable represents the measured phenomenon (with margins of error).
5. There is evidence that national culture influences how people answer surveys (Johnson et al. 2005; van Herk et al., 2004). Comparing scores between organizations based in different countries can thus lead to wrong conclusions as differences are due to the nationality of respondents rather than the cultures of the given organizations.
6. Antonsen (2009b) found in his study of an offshore drilling platform before and after an accident happened there that a pre-accident safety culture questionnaire failed to detect safety issues that were identified after the accident.
7. Interviewees may also lie for one reason or another, or deliberately fail to disclose certain issues. For this and the above reasons, the key is to identify repeated observations, rather than those of one individual.
8. Le Coze (2019a) illustrates how both theories actually developed in parallel, but did not interact very much in the early phases.

12 Creating a Culture for Safety

Creating a culture for safety in complex organizations requires that we reflect on how we understand safety and how we see organizations and the human role in assuring safety. Assuring safety in complex systems necessitates a more nuanced approach than seeing safety merely as following the safety rules or avoiding errors. We need to recognize the inherent complexity of our organizations and their natural tendencies to evolve and adapt. We must also be mindful of the assumptions and simplifications that members of these organizations form about their environment and themselves.

Safety in complex organizations is not just the absence of incidents but the presence of adaptive capacity and proactive processes. For organizations, this means that the focus of management should be on the entire system, not on safety as a separate or independent issue. The focus needs to be placed on supporting humans at work by creating better preconditions with structures, technology, and organizing and facilitating everyday work with adaptive leadership. It means building on and enhancing the strengths of humans while organizing to prevent weaknesses. Furthermore, it has to be based on a solid understanding of the organizational core task and the associated hazards. This chapter summarizes the main lessons learned and proposes a shift in the perspective from which we approach safety in complex organizations.

The cultural approach is more a way of looking at organizations than a specific method of managing, assessing, or developing them. A cultural approach is very much about indirect effects and indirect influences. Culture cannot be directly changed but it can be influenced indirectly and with time. Understanding these influences and delayed effects is at the root of systems thinking. It is also at the root of understanding change in complex organizations.

12.1 CULTURE CHANGE

Culture change is often proposed as a solution to almost any problem organizations face, be it deficient quality, lack of customer focus, inefficiency, or apparent neglect of safety issues. This has become such a standard response that "culture change" nowadays does not refer to any specific type of change, as it can mean almost anything in practice. And it often also leads to nothing.

Often organizations would be better off without these superficial "culture change" initiatives. In fact, instead of "culture change" campaigns and development projects, it would often be enough to just view the problem *from a cultural perspective*. This means understanding the dynamics that have created the problem and designing practical solutions that take the system and its dynamics into account. It can also mean revising the understanding of what the actual problem is (e.g. moving from individual and linear explanations to systems-level understanding).

Only when the problem is rooted in the habitual ways of the organization, in the shared meanings attributed to various issues or in the underlying logics of action, is a change in the deeper elements of culture necessary. Even then, the organization needs to understand what culture change means, and what it requires from the members of the organization.

What does it mean to change culture? The answer to this depends on whether we view culture as a process or a structure (see Chapter 1 for a description of these views). From a process view, changing culture means changing how people interact and make sense of events, how they rationalize and explain what has happened, and how they attribute meaning to structures, events, and behaviors.

From a structural perspective, culture change means that shared assumptions or underlying logics change. As White (2021) put it, "to change culture means changing logics". To ensure safety, often it would be enough to be aware of and understand the logics rather than seek to change them.

DOI: 10.1201/9781003491156-12

And when the organization seeks to change the logics, or underlying assumptions, it is best done by developing the practices around the core task and internal integration (White, 2021).

Culture change takes place on multiple levels of the system. White (2017, p. 284) has summarized:

> Successful and sustained culture change is a complex, multifold two-dimensional process involving, on one axis, interventions at every level of an organizational system: individual, group and whole system. At the other [dimension], it entails changes in language and other semiotic structures (symbols, brands, logos) as well as physical layouts, but, most of all, changes to prevailing practices. But it begins with exposing and making sense of underlying schemas.

I would add technology and tools to the list of things that require changes when changing culture. Sometimes technological changes fail due to the lack of cultural considerations. For example, new technology changes work, ideas of professionalism, and power relations in the organization. These can cause change resistance the reasons of which are difficult for management to understand without adopting the cultural perspective.

A typical pitfall in considering culture change is to focus on surface level "changes" such as climate or official practices. Both can, in the long term, affect the deeper levels of culture but as such do not yet indicate cultural change. Culture forms by people interacting around a shared task. To change culture, interaction, work, and work practices should be the targets. This change should support the core task and consider the context in which the organization operates (see Chapter 10). Additionally, any change should be informed by a thorough understanding of the prevailing culture (Chapter 11).

There is often an underlying assumption that to improve organizational performance or safety we always need to implement new things or come up with novel solutions. In fact, as Collins (2009) has also noted in his analysis of major companies that failed financially, one symptom of a path to decline is frantically trying new things in search of the saving silver bullet. In a crisis, as in any learning situation, the organization should investigate its existing practices and ask whether these are up to date, or if they require change. The organization should also ask what obsolete ideas, assumptions, and routines they have that need to be eliminated. Maybe some of the safety initiatives have met their due date and need to go. Or maybe they have never been beneficial to safety, but have produced the right numbers for the management, at least until recently. Thus, change should be about changing the old, not just adding new (see maturity models of safety culture in Chapter 5). Changing the old is demanding and requires time and practice. Russell Ackoff, one of the systems thinkers introduced in Chapter 1, has reminded us that the only thing more difficult than starting something new is stopping something old.

When considering changes in practices, the following list of basic action types based on Georg Henrik von Wright's theory of action can help to identify the needed actions:

- Constructive or producing actions, such as a completely new practice or tool;
- Transformative or changing actions, such as changing an existing practice or modifying an existing tool;
- Preservative or maintaining actions, such holding on to good practices and well-functioning structures;
- Destructive or stopping actions, such as getting rid of existing practices or structures.

A typical mistake in development is to look at issues in isolation. Systems thinking (see Chapter 1) demonstrates how the focus should be on interactions and connections in addition to individual elements. Thus, when developing a certain practice, it may not be that practice that requires change but some related practices and structures that will indirectly affect the practice that needs to change. Change happens best by creating environments that support the kind of behavior and practices that (we believe) will contribute to safety.

12.2 CORNERSTONES OF CULTURE DEVELOPMENT

I have summarized the following cornerstones of culture development based on my previous research (Reiman et al., 2015; Reiman & Rollenhagen, 2018; Viitanen et al., 2018) and inspired by Carsten Busch (2021), and Alvesson and Sveningsson (2024).

1. *Do not focus only on "culture"*

Paradoxically, the first principle of developing culture for safety is to not focus directly on culture. Instead, develop the organizational prerequisites, working practices, and structures as part of the overall development of the organization. Also, do not separate "safety culture" development from development of the company culture. As I have argued throughout this book, "safety culture" and company culture refer to the same phenomenon. Develop a culture for safety, not a "safety culture". And when developing culture, do not announce that you will "develop a safety culture", but rather tell in as concrete details as possible what exactly needs to be developed and why. Avoid proclaiming that "in a good safety culture, people do not do X or Y". Rather talk about your concrete expectations and things that are important to you.[1] When discussing organizational issues, use the term "we", as in "our organization" and "we need to change". The concept of culture should be in your mind, and culture should be taken into account in the development plans, but when you communicate with personnel, adjust the message accordingly. The overall approach should be cultural, even if the actions are concrete.

The point is not to deny the importance of shared values, beliefs, and assumptions. On the contrary, their force is so strong that it is not possible to change them solely by talking about them. A leader needs to show what the values, beliefs, and assumptions look like in practice, and enable another way of thinking, feeling, and acting for employees. This needs consistent leadership as described in Chapter 7. A focus on abstract values often leads to an emphasis on the espoused values, or the official values and slogans, without a deeper effect on the internalized values, beliefs, and assumptions that in reality drive behavior.

Culture is developed best by simultaneously working with both the concrete and the abstract. This means that the focus of development should be equally on organizational structures, work practices, employees' understanding, and the collective mindset (attitudes, values, norms). The trick is in connecting the abstract and the concrete, in giving people the vocabulary to discuss their changing work and to view it in a new light, to attach new meanings to the structures, processes, and practices. Personnel need tools not only for dealing with the concrete, but also for sensemaking and learning (and unlearning).

2. *Consider integration, differentiation, and fragmentation*

Whatever the official goals of the change, organizational change is always interpreted by the employees. Culture affects these interpretations, but there is also large individual variance as some people dislike any change and some embrace it.

Culture is typically either one aspect that the change initiative aims to address or an aspect that needs to be considered when changing the organization. Alvesson and Sveningsson (2024) note that few if any organizational changes are "culture-free".

When developing organizational culture, one of the dilemmas concerns the issue of whether to aim at creating one shared culture or multiple co-existing subcultures. Probably both are needed. The important issues to discuss in the organizations are "what are the issues that need to be shared?" and "what issues can be decided locally and what issues need centralized management?"

When aiming to create a shared culture, it is good to consider what are the mechanisms that can create this "sharedness". Interaction, communication, and collaboration are the ways to learn together, but many organizations are limited in the capacity to interact due to size, different geographical location, or specialization in more or less independent functions. A strong shared culture is most likely impossible for these organizations. This does not mean that nothing can be shared, but that subcultures will form and attempts to develop culture centrally are always interpreted locally.

The existence of subcultures has implications for culture change; attempting to create a uniform culture might not be possible or even a desirable goal. Cultural heterogeneity implies that such attempts at influencing culture, which assume that culture is uniform, may result in unexpected outcomes because each subculture may respond to the culture change intervention differently (Viitanen et al., 2018). For example, how the various subcultures view safety and hazards in their work affects how they interpret and react to the change efforts. Cultural heterogeneity calls for a tailored or at least subculture-aware approach to culture development (Viitanen et al., 2018).

Very often subcultures are in fact necessary for the long-term performance of the organization. It is very typical for operating functions to be somewhat conservative, whereas engineering or R&D functions can and should be much more innovative. Consequently, managing cultural change often becomes a matter of achieving effective cultural integration.

The existence of conflicts and paradoxes in safety-critical organizations is not a negative phenomenon per se; rather, it is an inherent characteristic of any complex system (Chapters 1 and 4). Diversity, conceptual slack, and even conflict can in many ways contribute to safety (Reiman et al., 2015; Schulman, 1993; Weick et al., 1999). Any organization requires sufficient variety (e.g., in terms of interpretations) to be able to manage safety-critical activities or to facilitate learning (Antonsen, 2009c; Weick, 1987).

All cultures ignore issues they do not consider relevant (Turner, 1978). Strong cultures strongly agree on those issues, whereas less homogenous cultures can incorporate multiple views on what is relevant and what is not. However, having multiple views inside the organization is not the same as utilizing multiple views in decision making. The organization needs an inclusive culture built on trust in order to surface the views and build adequate consensus on topics where consensus can be reached and agreeing to disagree on the other topics – and do all this without silencing the opposition. Leadership (see Chapter 7) is in a key role in achieving this.

3. *Develop in a goal-oriented, contextual, and systematic way*

Culture should be developed in a goal-oriented and systematic way. A systematic approach to development is needed because changing the deeper elements of culture takes time. One of the challenges in organizational development is that different development activities have different effects on culture – sometimes even cancelling each other out. Seeking a development of only one component (e.g., attitudes) can lead to a misbalance in the organization between the new elements and the old structures and elements. This in turn will most likely lead to the organization reverting to the old ways and learning only a negative attitude towards future improvement efforts ("not another safety campaign"). Some commercial safety culture development programs have a narrow view on safety culture as either individual behaviors or attitudes. These programs should always be complemented with other more systems-oriented ways of developing culture.

Being goal-oriented means also that the organization needs a vision of its future state. This vision and the development that follows needs to be based on an adequate view of the current culture, including its strengths and weaknesses. Systematic development is planned and resourced appropriately, and its effectiveness is monitored. Adjustments are made, if necessary, either to the development approach itself or to the desired culture. Discussions are needed to form a shared view on what good culture for safety means for the organization. Organizations must ease the unlearning of the old ways of thinking and acting and offer a safe space for practicing the new.

A goal-oriented approach does not mean that it needs to be quantitative. When developing culture, it is important to avoid seeking to optimize quantitative indicators (see Chapters 6 and 9). Goals need to be related to the organization's core task and strategy, and culture should support achieving those, not scoring X (or better than Y) on some generic (or very specific) questions.

Behavioral or organizational changes can be implemented within a short period of time, but a corresponding change in values and assumptions takes longer (Viitanen et al., 2018). When successfully implementing culture change tools or methods, their direct effects should result in behavior or structure change, but most likely it does not lead to immediate changes in values and assumptions. For this reason, culture development needs to be a systematic and long-term effort rather than a quick, one-off project. Time and consistency of actions are required to achieve cultural change.

The development of a culture for safety requires competence and human resources. In addition to these, it must be ensured that the development is in line with the organization's strategy, and not, for example, a "counter-campaign" by the safety department or another program separate from the general development of operations. Again, consistency and systematic approach on the level of the entire organization level is the key.

Sometimes organizations launch campaigns to change culture without giving the people in charge of them any authority to actually make changes. And even when there is some authority to, for example, propose changes, these changes are typically thought to target the front-line staff, not senior managers (or even middle managers). If the senior management is not prepared to reflect their leadership style and its influence on culture, culture change happens under heavy constraints. In this situation, a real and lasting change in culture is typically not possible, and the task of the culture development initiative must shift from "culture change", as the leadership perceives it, to gradually changing the leadership perceptions. This challenging task falls on the shoulders of safety professionals, occasionally backed up by HR and a few enlightened senior managers. Expect a long and winding road ahead in that case.

4. *Develop the system, not only individuals*

Culture should be developed holistically, with a focus on structures as well as practices, interaction, leadership, competence, and technology. Do not focus just on visible behavior, which is only an artifact of culture. We discussed in Chapter 5 how safety culture has often been misused in the industry to denote individual attitudes instead of organizational phenomena. James Reason (1997) already noted in reference to human errors that "we cannot change the human condition, but we can change the conditions under which humans work" (Reason, 2000, p. 769). Hopkins (2019) further notes that the culture of an organization is largely determined by its structure. Structures have a significant influence on both behavior and culture, and changing structures will affect culture. The trick is in changing them so that the effect on culture is positive rather than negative (e.g. hiding some risks, contributing to normalization and drift, oversimplifying the nature of safety; see Chapters 3 and 6).

Development of culture should build on people's strengths and weaknesses and use technical solutions when appropriate. Staff shouldn't needlessly compensate for poor planning or insufficient investments. Culture development should not be an excuse to avoid redesign of structures or technology (Rollenhagen, 2010).[2]

A common mistake when developing culture is to focus directly on individual behavior, rather than the conditions and structures influencing the behavior. This is due to a misunderstanding of what the emphasis on humans, or placing humans in the center of safety, means. Focusing directly on individuals and their behavior may end up putting extra pressure on them to perform in a dysfunctional system – making the overall human–organization–technology system even more dysfunctional. Thus, to achieve a lasting culture change, leaders need to change the work conditions, structures, interaction, tools, and maybe even technology. And they need to change the way they lead. One could argue that to change a culture, leaders rather than "followers" need to change – at least they have to change first.

It is also important to consider that one cannot target only one element of the sociotechnical system (e.g. management system, competence, or leadership skills) when leading culture change

(Viitanen et al., 2018). Organizations are adaptive entities, and change in one element typically does not lead to a lasting change. To change an organization, typically several things must do done, often on various levels of the system. On the other hand, trying to do too much often leads to different type of failure as resources do not permit enough focus on each development initiative. Thus, a balance should be found between attempting to change too much versus changing only few things.

5. *Acknowledge that culture development is a dynamic process*

The development of a culture cannot ever be fully predicted in advance. Culture is an emergent property of the sociotechnical system. It should not be oversimplified. Culture can be influenced by systematic work, but it cannot be measured quantitatively or controlled exactly. Culture evolves partially by chance, making its development and transformation inherently unpredictable and uncontrollable. Minor adjustments or random events can subtly influence cultural norms over time. For instance, modifying workplace seating arrangements due to the sale of office space under financial constraints (or other reasons) might cause new interactions among different groups while distancing others more than before.

It is important to work with culture, not against culture – unless absolutely necessary. Focusing on the strengths of the current culture and considering how to better attain the organizational goals with the current cultural values and assumptions may in many cases be more efficient than striving to change the deeper layers of culture. This also means that there will be something familiar left for the employees; if everything changes at the same time, it is very anxiety provoking, and as people have a need for clarity (Chapter 2), routines can form too early for many issues. A safe atmosphere and the opportunity to practice (and fail) are important in any change project, as it takes time and effort both to unlearn the old and learn the new. Practices do not change overnight, and cultures change even slower.

Culture development is always more or less an uncertain and unpredictable process, and no amount of planning and preparation will completely remove the uncertainties. There are several sources of uncertainty, starting with the complexity of the organization and its culture. Any view of culture is always incomplete, and as has been emphasized in this book, any attempt to change culture is also an opportunity to learn about the culture. Likewise for the ideal culture: the view of the wanted cultural dimensions, meanings, values, beliefs, and practices changes as understanding of the current culture improves. Finally, unexpected events and coincidences may offer opportunities to build culture or setbacks to be recovered from.

An iterative approach means also being sensitive to employee interpretations and general reception of the core message. It requires reflecting on the reasons for employee change resistance and adapting either the approach or the content depending on the need. Despite being iterative, there must be enough consistency in the approach that people recognize the direction. There also needs to be repetition and reinforcement of the core message, with small variations to maintain interest.

Kurt Lewin originally proposed that "one can best understand a system by trying to change it". This understanding does not come automatically, but it requires active reflection on how and why the organization responds to the various change initiatives. When developing culture, there are always surprises and unanticipated side-effects. Thus, it is important to actively monitor culture and maintain readiness to react to surprises, and to use these findings for further increasing understanding of the current culture. Some of the surprises are due to the limited understanding of the current culture, whereas others stem from the complexity and inherent unpredictability of social systems.

As was illustrated in Chapter 11, conducting a thorough culture assessment provides a snapshot of the existing culture. Nonetheless, culture is always developing, and the snapshot needs to be updated regularly. Furthermore, assessments always include a degree of uncertainty concerning whether all essential aspects of culture have been uncovered. Active monitoring of potential changes, overlooked aspects, new risks, and developments is needed, whether the culture is intentionally developed or not. Conducting comprehensive cultural assessments every year is typically unnecessary unless active monitoring indicates the need for more detailed analysis.

6. *Realize that you are part of the problem (as well as part of the solution)*

Some leaders may expect the culture to change just by modifying a few of the visible artifacts such as policy, procedures, or processes. Such actions rarely lead to long-term change alone, because the underlying values and basic assumptions remain the same. Even more, these actions miss the fact that as has been highlighted in this book, leaders are part of the system they are trying to change. This top-down approach to culture change may manifest as managers talking about the workers as needing change: "they" need to change "their" culture. Senior management can also talk about culture as an object out there in the field. The paradox of culture change is that leaders are part of the system they are supposed to change, but being part of the system, they actually have the best possibility to do it. For this reason, culture change should never be left for consultants or even the HR department alone, without ownership and direction from the line organization.

Culture change should be an effort that the entire organization takes part in. This includes the leaders and senior management. There is a big difference between a manager saying "you need to change" and saying "we need to change". Similarly, if managers refer to culture as an external force, "it", this distances people from the phenomena and may hide the fact that they are the ones who produce and reproduce culture. As in social identity in general, the feeling of "we" is essential to culture development.

As we discussed in Chapter 7, one of the dilemmas of leadership is that in order to gain (ideological) influence in a group, the group must perceive the person as "one of us", a manifestation of the kind of a person they are or want to be. For this reason, the personal example of a leader is so important especially when aiming at culture change. This also means that culture change cannot be driven by consultants (who will never be "one of us") or even HR (who are "one of us" only in terms of the organizational level social identity, not department or function level), but rather by the leaders of the various functions and departments, all the way down to the supervisors and informal leaders in teams.

Due to the paradox that culture can best be changed by insiders who may not be in the position of having the adequate requisite variety (of imagination and competence) to accomplish the needed change, outside help is often needed. Outsiders can bring in diversity and reflection concerning current cultural assumptions as well as systematic development methods. They should not, however, dictate the direction of the culture change.

To change culture, the deeper levels of culture need to be addressed in parallel with the surface levels. This takes time and requires consistency in the actions and words of leaders at all organizational levels. To change a culture, typically a change in leadership practices is also needed.

7. *Realize what change means to professional identities in the organization*

Professional identity (a type of social identity, see Chapter 2) is an important part of who we are at work. Any changes seeming to threaten our possibilities to carry out our work as well as we wish, and in a manner that we think is best, is seen as threatening. Professional identity is often connected to the organization and the professions at large rather than the specific unit or department where one is working (Chapter 2). Local change efforts may thus be resisted if these are seen as contrary to the best practices of the profession (cf. Sennett, 1998). To understand resistance to change, it is essential to understand the process by which professionals within your organization establish their professional identities.

Professional identities are frequently constructed around specific tools, methodologies, and practices. Efforts to modify or challenge these established practices may be interpreted as a threat to one's professional identity, often resulting in strong emotional resistance. This phenomenon is visible also in the practice of safety, where challenging certain methods or models (Behavior Based Safety and Safety Culture come to mind) raise emotional reactions from those that have built their careers and identities around them. This is an understandable reaction to a threat, and the countermeasure is not to increase the strength of the attack but rather help people cope. A perceived threat to one's own identity is highly stressful (George et al., 2023).

Very often, senior managers plan the new strategy and decide on the official values and the needed culture change without involving the personnel in this process. This may work if the senior leaders implement the change themselves first, but often the assumption seems to be that it is the worker level that needs to change, not the management level. In fact, culture for safety should support the development of a healthy professional identity of the practitioners at the various levels of the organization. This is very difficult to achieve without involving the professionals themselves in the change process.

Some culture-change efforts suffer from a compartmentalized approach. This means an approach to culture change where one department or one professional group designs and implements the culture change. This change may be effective within the group, but if other groups are not involved, the results on the system level may not be as positive as those on the group-level were.

Various silos and communication problems are typical findings in any assessment of (safety) culture. These issues need to be set in context, as there is a danger of relying on too easy solutions if we do not understand the cultural, task-related, and profession-related factors that contribute to silos. Silos need to be understood as a natural consequence of how we organize our world and interactions, as well as how our professional structures have been built. Thus, we will probably never get rid of silos completely, and in fact we should not. But we need to make sure they are not closed from the outside world, that information is shared, collaboration is possible, and there is sufficient common language.

Most people inherently resist change that they have no control over, or that they have not decided themselves. This is a natural and perfectly understandable response and highlights the importance of involvement and communication in any change effort. The "not invented here" syndrome works on a local scale as well. There are also many other reasons to resist change. Organizational politics is one of them. Changes can threaten the existing power structures in the organization, and those with more power will resist the change whereas those with less power will welcome the change – but they lack the power to positively influence it.

8. *Be mindful of the tools and models that you use*

According to Viitanen et al. (2018), some of the most common types of tools that are used for "safety culture change" include trainings and seminars; influencing behavior using, for example, Human Performance Programs; increasing the awareness of safety or safety culture through promotional materials; building employee involvement and participation; and developing organizational structures with the intent of ensuring a good safety culture. They (ibid.) note that these tools are designed and implemented based on specific assumptions about their effectiveness and impact.

Developers' assumptions about safety, culture, organizations, and leadership can influence tool development. These assumptions, whether explicitly stated or not, become embedded in the tools (see Chapter 1) making the tool an artifact of the developer's culture. However, there is always a translation process when the tools are applied by others. As we have previously noted (Reiman & Viitanen, 2019b), the assumptions embedded in the tools and methods gradually change as practitioners adapt them to their particular context. This can result in a change of the original message embedded in the tool. Gradually, new assumptions are formed and embedded into the community's culture of how the tool is used and understood. The tool starts to mean something else to its users than the designer intended.

This can lead to a gap between the developers' intentions and the end-users' understanding, possibly resulting in improper use or ineffectiveness of the tools (Viitanen et al., 2018) but also innovative and new ways of using the old tools. Thus, it is important to understand assumptions embedded in our tools and methods. When aiming to develop culture, one should carefully consider the benefits and limitations of various development methods and tools. For example, a focus on concrete behaviors works better with local hazards that are easy to perceive on the spot, whereas a focus on more abstract culture principles may be beneficial when hazards require integration of multiple sources of information to monitor and review, as most core task–related hazards do (see Chapter 4).

The concepts, methods, and tools very seldom contain the technical specifications for their use, or the "safety instruction" warning about their misuse. The translation of those tools into practice is very often left to individual practitioners or to consultants. This may lead to concepts that are originally meant to be used as analysis/modelling tools to be used in improvement, causal explanation, and so on. The original assumptions embedded in the tools and methods get lost, and gradually new assumptions are formed and embedded into the practitioners' culture of tool using. Questionnaires are a good example of this, for example, the use of mean scores, simplified questions, or non-representative samples. Another example is the entire concept of (safety) culture. Originally, culture in a sense of how Schein (1985) defines it referred to shared basic assumptions. Nowadays "culture" can be used in a multitude of different ways, from a management philosophy (a tool) or a causal factor in an accident investigation (see Chapter 5) to a dynamic social process (Figure 1.2) or an emergent structure (Figure 1.1).

The normative models of safety culture (see Chapter 5) need to be utilized correctly to benefit from their normative nature while retaining the descriptive nature of culture. Sometimes culture assessments are conducted using a "checklist approach", with assessors going through documents and practices guided by the normative model. This likens the assessment to a compliance audit, and that loses the point of culture assessment as a complementary approach to the existing safety management methods such as audits and inspections. Normative models should thus not be used as checklists or quantitative indicators.

When development is too strongly guided by normative models, it tends to reinforce the gap between Work-as-Imagined and Work-as-Done. This is due to the superficial approach of implementing ready-made solutions to an organization without considering the existing culture and current ways of working. The added value of cultural approach is that it can offer us a deeper and more systemic understanding of organizational reality. If we rush to normative statements and easy solutions, we risk losing this understanding. With an overly normative approach, we also risk missing the local adaptations that may in fact be safer than the non-contextual normative approach would dictate.

As discussed in Chapter 5, normative safety culture models typically include dimensions that are qualitatively different from each other and that focus on different levels of the sociotechnical system. Without understanding the differences and the interrelationships between the dimensions, interventions to the organization can be misaligned.

The normative models of culture also tend to emphasize the soft aspects of culture and miss the structural and technical manifestations. Focusing too strongly on the normative models can thus lead to using culture as an excuse for not dealing with technical or structural problems (Rollenhagen, 2010). Whatever the model of culture that is being used, it is imperative to ask *what else there is in addition to culture.* Some models of culture do not consider behavior an aspect of culture, but that does not mean that we should not pay attention to behavior at all when assuring safety.

The original "ladder model" or typology of safety culture (see Chapter 5) was developed to facilitate discussion and deepen the given organization's insights into safety (Westrum, 1993). However, it was quickly transformed into a tool for development and benchmarking (Hudson, 2007). When culture turns into an indicator to be improved and compared, the nuances of its unique nature disappear. One could argue that the "highest level" of safety culture is unique to each organization, not something that can be implemented from a checklist and audited afterwards.

Transforming phenomena into tools means reifying abstract concepts into concrete management actions. If the phenomenon in question is converted into a tool and development focuses on implementing that tool, the connection to the original concept may be lost. Concepts of safety culture, just culture, and psychological safety are examples of these dual-use items, which may refer to either an underlying phenomenon or to a management approach. As with dual-use items generally,[3] one has to be careful that these are not misused.

9. *Focus on meanings and assumptions as the core of culture*

Alvesson and Sveningsson (2024) propose that when changing culture, it is better to focus on meanings instead of values. Values are often too abstract and too positive to be utilized when discussing the daily trade-offs and conflicting priorities at work. More valuable than the values themselves are the meanings that employees attribute to them. In terms of safety, this would mean that instead of advocating safety as the priority value, culture development should seek to make safety and the associated practices meaningful for the personnel. This in turn requires becoming aware of the current meanings and assumptions that people hold about safety in the organization. Similarly, if the topic of development is leadership, a culturally oriented development program would investigate the current meanings attributed to leadership and take these as a starting point for development.

This does not mean that organizations should not talk about values. On the contrary, values are a good starting point for discussions about what is meaningful and important in work. Safety, teamwork, collaboration, openness, customer service, quality, reliability, and so on can all be values in an organization, but these concepts are given multiple meanings by employees based on their local cultures, as well as by the various stakeholders. Without localizing these values, there is a risk of miscommunication. Localization also shifts the focus to practices in addition to meanings. Practices are the way values can manifest in daily life, and thus if values give the direction, practices give the means.

Sharman (2019) notes that a key aspect of enabling people to support a change is for the change leaders to consistently and continually explain why the operations, task, or goals have changed – and ensure that all workers are fully cognizant to the reasons why those changes are taking place. Similarly, White emphasizes that

> leaders must connect to what is already meaningful, and do so in meaningful ways. And what is meaningful is usually rooted in the dominant, but often tacitly assumed professional orientation or strategic task focus of the collective.
>
> **(White, 2017, p. 276)**

Leaders need to make safety a clear part of the core task of the organization and an aspect of professionalism among the personnel.

The reciprocal causality of technology and the human elements of the system (Reiman & Rollenhagen, 2011) need to be considered when considering safety from a cultural perspective. People create technology, structures, and processes, which in turn influence how they, and others, think, feel, and act. The members of an organization assign meanings and beliefs to organizational elements (structures, systems, tools, and others' behavior) and these assigned meanings in turn influence the ways in which the members behave (Alvesson, 2002; Hatch & Schultz, 2002; Weick, 1995).

10. *Development must be connected to the organizational core task and the actual work*

Developing culture, or change in general, is not an end in itself. If safety is not a genuine goal in the organization, it does not make any sense to try to develop a culture for safety only because some industry standard or best practice mentions "safety culture". The organization first needs to realize why culture matters and why safety is an issue worth pursuing, and how it all relates to the organization's strategy. After that, the organization needs to realize that creating a culture for safety means that also the senior management must change their practices and their assumptions.

Consider what are the other strategic objectives of the organization and what constraints and requirements the organizational core task sets for the organization. How much innovation and flexibility are needed in carrying out the core task? How much uncertainties are there in the operational environment and in the core task itself? Is the core technology mature? What is the life cycle of your sociotechnical system, if that is applicable (e.g. design, construction, operation, decommissioning)? These factors imply both what the specific safety challenges are and what are the operational and strategic realities that safety needs to be integrated into (see Chapter 10 for more details).

Change is not an end in itself. When creating a culture for safety, there is a danger, as in leadership in general (Haslam et al., 2024), of overemphasizing the need for change. When we are dealing with safety and the associated control measures and barriers, as important as it is to continuously develop, we also need to maintain the issues that make work possible and safe now. We need to make sure that the attempts to develop culture do not work counter to the daily attempts to conduct the work. Sometimes we as leaders and safety professionals need to step aside and let people work.

Only change what needs to be changed. Again, change is not an end in itself, and if something is working well, perhaps it is better to leave it alone – provided that there is adequate proof it is actually working and will be working in future as well. As important as changing something is maintaining practices and structures that are currently working well. But to be able to maintain what is working, it first needs to be identified as a strength.

When developing culture, it is important to consider the current culture in the organization and how it has been formed, and how it currently contributes to safety. The challenges faced by development initiatives are different in a well-developed culture for safety than in an immature culture. For example, the complexity of the safety management system itself can hide certain risks and create a false sense of safety even though the risks are in good control at the moment. In a low-safety-level organization, the main challenges are overcoming the excessive production focus and unsystematic nature of organizational practices (Amalberti et al., 2005).

As Viitanen et al. (2018) note, change leaders should not distance themselves from the organizational members, but instead they should try to understand how the work is conducted in practice and how the various individuals and groups interact with each other. This can give insight into what are the driving forces that have temporarily resulted in the current values and assumptions. Consequently, this insight can help identify the leverage points for culture change.

11. *Mind your language when developing culture*

The role of language has already been discussed in relation to learning, leadership and rules (Chapters 6–9). When changing culture, one aspect is to learn a new language, new concepts, and metaphors (Anand, 2024; Alvesson & Sveningsson, 2024; Marquet, 2020). However, care must be taken that the language will eventually line with practice as well as with the deeper levels of culture, meanings, beliefs, and assumptions.

Organizations typically have slogans, and safety campaigns seem to love statements such as "all accidents are preventable" or "you choose to be safe", or "safety is up to you". Geller (2001) noted that "all injuries are preventable" almost seems like a slogan that safety professionals use to keep themselves motivated. He notes that if organizations combine this slogan with a *goal* of zero injuries and a *reward* for not having an injury, human nature will dictate covering up an injury if possible (see Chapter 6). As we established in Chapter 2, language, thinking, emotion, and action are all connected.

According to Nippin Anand (2024), the most powerful way to change a culture is to change the language that you use in your organization, for example, the words we use when we talk about safety, accidents, leadership, or risk. Credible and pedagogical examples (stories) that capture something in a way that appeals to experience are an important tool in culture change. These stories need to be locally relevant and connected to practice, otherwise they may suffer the same fate as many slogans and official values (Alvesson & Sveningsson, 2024).

Different interpretations of key terms and concepts are quite typical in any organization. These cause communication problems and may even lead to conflicts and frustrations (see Chapter 3). Among these key issues that seldom have a unified interpretation are safety, quality, human error, and accident. Culture change should challenge the existing interpretations and work toward a unified understanding of the key concepts. At the same time, it should be acknowledged that exact definitions and unified views will never be perfectly accomplished (and are not even the goal). As in cultural approach in general, the process of interpretation and reinterpretation is what matters. This process should be transparent, inclusive, and generative.

12. *Avoid an oversimplified and/or shallow approach to culture*

Culture development should not be a tick-the-box exercise that is run once and then forgotten. It needs to be a continuous practice in the organization. Values and culture need to be considered from recruitment to promotion and termination. They need to be discussed together in teams and working groups from time to time. Leaders need to reflect on culture and their leadership style regularly. Warning signs of a degrading culture need to be monitored (see Chapter 11).

A shallow approach to culture can manifest as, for example, development programs consisting of either training individuals at the sharp end or writing documents and polishing the management system at the blunt end. Paradoxically, a shallow change project that fails to consider the basic assumptions may not only have a limited effect on the current culture, but may also end up reinforcing the existing culture (Alvesson & Sveningsson, 2024, p. 160). For example, technology and engineering-driven organizations may have skepticism towards "soft" approaches to safety and see them as providing little added value. A superficial "safety culture" campaign focusing on safety values without a clear link to structures or practices may be interpreted as further proof of the futility of the safety culture approach.[4] Another very easily reinforced assumption concerns senior management's understanding (or lack thereof) of the workplace reality. An abstract development project implemented top-down without adequate employee involvement most likely confirms employee suspicions that senior management does not know the realities of the field, whatever new values and slogans the change project advertises. This example can also be seen as an example of a compartmentalized approach, if the senior management subculture is the only group involved in planning the change.

When considering problems or issues at the organization, they typically are influenced by several factors. For example, an organization that has difficulties in getting people to report near-misses should consider a) the usability of the reporting tool itself; b) practices of reporting, for example, while working or after the work has been conducted; c) employees' skills at observing work and the work conditions; d) employees' skills at making the report; e) employees' willingness to observe their work environment from a safety perspective; f) the climate of reporting, that is, whether employees feel they can report openly or do they fear consequences; g) how leaders at different levels intentionally and unintentionally affect employees willingness to report; h) how the reports are used in making changes at the organization, for example, do they lead to effective corrective actions; and i) whether employees are measured (and rewarded) on the quantity of reports or is the focus on the quality of reports. These all affect and are affected by the rationalizations that employees have made about the reporting process and the beliefs and assumptions they have concerning, for example, the importance of reporting, leadership, and its objectives as well as their beliefs about safety and risks in general. Before understanding the issue in more detail, corrective actions are like fumbling in the dark.

Often organizations react to various other findings by seeking to directly address the individual findings, without considering the entire system. This is the same problem as with indicators (see Chapters 6 and 9). For example, let's consider an organization that gets a low mean score on an employee wellbeing survey. Senior management decides that they need to devise a plan to "raise the scores". This is where things have already taken a wrong turn. One should not seek to raise the scores, but rather to improve employee wellbeing. If the measures do in fact improve wellbeing, the next scores should indicate that by being higher. But by seeking to merely raise the scores, the organization will most likely get at least moderately higher scores next year (e.g. by rewarding for higher scores or implicitly indicating that the next measurement needs to be better or there will be consequences), but the scores and the phenomena they are used to measure (employee wellbeing) have become disconnected. Similar phenomena take place when the management wants to lower lost-time incidents or increase the number of observations among personnel.

Schein (1985, p. 47) noted that just because the manifest environment of a company feels informal and open does not mean that the actual culture is informal and open. To understand culture and separate it from artifacts and temporary feelings, it is necessary to uncover the more stable (yet also slowly changing) aspects of the organization, the meanings, underlying logics, and assumptions. The deeper elements of culture are best revealed by a joint effort of knowledgeable outsiders and insiders (Schein, 1985) or a comprehensive self-assessment conducted by competent group of insiders (see Chapter 11).

In order to think culturally, we have to keep an open mind. This means being aware of our own biases and mindset, and respecting that others have their views, which are as real to them as our experiences are to us (Tett, 2015, pp. 311–312). We need to listen with an intent to learn.

12.3 SAFETY IN COMPLEX ORGANIZATIONS – A CHANGE IN PERSPECTIVE

This book proposed a cultural approach to safety in complex organizations. It differs from a traditional, mechanistic approach to safety and organizations in some important respects. The cultural approach pays attention to subjective issues such as meanings and beliefs, whether these are shared, individual, or something between. It facilitates and acknowledges both the diversity of views in organizations and sometimes their integration and homogeneity. The cultural approach further recognizes the role of artifacts such as management systems, official procedures, plans, tools, and technology in shaping interactions and influencing sensemaking within organizations. It is a systems approach, and the system includes people, technology, and their interaction and organizing. It advocates seeing culture not as a tool or a causal explanation (see Chapter 5) but rather as a dynamic social process (see Figure 1.2) and an emergent structure (Figure 1.1).

The following is a summary of key changes needed in thinking to transition from a component-based approach to a cultural and systems-based approach to safety. I propose a change in focus, not a complete abandonment of the old. In most issues, the question is not an either/or but rather yes/and.

From Individual Behaviors to Shared Practices

Culture is a property of groups, not an individual trait. It is expressed in behavior, and it can be seen in actions, events, and other physical material aspects but it does not refer to these elements as such. Rather culture refers to the meaning and beliefs these have for people (Alvesson & Sveningsson, 2024) and the underlying assumptions that the group has learned while solving its task and group-related challenges (Schein, 1985).

This also means that deviations and errors need to be looked at from a system perspective. Blaming individuals for human errors does not promote learning. Mistakes are typically a symptom of a lack of prerequisites for interaction and work, not just the cause of an individual's behavior. Simultaneously, organizations need to hold people accountable for doing their best and not tolerate intentional negligence or wrongdoing.

The human mind best perceives surprising and extraordinary things – the kind that don't live up to our expectations. If, on the other hand, our daily routines go as we anticipate, we can significantly save our precious cognitive capacity. However, we are bad at noticing incremental small changes in our environment. Our expectations are also the result shared learning – they are influenced by culture. In fact, culture is created and shaped when we act and communicate in our work communities in ways that seem natural, normal, and irrefutable to us.

The cultural approach means focusing on interactions, meanings, and preconditions for behavior rather than the actions and behaviors as such. Behaviors are naturally important by either reflecting culture or reflecting on culture, and sometimes there is a need to single out certain behaviors as dangerous or reckless. However, this should not be the main focus of the cultural approach.

From Hierarchical Levels to a System of Systems

Many safety management approaches, including some safety culture approaches, have an overly strong emphasis on the front-line workers (the so-called sharp end) instead of the so-called blunt end, including designers, developers, and senior management. On the other hand, some approaches have been criticized for the exact opposite issue, for example, for an overemphasis on organizational issues and the role of senior management in safety (and accidents). These different levels of the sociotechnical system are often considered in hierarchical terms, that is, senior management is above workers, and designers and technical people are somewhere between. Furthermore, an additional layer of middle managers and supervisors acts as a barrier and a potential latent weakness (cf. the Swiss Cheese model introduced in Chapter 4). This hierarchical view of an organization is challenged in systems thinking (see Chapter 1).

Generic cultural approaches are sometimes criticized for focusing too much on organizational level issues and neglecting the individual and the particular. This is a valid point. In addition to wider learning and development, individual cases should not be forgotten. There is always a need to attend to the particular, whether the "case" is hazardous machinery, a burned-out individual, or an unsafe method of work. A cultural approach needs to balance between the system-level interventions that take time but have a wider effect and more precise corrective actions that may be necessary in the short term.

Looking at organizations from a systemic perspective, the hierarchical separation loses significance. An organization can be considered a system of systems, with each work group forming their own system (and subculture). Senior management and all other members of the organization belong to the overall system, and as such they are not independent of the sharp-end actors. This also means that a focus on managers, designers, or any other specialist group does not yet make the approach a systems approach. The systems approach is not about where in the system the focus is, but rather in how the system is looked at and analyzed.

From Prescribing Routines to Mindful Organizing

Mindful organizations keep the discussion on risk alive even when everything looks safe. They do not take past success as a guarantee of future safety, but they acknowledge that their model of risk may be outdated or wrong (Dekker, 2019, p. 395). Routines do not need to be mindless, but they often drift into mindlessness. Mindless acts are automatic, routine ones that are merely customary (Weick & Sutcliffe, 2007, p. 61). Mindful application of routines allows people to fit the routine into its context and learn from it, if there is something to be learned. Organizations must continuously work to override the typical human tendencies (see Chapter 2) to seek confirmation and avoid disconfirmation (Weick & Sutcliffe, 2007). Culture should promote adaptiveness and reflection, not closed-minded rule following.

In a complex organization, no-one can understand all the interconnections and all the consequences of actions. This impossibility of understanding applies to both managers and experts. Humility instead of hubris is needed to manage a complex organization. Managers as well as experts need to admit that they depend on other people. There is a practical need for deference to expertise (see HRO in Chapter 5). At the same time, there is a need for understanding what expertise entails, and how different experts perceive the organization differently. We introduced the difference between Work-as-Imagined and Work-as-Done in Chapter 3. People who operate the technology or deal daily with the hazards have expertise in the Work-as-Done, which is as important as the more theoretical knowledge of those typically in charge of organizing and creating structures. Mindful organizing requires bridging this gap between Work-as-Done and Work-as-Imagined (or work as designed and structured).

One important way to develop a culture is to learn from successful and unsuccessful work performance. Learning does not always mean incident investigations; rather it is more about the continuous development of operating methods and culture at work. It is essential to try to understand the current practices and how these have formed. A blame-free and systemic approach is key to this. This also applies to the attitude towards various mistakes. From the point of view of learning and development, it is useless to blame individuals for mistakes, the real causes of which can usually be found in the activities and operating conditions of the entire work community.

To ensure safety, organizations need ways to monitor both the usual and the unusual. They also need to monitor the process in which the unusual changes into the usual, and vice versa. Mindful organizations pay attention not only to new or surprising events but also to the mundane, and the issues that are considered normal.

From Separate Dimensions to an Integrated Approach

It is necessary to shift from looking at failures of individual components to a systems view that looks at how the entire system functions. The safety of an organization is a system property, and it cannot be deciphered only from past outcomes (e.g., the number of incidents) or past behavior (e.g., rule following), but from its potential for future behavior. This means that management needs to focus on building organizational capability for recognizing and dealing with hazards on the system level. This includes competence, tools, systems, and practices as well as a culture supporting safety and quality. Safety managers in turn are not only rule setters and auditors but also change agents and organization developers, as was illustrated in Chapter 8. Moving the focus of safety management on organizational capabilities brings safety function closer to quality and human resources functions, in addition to bringing these closer to the actual conduct of work. Safety should be integrated into the daily operations of the organization. This does not mean that there is no longer a need for safety specialists or a strong safety department. Depending on the development stage of the company's safety culture, safety professionals' role is to facilitate the fulfillment of legal requirements, assist in perceiving and controlling risks, and acting as change agents in the continuous development of safe operations (see Chapter 8).

Having an integrated approach to safety means that instead of focusing on individual factors, we should look at their interactions. For example, power, culture, and leadership cannot be treated as independent dimensions with separate interventions or corrective actions for each. Linear, factor-based approaches need to be complemented with a view of how the factors interact and what emergent properties this interaction creates. As the focus of attention is on interactions and emergent properties, safety assurance should also look at the whole system instead of each component individually.

Understanding safety in complex organizations requires looking at the big picture as more than the sum of its parts. Often, after events or mistakes, we try to understand the situation by delving deeper and deeper into the details. Details are important, but even more important is understanding the big picture and connections. When we look at work and the organization from the cultural perspective, instead of delving deeper, we raise the perspective and try to see things in context. In other words, we try to perceive the context in which work is done, rather than individual work performances. To see culturally, we need to apply the macroscope in addition to the microscope.

Since culture shows up in practice, is embedded in structures, and depends on interaction, all these elements must be further analyzed to really understand safety in the given sociotechnical system. What is important is to understand and analyze them from a cultural perspective:

- What meanings, beliefs, and assumptions currently exist in our culture? How widely are they shared? How consistent are they with each other?
- What structures, technologies, and tools currently maintain the meanings, beliefs, and assumptions?
- What historical events could explain the current practices?
- How does a particular change in structure influence interaction and behavior, and subsequently culture?
- What behavior is most likely in the context, given the structures and prevalent interaction patterns? Do we need to change them before behavior can change?
- What type of safety interventions work best in our culture? What kind of side-effects may the various interventions have?

A cultural approach is not only about changing culture but also about becoming aware of the cultural nuances in one's own organization and using that awareness to tackle concrete daily problems (and changing culture as a side-effect). It is about seeing the culture and systems in our ordinary everyday work.

From Strong to Healthy Cultures

A cultural approach to organizations considers the potential negative effects of culture formation in groups, which often lead to simplistic, categorical (yes/no) solutions, denial of uncertainties, and narrow attention. The cultural approach needs to acknowledge the paradox that culture often hides the dynamics of everyday life and the inherent uncertainties in it – the very things that the cultural approach is interested in. This may be beneficial for the fluency of everyday life in the short-term, but in safety-critical contexts, these simplifications may have fatal effects.

As has been argued in this book, a strong culture is not the same as a healthy culture, or a culture that contributes positively to safety. A strong culture can as easily mean conformity and blindness as it can mean diversity and openness. A healthy culture means an inclusive culture built on trust and an adequate consensus on topics where consensus can be reached. In addition, personnel need to maintain a constructive criticism towards each other's ideas but also acknowledge that there will always be disagreements and differences of opinion. Mutual respect, fairness, and trust need to be the basis of operations and management.

A healthy organizational culture should support the generic core task requirements of anticipating, monitoring, and reacting as well as the instrumental core task requirements of adaptability, systematic response, and learning (see Chapter 10).

From Preventing Failure to Facilitating Success

When safety is seen as an organization's ability to succeed in varying circumstances, development focuses on developing practices, structures, technology, and culture in such a way that the organization's ability to manage the daily work, and prepare for various contingencies and recover from both anticipated and non-anticipated events is emphasized.

The development of culture (and therefore safety) must be active and proactive. The level of safety will not be maintained unless it is actively developed. The organization and different teams must maintain an understanding of the risk profile and its changes and react (learn) if changes occur in the environment, work, or organization or new risks arise. However, the key to a proactive approach is that safety is developed without its noticeable deterioration. Organizations need to draw attention to weak signals and anticipate how they may develop into bigger problems.

Complex organizations require adaptive management whereby the system is guided, nurtured, and influenced rather than controlled (see Chapter 7). Developing safety in a complex organization entails establishing the conditions for the system to adapt and self-organize rather than defining the ideal state in advance and implementing it. It is important to remember that learning and development should not lead to more elaborate defenses, such as new rules and restrictions, but more elaborate response capabilities (Weick & Sutcliffe, 2007, p. 72).

From Simple Solutions to Better Questions

In rushing to develop a "safety culture" and implement various normative criteria, organizations often overlook the potential added value of the concept of safety culture, that is, that it helps to understand what is happening in organizations, why people react the way they do, what they consider relevant ,and what is less meaningful. If these organizations stop and reflect on their culture for a minute or two, they will probably realize that the solutions they are advocating are too simplistic. They would understand that it all starts with asking the right questions. By identifying organizational and cultural phenomena and asking where they originate, what sustains them, and what prevents them from being changed, we get rid of overly simplistic solutions that do not actually solve organizational problems.

Development of a culture for safety should involve a qualitative increase in understanding rather than a quantitative increase in the number of tools at the organization's disposal. Sometimes this quantitative increase is needed too, but as all meaning is qualitative, all change has a qualitative component.

We can never truly understand a culture if we do not let its inhabitants speak freely. But we also cannot ever completely assess the culture as long as we are inside it. Thus, a balance needs to be struck between relying on the outside view of cultural strengths and weaknesses and an insider understanding of the workplace reality. These offer complementary perspectives on the challenges of mindful organizing.

From Overseeing Results to Reviewing Process And Effort

It is very typical for organizations to focus on numerical results and indicators. Often there are rewards for making safety observations, or for the low number (or lack of reporting) of incidents, accidents, and adverse events. Staff quickly learn that what matters is not the work itself, but the numbers. When a culture becomes an indicator for improvement and comparison, the nuances of its unique character disappear. When examining the organizational culture, attention must be paid not to the end results but to the process and effort of the work. Remuneration should also focus on processes and effort in addition to the end results.

The gap between results and process/effort is one of many reasons to move safety management towards Safety II thinking (cf. Hollnagel, 2014). At the same time, this property of complex systems (disconnect between inputs and outputs, between cause and effect) reveals one of the practical challenges of Safety II thinking: we traditionally identify successes and failures precisely on the basis of consequences – success leads to positive outcomes whereas failure leads to negative ones. The "successes" advocated by Safety II thinking may mislead the organization; great successes are almost as rare as major failures. They also carry the risk that success has been achieved with enormous effort under pressure. In this case, the lessons to be learned from this success are very different than if we were to identify the prerequisites for smooth everyday work.

When we look at the way an organization creates safety, the effort it puts into it, and the strain this has on employees, we identify the daily adaptations, improvisations, and heuristics that are needed to get the job done. Many of these would count as human error or transgressions if the outcomes were negative. However, systems thinking reminds us that some of these adaptations are due to shortcomings in the functioning of the "higher" level of the system. These adaptations should therefore not be promoted as anything other than the necessary improvisation of an imperfect organization, while at the same time trying to influence top-level processes and improve the entire system. Thus, instead of supporting these actual adaptations, we should support the processes that enable necessary adaptations in the future. These can be, for example, different interaction practices, ways of acquiring information, management practices, or decision-making practices. In the same way as with mistakes, where we need to understand the organizational processes that contribute to them, we need to understand the processes that promote the fluency and quality of work, enabling situational adaptations and improvisations when necessary. These are, of course, to a large extent, the same processes as those contributing to failures, as illustrated in Chapter 4 (see Safety as Capability).

Looking at effort in addition to successful outcomes is important precisely because adaptation and improvisation are required in complex organizations. In many contexts, individuals compensate with their commitment and sense of responsibility for a dysfunctional system. This is not a viable long-term strategy as it leads to side-effects, such as burnout, stress-related deviations, and local policies and solutions that are not transparent to other actors in the system. It contributes to drift and normalization (see Chapter 3) and may create new or reinforce existing psychosocial hazards (see Chapter 4).

Another kind of vicious circle arises when personnel gradually, and often under duress, specialize in flexibility and situational adaptation (Smaggus, 2019). Personnel can become so good at compensating for the shortcomings of the system that these system-level deficiencies are no longer detected, at least not as long as the staff can put in the extra effort.

People tend to see causal connections between things that happen sequentially in time (Chapter 2). For this reason, we often consider the person who is physically closest to the accident or adverse event to be the cause of the event in question, or at least partly to blame. However, this is a feature of human thinking that rarely matches the diversity of the real world. In organizations and other complex environments, many things happen with a delay, and many interactions are indirect.

Since complex organizations evolve over time through learning, feedback, and adaptation, today's strengths can be tomorrow's weaknesses, and vice versa. Adaptive learning is both more difficult but also more important to ensure safety.

From Either/Or Thinking to Yes/And Thinking

People tend to categorize things into mutually exclusive categories. Many safety assurance practices are based on either individual-centered or organization-centered thinking. In the event of an accident, we look for either a human or a technical cause. Root cause analysis often tries to find one cause for any event. However, reality is more diverse than the categories with which we try to simplify and manage it.

Similar either/or thinking also occurs in safety research, where it is debated, for example, whether solutions based on harmonization and guidance are better for safety than solutions based on situational decision-making and skills of the professionals. These extremes are sometimes referred to as Safety I and Safety II thinking, as has been discussed in the book several times. In reality, both approaches are needed, and one might think that obsessing over finding the one right solution, whether you call it Safety I, II, or III, is old-fashioned thinking.

Yes/and thinking involves acknowledging that organizations always contain contradictory elements and issues. The cultural perspective does not assume that the organization is fully cohesive, but the organization is approached openly, examining which issues are shared and which issues do not have a common understanding.

The principles of adaptive safety management introduced in Chapter 7 build on the idea that safety in complex organizations requires stability *and* change, uniformity *and* diversity, participation *and* direction, and finally, consideration of local goals *and* system goals (see also Figure 4.1).

From Concrete to Abstract, and Back to Concrete

Taking culture into account does not mean focusing only on abstract values and assumptions, or, for example, on language and speech. On the contrary, a cultural approach should help to understand the role of technology in creating opportunities and constraints. A cultural approach should consider the materiality of the system and its role in the construction of organizational activities (Pettersen Gould & Macrae, 2021, p. 12) in conjunction with the cultural features. The cultural approach should help the organization, and designers, to realize the importance of technology for creating opportunities and constraints for action.

One of the most difficult things about the cultural approach is this need for dualistic thinking: it is not about abstract or concrete but about understanding the abstract in the concrete and understanding how the physical and the concrete shapes and transforms the abstract. Since culture is always more or less rooted in organizational structures, tools, and technology, structures must also be analyzed if culture is to be understood and changed. Analyzing language and speech is important, but simply changing it will not change the culture (at least not in the desired/intended direction).

A culture is best developed by simultaneously working with both the concrete and the abstract. This means that the focus of development should be equally on organizational structures, work practices, understanding of employees, and collective mindset (attitudes, values, norms). Although cultural development must focus on both the abstract and tangible, it must be understood that all development of organizations and work is cultural, intentionally or unintentionally. This is one of the reasons why so many development projects and change initiatives fail; they neglect the cultural dimension, the importance of human interaction, and our need to belong, be heard, and be involved. At the same time, too little attention is paid to the impact of beliefs and assumptions on our behavior.

Assumptions about their use, rooted in technology and tools, shape operating methods by creating affordances, that is, obvious measures, in the environment. The old saying that for a person with only a hammer, everything looks like nails, is very true. For this reason, technical change is never just a technical change but always includes a cultural dimension. New technology offers new ways of conducting work – and not only new ways of working, but also of thinking about work and conceptualizing it. At the same time, technology changes power relations in the organization and thus can threaten the structures that sustain culture.

The cultural perspective on safety considers how technology and structures shape interaction and slowly alter the underlying logics of working. It also considers how safety is perceived and how structures and interaction support creating and maintaining an adequate view on safety. Development of this culture for safety takes place by addressing how people work and interact, as well as the structures and technology they use. In addition, it involves integrating (the development of) practices, structures, and culture.

This principle applies also to safety research, which is also about understanding the abstract in the concrete (and vice versa). Research in safety should be based on the realities and challenges of people doing safety critical work, be they the front-line operators, maintenance personnel, safety professionals, or senior managers. The practical problems and challenges they face should be looked at from the theoretical perspective of safety science and put in the wider context of assuring safety in complex organizations. However, there is always a need to go back to the concrete: what can we say about the particular work and way of organizing it based on theoretical analysis?

12.4 WHERE TO START?

A good starting point for the cultural approach is to ask yourself the following questions, and if/when you do not know the answer, consider how to gain that knowledge from your organization – and consider who you need to talk to, and who you need to listen to, in order to gain a better understanding of safety in your organization. Bear in mind, however, that more important than finding answers to these exact questions is to learn to ask more questions and teach others to be as mindful and curious as they need to be to manage safety in a complex organization.

1. Who or what are the functions, units or individuals with most power in your organization? How do they accumulate and use their power? Who is really in charge and what are their guiding values?
2. What is the strategy for assuring safety in your organization? Do you employ the same strategy for all types of safety in your organization? How have these strategies been chosen and implemented?
3. How is the strategy implemented in practice? How do managers show visible commitment to it? What is the connection between your core business and safety?
4. In addition to history and records such as lost-time incidents, what other evidence of safety do you have in your organization? How about evidence of vulnerability and risks?
5. How is safety being monitored or measured? How are the measures of safety and safety targets differentiated? What are the side-effects of the current ways of measuring safety?
6. How does your organization set safety targets? How are the targets implemented in daily work? Do they facilitate safety in practice, and what side-effects may they have?
7. How have the vulnerabilities and risks in your organization changed during the past years? What safety issues do you consider "solved" and what is the evidence they actually have been solved instead of temporarily tamed?
8. What are the values and norms guiding behavior at the various levels and functions of the organization? How much do these differ between groups?
9. How are the values and norms that guide the actual work related to the official values of the organization?
10. Is there adequate diversity of views in your organization? How do you get a fresh perspective on various issues? Whose voices are typically silenced?
11. On what issues are there contradictory views in your organization? Can these views be beneficial for the organization and safety?
12. How is safety viewed at the various levels and functions of the organization? How is it understood and how is it prioritized? How is the role of humans seen in safety?
13. How are failures, errors, and accidents viewed at the organization? Is the view systemic enough to support organizational learning?
14. How well does the organization know the hazards and the associated risks? What is done to ensure and develop this understanding?
15. How does the organization deal with negative information or "bad news"? What is done to resist drifting into a "good news" culture?
16. Are people willing to report issues they notice at work, or their safety-related concerns? How well are they able to identify various weak signals?
17. How well are you aware of the workplace reality in different functions and departments of your organization? How do you maintain an understanding of how and under what conditions work is conducted daily?
18. How is the gap between Work-as-Imagined and Work-as-Done monitored, and what actions are taken to manage it? Are they both changing in the same direction?

19. Do you know what aspects of working in your organization employees find most important, most frustrating, and most demanding? What are the reasons for the employees' opinions? How can this information be utilized in safety management?
20. How does your organization approach uncertainty and complexity? How is uncertainty quantified in risk assessments, and how are the unquantifiable aspects treated? How does the organization balance between reducing and coping with uncertainty? When is uncertainty beneficial or necessary?
21. Are people, including you, mindful of what is happening now and what may happen in future? Are they questioning routines and potentially dangerous conditions? How is this mindfulness maintained and developed?
22. How does your organization make sure decisions, activities, and analyses can be independently assessed and, if needed, challenged?
23. What elements of the organization are currently maintaining the values, beliefs, and norms in the organization, and what elements are changing them?
24. How does the organizational structure support safety and culture for safety in general?
25. How are gradual changes considered in your organization? What once-exceptional problems are now considered part of everyday work?
26. How does your organization deal with pressures from the environment? How are competition, customer feedback, owners' interests, regulators' observations, and public opinion interpreted and acted upon in the organization?
27. How do you maintain a discussion of risks and hazards yourself also when things are going seemingly well?
28. How aware are you of your leadership style and its influence on the culture of your organization? What aspects of leadership do you need to develop? What aspects of leadership do you need to emphasize more and what aspects less?
29. How does your organization analyze internal incidents for learning purposes?
30. How does your organization seek out information and lessons learned from other organizations?
31. How does your organization involve employees to develop their work, safety, and the organization in general?

NOTES

1. Your values, that is, but in concrete language that is localized to the particular context and not rehearsed from management literature or official policies.
2. Naturally, it is seldom practical to invest with the aim of zero risk. Some residual risk will always be left for the employees to manage, but reliance on human performance when a technical solution is available and affordable is an example of a poor culture for safety.
3. A dual-use item refers to an item (system, equipment, document, etc.) that can be used for both civil and military purposes.
4. This happens in some empirical studies too. Sometimes safety culture is defined and measured superficially, results indicate little relation to anything significant, and consequently the entire concept (as defined by the researchers) is denounced as useless.

References

Accou, B., & Reniers, G. (2020). Introducing the extended safety fractal: Reusing the concept of safety management systems to organize resilient organizations. *International Journal of Environmental Research and Public Health, 17*, 5478.

Ackoff, R. L. (1979). The future of operational research is past. *Journal of the Operational Research Society, 30*(2), 93–104. https://doi.org/10.1057/jors.1979.22

Acosta, M. (2024). *Safety capacity. Leadership practices for failing safely.* Pre-Accident Investigation Media.

Almas, I., & Lordos, A. (2025). A narrative review of psychopathy research: Current advances and the argument for a qualitative approach. *The Journal of Forensic Psychiatry & Psychology, 36*(3), 356–406. https://doi.org/10.1080/14789949.2025.2456208

Almklov, P. G., Rosness, R., & Størkersen, K. (2014). When safety science meets the practitioners: Does safety science contribute to marginalization of practical knowledge? *Safety Science, 67*, 25–36. https://doi.org/10.1016/j.ssci.2013.08.025

Alvesson, M. (2002). *Understanding organizational culture.* SAGE Publications Ltd.

Alvesson, M., & Spicer, A. (2012). A stupidity-based theory of organizations. *Journal of Management Studies, 49*(7), 1194–1220. https://doi.org/10.1111/j.1467-6486.2012.01072.x

Alvesson, M., & Sveningsson, S. (2024). *Changing organizational culture. Cultural change work in progress* (3rd ed.). Routledge.

Amalberti, R. (2001). The paradoxes of almost totally safe transportation systems. *Safety Science, 37*, 109–126.

Amalberti, R. (2013). *Navigating safety: Necessary compromises and trade-offs - Theory and practice.* Springer. https://doi.org/10.1007/978-94-007-6549-8

Amalberti, R., Auroy, Y., Berwick, D., & Barach, P. (2005). Five system barriers to achieving ultrasafe health care. *Annals of Internal Medicine, 142*(9), 756–764. https://doi.org/10.7326/0003-4819-142-9-200505030-00012

Anand, N. (2024). *Are we learning from accidents? A quandary, a question, and a way forward.* Novellus.

Anand, V., Ashforth, B. E., & Joshi, M. (2004). Business as usual: The acceptance and perpetuation of corruption in organizations. *Academy of Management Perspectives, 18*(3), 39–53. https://doi.org/10.5465/ame.2004.13837437

Andersen, M. F., Svendsen, P. A., Nielsen, K., Brinkmann, S., Rugulies, R., & Madsen, I. E. H. (2022). Influence at work is a key factor for mental health – But what do contemporary employees in knowledge and relational work mean by "influence at work"? *International Journal of Qualitative Studies on Health and wellbeing, 17*(1), 2054513. https://doi.org/10.1080/17482631.2022.2054513

Antonsen, S. (2009a). Safety culture and the issue of power. *Safety Science, 47*(2), 183–191. https://doi.org/10.1016/j.ssci.2008.02.004

Antonsen, S. (2009b). Safety culture assessment: A mission impossible? *Journal of Contingencies and Crisis Management, 17*(4), 242–254. https://doi.org/10.1111/j.1468-5973.2009.00585.x

Antonsen, S. (2009c). *Safety culture: Theory, method and improvement.* CRC Press.

Antonsen, S., & Almklov, P. (2020). Revisiting the issue of power in safety research. In J.-C. Le Coze (Ed.), *Safety science research. Evolution, challenges and new directions.* CRC Press.

Antonsen, S., Almklov, P., & Fenstad, J. (2008). Reducing the gap between procedures and practice – Lessons from a successful safety intervention. *Safety Science Monitor, 12*(1), 1–16.

Arditi, D., & Chotibhongs, R. (2005). Issues in subcontracting practice. *Journal of Construction Engineering and Management, 131*(8), 866–876. https://doi.org/10.1061/(ASCE)0733-9364(2005)131:8(866)

Argyris, C. (1999). *On organizational learning.* Blackwell.

Arnold, J. A., Arad, S., Rhoades, J. A., & Drasgow, F. (2000). The empowering leadership questionnaire: The construction and validation of a new scale for measuring leader behaviors. *Journal of Organizational Behavior, 21*(3), 249–269. https://doi.org/10.1002/(SICI)1099-1379(200005)21:3<249::AID-JOB10>3.0.CO;2-#

Asch, S. E. (1956). Studies of independence and conformity: I. A minority of one against a unanimous majority. *Psychological Monographs: General and Applied, 70*(9), 1–70.

Ashby, W. R. (1956). *An introduction to cybernetics.* John Wiley & Sons.

Avolio, B. J., Gardner, W. L., Walumbwa, F. O., Luthans, F., & May, D. R. (2004). Unlocking the mask: A look at the process by which authentic leaders impact follower attitudes and behaviors. *The Leadership Quarterly, 15*(6), 801–823. https://doi.org/10.1016/j.leaqua.2004.09.003

Bainbridge, L. (1983). Ironies of automation. *Automatica, 19*(6), 775–779.
Baker III, J. A. (2007). *The report of the BP U.S. refineries independent safety review panel.* The BP US Refineries Independent Safety Review Panel.
Baron, J. (2008). *Thinking and deciding* (4th ed.). Cambridge University Press.
Baron, J. & Hershey, J.C. (1988). Outcome bias in decision evaluation. *Journal of Personality and Social Psychology,* 54, 569-579.
Barondes, A. D. (2014). Exxon Valdez: Human error, plain and simple. *Journal of System Safety, Spring/Summer 2014*, 22–30.
Bass, B. M. (1985a). Leadership and organizational culture: New perspectives on administrative theory and practice. *The Journal of Higher Education, 56*(5), 592–595. https://doi.org/10.1080/00221546.1985.11778728
Bass, B. M. (1985b). *Leadership and performance beyond expectations.* The Free Press.
Bass, B. M. (1990). From transactional to transformational leadership: Learning to share the vision. *Organizational Dynamics, 18*(3), 19–31.
Bass, B. M., & Avolio, B. J. (1994). *Improving organizational effectiveness through transformational leadership.* SAGE Publications Ltd.
Baumard, P., & Starbuck, W. H. (2005). Learning from failures: Why it may not happen. *Long Range Planning, 38*(3), 281–298. https://doi.org/10.1016/j.lrp.2005.03.004
Bazerman, M.H. (2022). *Complicit. How We Enable the Unethical and How to Stop.* Princeton University Press.
Bazerman, M. H. & Tenbrunsel, A.E. (2011). *Blind Spots. Why We Fail to Do What's Right and What to Do about It.* Princeton University Press.
Beer, S. (1984). The viable system model: Its provenance, development, methodology and pathology. *Journal of the Operational Research Society, 35*(1), 7–25.
Bella, D. A. (1987). Engineering and erosion of trust. *Journal of Professional Issues in Engineering, 113*(2), 117–129. https://doi.org/10.1061/(ASCE)1052-3928(1987)113:2(117)
Bellamy, L. J. (2015). Exploring the relationship between major hazard, fatal and non-fatal accidents through outcomes and causes. *Safety Science, 71*, 93–103. https://doi.org/10.1016/j.ssci.2014.02.009
Bergman, D., & Nelson, D. (2013, May 7). Bangladesh building collapse: Dhaka buildings not given final safety clearance. *The Daily Telegraph.*
Bieder, C., & Bourrier, M. (Eds.). (2013). *Trapping safety into rules. How desirable or avoidable is proceduralization*? Ashgate.
Bieder, C., Amalberti, R., Pariès, J., Laroche, H., Marsden, E., & Kamaté, C. (2024). *Industrial safety in the age of "living with": Uncertainty, complexity and rising expectations.* Fondation pour une Culture de Sécurité Industrielle (Foncsi). https://doi.org/10.57071/420lwu
Bisbey, T. M., Kilcullen, M. P., Thomas, E. J., Ottosen, M. J., Tsao, K., & Salas, E. (2021). Safety culture: An integration of existing models and a framework for understanding its development. *Human Factors: The Journal of the Human Factors and Ergonomics Society, 63*(1), 88–110. https://doi.org/10.1177/0018720819868878
Bonjeer, T., & Vonkova, H. (2024). Relationships between response styles and the Hofstede and GLOBE dimensions of culture in a sample of adolescents from 33 countries. *Cross-Cultural Research, 58*(2–3), 180–207. https://doi.org/10.1177/10693971231203759
Borys, D., Else, D., & Leggett, S. (2009). The fifth age of safety: The adaptive age. *Journal of Health Services Research & Policy, 1*(1), 19–27.
Bourrier, M. (1996). Organizing maintenance work at two American nuclear power plants. *Journal of Contingencies and Crisis Management, 4*, 104–112.
Bozeman, B. (1993). A theory of government 'Red Tape'. *Journal of Public Administration Research and Theory: J-PART, 3*(3), 273–303.
Brooks, N., Fritzon, K., & Watt, B. (2020). Overview of theories and empirical findings relevant to psychopathic personality characteristics amongst high-functioning populations. In K. Fritzon, N. Brooks, & S. Croom (Eds.), *Corporate psychopathy: Investigating destructive personalities in the workplace.* Springer International Publishing. https://doi.org/10.1007/978-3-030-27188-6
Burns, J.M. (1978). *Leadership.* Harper & Row.
Buch, R., Martinsen, Ø. L., & Kuvaas, B. (2014). The destructiveness of laissez-faire leadership behavior: The mediating role of economic leader–member exchange relationships. *Journal of Leadership & Organizational Studies, 22*(1), 1–10.
Busch, C. (2016). *Safety myth 101: Musings on myths, misunderstandings and more.* Mind the Risk.

Busch, C. (2020). *Preventing industrial accidents: Reappraising H. W. Heinrich – More than triangles and dominoes* (1st ed.). Routledge. https://doi.org/10.4324/9780429325397
Busch, C. (2021). *The first rule of safety culture. A counter-C-word manifesto.* Mind the Risk.
CAIB. (2003). *Report volume 1.* Columbia Accident Investigation Board.
Caldwell, C. L. (2018). *Safety culture and high-risk environments. A leadership perspective.* CRC Press.
Callahan, P. (2004, February 29). So why does Harry Stonecipher think he can turn around Boeing? *Chicago Tribune.* https://www.chicagotribune.com/chi-0402290256feb29-story.html
Cameron, K. S. (2021). *Positively energizing leadership.* Berrett-Koehler.
Cameron, K. S., & Quinn, R. E. (1999). *Diagnosing and changing organisational culture: Based on the competing values framework.* Addison-Wesley.
Cameron, K. S., & Quinn, R. E. (2011). *Diagnosing and changing organizational culture. Based on the competing values framework* (3rd ed.). Jossey-Bass.
Campbell, J. P. (1977). On the nature of organizational effectiveness. In P. S. Goodman & J. M. Pennings (Eds.), *New perspectives on organizational effectiveness* (pp. 13–55). Jossey-Bass.
Cannon, J. (2023). *Toxic cultures at work. The eight drivers of a toxic culture and a process for change.* Routledge.
Cantu, J., Gharehyakheh, A., Fritts, S., & Tolk, J. (2021). Assessing the HRO: Tools and techniques to determine the high-reliability state of an organization. *Safety Science, 134*, 105082. https://doi.org/10.1016/j.ssci.2020.105082
Cantu, J., Tolk, J., Fritts, S., & Gharehyakheh, A. (2020). High Reliability Organization (HRO) systematic literature review: Discovery of culture as a foundational hallmark. *Journal of Contingencies and Crisis Management, 28*(4), 399–410. https://doi.org/10.1111/1468-5973.12293
Carrillo, R. A. (2020). *The relationship factor in safety leadership. Achieving success through employee engagement.* Routledge.
Carroll, J. S., Pfeiffer, Y., Nowak, H., & Friis, S. (2022). The variety of beliefs about the causes of safety among safety practitioners. *Safety Science, 148*, 105641. https://doi.org/10.1016/j.ssci.2021.105641
Chapman, J. (2004). *System failure: Why governments must learn to think differently* (2nd ed.). Demos.
Chatman, J. A., & Jehn, K. A. (1994). Assessing the relationship between industry characteristics and organizational culture: How different can you be? *Academy of Management Journal, 37*(3), 522–553.
Checkland, P. (1981). *Systems thinking, systems practice.* John Wiley & Sons.
Chemers, M. M. (1997). *An integrative theory of leadership.* Lawrence Erlbaum Associates.
Chernov, D., & Sornette, D. (2016). *Man-made catastrophes and risk information concealment. Case studies of major disasters and human fallibility.* Springer.
Chhokar, J. S., Brodbeck, F. C., & House, R. J. (Eds.). (2007). *Culture and leadership across the world: The GLOBE book of in-depth studies of 25 societies.* Lawrence Erlbaum Associates. http://www.tandfonline.com/doi/abs/10.1080/13602381.2011.559082
Chiles, J. R. (2002). *Inviting disaster: Lessons from the edge of technology.* HarperCollins.
Choudhry, R. M., & Fang, D. (2008). Why operatives engage in unsafe work behavior: Investigating factors on construction sites. *Safety Science, 46*(4), 566–584. https://doi.org/10.1016/j.ssci.2007.06.027
Choudhury, P., Khanna, T., Makridis, C. A., & Schirmann, K. (2022). *Is hybrid work the best of both worlds? Evidence from a field experiment* (Working Paper 22-063). Harvard Business School.
Cilliers, P. (1998). *Complexity and postmodernism. Understanding complex systems.* Routledge.
Cilliers, P. (2010). Difference, identity and complexity. *Philosophy Today, 54*(1), 55–65.
Clarke, S. (2013). Safety leadership: A meta-analytic review of transformational and transactional leadership styles as antecedents of safety behaviours. *Journal of Occupational and Organizational Psychology, 86*(1), 22–49. https://doi.org/10.1111/j.2044-8325.2012.02064.x
Collins, J. (2009). *How the mighty fall. And why some companies never give in.* CollinsBusiness.
Collins, J., & Hansen, M. T. (2011). *Great by choice. Uncertainty, chaos, and luck—why some thrive despite them all.* HarperBusiness.
Conant, R. C., & Ashby, R. W. (1970). Every good regulator of a system must be a model of that system. *International Journal of Systems Science, 1*(2), 89–97. https://doi.org/10.1080/00207727008920220
Conchie, S. M., Moon, S., & Duncan, M. (2013). Supervisors' engagement in safety leadership: Factors that help and hinder. *Safety Science, 51*(1), 109–117. https://doi.org/10.1016/j.ssci.2012.05.020
Cook, R. I., & Woods, D. D. (2006). Distancing through differencing: An obstacle to organizational learning following accidents. In E. Hollnagel, D. D. Woods, & N. Leveson (Eds.), *Resilience engineering. Concepts and precepts* (1st ed., pp. 329–338). Ashgate. https://doi.org/10.1201/9781315605685-28
Cooper, M. D. (2000). Towards a model of safety culture. *Safety Science, 36*(2), 111–136. https://doi.org/10.1016/S0925-7535(00)00035-7

Cooper, M. D. (2009). Behavioral safety interventions. A review of process design factors. *Professional Safety, February 2009*, 36–45.

Cooper, S., & Wakelam, A. (1999). Leadership of resuscitation teams: 'Lighthouse leadership'. *Resuscitation, 42*(1), 27–45. https://doi.org/10.1016/S0300-9572(99)00080-5

CSB. (2007). *Investigation report. Refinery explosion and fire. BP Texas City, Texas. March 23, 2005* (REPORT NO. 2005-04-I-TX). US Chemical Safety and Hazard Investigation Board (CSB).

CSB. (2014). *Investigation report. Explosion and fire at the Macondo well* (REPORT NO. 2010-10-I-OS). US Chemical Safety and Hazard Investigation Board (CSB).

Cullen, W. D. (1990). *The public inquiry into the Piper Alpha disaster.* HMSO.

Czarniawska-Joerges, B. (1992). *Exploring complex organizations. A cultural approach.* SAGE Publications Ltd.

Daudigeos, T. (2013). In their profession's service: How staff professionals exert influence in their organization. *Journal of Management Studies, 50*(5), 722–749. https://doi.org/10.1111/joms.12021

de Boer, R. J. (2021). *Safety leadership: A different, doable and directed approach to operational improvements.* CRC Press.

De Bono, E. (1985). *Six thinking hats.* Little, Brown and Company.

De Meuse, K. P., Dai, G., & Wu, J. (2011). Leadership skills across organizational levels: A closer examination. *The Psychologist-Manager Journal, 14*(2), 120–139. https://doi.org/10.1080/10887156.2011.570143

Deal, T. E., & Kennedy, A. A. (1982). *Corporate cultures: The rites and rituals of corporate life.* Addison-Wesley.

Dechy, N., & Largier, A. (2022). Workload planning management of maintenance activities in nuclear power plants: Compensation mechanisms at the contractor interface. In J. Hayes & S. Tillement (Eds.), *Contracting and safety: Exploring outsourcing practices in high-hazard industries.* Springer International Publishing. https://doi.org/10.1007/978-3-030-89792-5

DeFazio, P. A., & Larsen, R. (2020). *Final committee report. The design, development, & certification of the Boeing 737 MAX.* The House Committee on Transportation & Infrastructure.

Dekker, S. W. A. (2005). *Ten questions about human error: A new view of human factors and system safety.* Lawrence Erlbaum.

Dekker, S. W. A. (2006). *The field guide to understanding human error.* Ashgate.

Dekker, S. W. A. (2011). *Drift into failure: From hunting broken components to understanding complex systems.* Ashgate.

Dekker, S. W. A. (2014). *The field guide to understanding 'human error'* (3rd ed.). Ashgate.

Dekker, S. W. A. (2015). *Safety differently: Human factors for a new era* (2nd ed.). CRC Press.

Dekker, S. W. A. (2017a). *Just culture: Restoring trust and accountability in your organization* (3rd ed.). CRC Press.

Dekker, S. W. A. (2017b). *The end of heaven: Disaster and suffering in a scientific age.* Routledge.

Dekker, S. W. A. (2018). *The safety anarchist. Relying on human expertise and innovation, reducing bureaucracy and compliance.* Routledge.

Dekker, S. W. A. (2019). *Foundations of safety science: A century of understanding accidents and disasters.* CRC Press.

Dekker, S. W. A. (2021). *Compliance capitalism: How free markets have led to unfree, overregulated workers* (1st ed.). Routledge. https://doi.org/10.4324/9781003177807

Dekker, S. W. A., & Nyce, J. M. (2014). There is safety in power, or power in safety. *Safety Science, 67*, 44–49. https://doi.org/10.1016/j.ssci.2013.10.013

Diehl, E., & Sterman, J. D. (1995). Effects of feedback complexity on dynamic decision making. *Organizational Behavior and Human Decision Processes, 62*(2), 198–215.

DOE. (2009a). *Human performance improvement handbook. Volume 1: Concepts and principles* (DOE-HDBK-1028-2009). U.S. Department of Energy.

DOE. (2009b). *Human performance improvement handbook. Volume 2: Human performance tools for individuals, work teams, and management* (DOE-HDBK-1028-2009). U.S. Department of Energy.

Donovan, S.-L., Salmon, P. M., & Lenné, M. G. (2016). Leading with style: A literature review of the influence of safety leadership on performance and outcomes. *Theoretical Issues in Ergonomics Science, 17*(4), 423–442. https://doi.org/10.1080/1463922X.2016.1143986

Dos Santos Grecco, C. H., Vidal, M. C. R., Cosenza, C. A. N., Dos Santos, I. J. A. L., & De Carvalho, P. V. R. (2014). Safety culture assessment: A fuzzy model for improving safety performance in a radioactive installation. *Progress in Nuclear Energy, 70*, 71–83. https://doi.org/10.1016/j.pnucene.2013.08.001

Dunning. (2011). The Dunning–Kruger effect: On being ignorant of one's own ignorance. *Advances in Experimental Social Psychology, 44*, 247–296.

Dutch Safety Board. (2024). *Railway accident in Voorschoten*. Dutch Safety Board.

Dyreborg, J., Lipscomb, H.J., Nielsen, K., Törner, M., Rasmussen, K., Frydendall, K.B., Bay, H., Gensby, U., Bengtsen, E., Guldenmund, F., Kines, P. (2022). Safety interventions for the prevention of accidents at work: A systematic review. *Campbell Systematic Reviews*, e1234. DOI: 10.1002/cl2.1234.

Edmondson, A. C. (1996). Learning from mistakes is easier said than done: Group and organizational influences on the detection and correction of human error. *The Journal of Applied Behavioral Science*, *32*(1), 5–28. https://doi.org/10.1177/0021886396321001

Edmondson, A. C. (2003). Speaking up in the operating room: How team leaders promote learning in interdisciplinary action teams. *Journal of Management Studies*, *40*(6), 1419–1452. https://doi.org/10.1111/1467-6486.00386

Edmondson, A. C. (2018). *The fearless organization: Creating psychological safety in the workplace for learning, innovation, and growth*. Wiley.

Edmondson, A. C., Roberto, M. A., Bohmer, R. M. J., Ferlins, E. M., & Feldman, L. R. (2005). *The recovery window: Organizational learning following ambiguous threats*. In W. H. Starbuck & M. Farjoun (Eds.), *Organization at the limit: Lessons from the Columbia disaster*. Blackwell.

Eichenwald, K. (2005). *Conspiracy of fools: A true story*. Broadway Books.

Einarsen, S., Aasland, M. S., & Skogstad, A. (2007). Destructive leadership behaviour: A definition and conceptual model. *The Leadership Quarterly*, *18*(3), 207–216. https://doi.org/10.1016/j.leaqua.2007.03.002

Enomoto, C. E., & Geisler, K. R. (2017). Culture and plane crashes: A cross-country test of the gladwell hypothesis. *Economics & Sociology*, *10*(3), 281–293. https://doi.org/10.14254/2071-789X.2017/10-3/20

Eoyang, G. H., & Holladay, R. J. (2013). *Adaptive action: Leveraging uncertainty in your organization*. Stanford University Press.

ERA. (2020). *Introduction to the European railway safety culture model*. European Union Agency for Railways.

Evan, W. M., & Manion, M. (2002). *Minding the machines. Preventing technological disasters*. Prentice Hall.

Evans, B. (2021). *The Space Shuttle: An Experimental Flying Machine*. Springer Cham.

Farjoun, M. (2005). History and policy at the space shuttle program. In W. H. Starbuck & M. Farjoun (Eds.), *Organization at the limit: Lessons from the Columbia disaster*. Blackwell.

Farjoun, M. (2010). Beyond dualism: Stability and change as a duality. *Academy of Management Review*, *35*(2), 202–225.

Farrington-Darby, T., Pickup, L., & Wilson, John. R. (2005). Safety culture in railway maintenance. *Safety Science*, *43*(1), 39–60. https://doi.org/10.1016/j.ssci.2004.09.003

Fennell, D. (1988). *Report on the investigation into the King's cross underground fire*. HMSO.

Feynman, R. P. (1988). *'What do you care what other people think?': Further adventures of a curious character*. W.W. Norton.

Filburn, T., & Bullard, S. (2016). *Three mile Island, Chernobyl and Fukushima. Curse of the nuclear genie*. Springer.

Fischhoff, B. (1975). Hindsight is not equal to foresight: The effect of outcome knowledge on judgment under uncertainty. *Journal of Experimental Psychology: Human Perception and Performance*, *1*(3), 288–299.

Fiske, S. T., & Taylor, S. E. (2008). *Social cognition: From brains to culture* (2nd ed.). Sage.

Fiske, S. T., & Taylor, S. E. (2017). *Social cognition: From brains to culture* (3rd ed.). Sage.

Fleming, P., & Spicer, A. (2014). Power in management and organization science. *The Academy of Management Annals*, *8*(1), 237–298.

Flin, R., & Martin, L. (2001). Behavioral markers for crew resource management: A review of current practice. *The International Journal of Aviation Psychology*, *11*(1), 95–118. https://doi.org/10.1207/S15327108IJAP1101_6

Flin, R., Mearns, K., O'Connor, P., & Bryden, R. (2000). Measuring safety climate: Identifying the common features. *Safety Science*, *34*(1–3), 177–192. https://doi.org/10.1016/S0925-7535(00)00012-6

Flin, R., O´Connor, P., & Crichton, M. (2008). *Safety at the sharp end. A guide to non-technical skills*. CRC Press.

Fransen, K., McEwan, D., & Sarkar, M. (2020). The impact of identity leadership on team functioning and wellbeing in team sport: Is psychological safety the missing link? *Psychology of Sport and Exercise*, *51*, 101763. https://doi.org/10.1016/j.psychsport.2020.101763

Fruhen, L. S., Flin, R. H., & McLeod, R. (2014). Chronic unease for safety in managers: A conceptualisation. *Journal of Risk Research*, *17*(8), 969–979. https://doi.org/10.1080/13669877.2013.822924

Gallup. (2025). *State of the global workplace: 2025 report. Understanding employees, informing leaders*. Gallup.AAAA

GAO. (2012). *Workplace safety and health. Better OSHA guidance needed on safety incentive programs* (GAO-12-329). United States Government Accountability Office.

Gardner, W. L., Avolio, B. J., Luthans, F., May, D. R., & Walumbwa, F. (2005). "Can you see the real me?" A self-based model of authentic leader and follower development. *The Leadership Quarterly, 16*(3), 343–372. https://doi.org/10.1016/j.leaqua.2005.03.003

Gawande, A. (2009). *The checklist manifesto: How to get things right.* Metropolitan Books.

Geller, E. S. (2001). *The psychology of safety handbook.* Lewis Publishers.

George, M. M., Strauss, K., Mell, J. N., & Vough, H. C. (2023). When "who I am" is under threat: Measures of threat to identity value, meanings, and enactment. *Journal of Applied Psychology, 108*(12), 1952–1978. https://doi.org/10.1037/apl0001114

Gibson, J. J. (1979). *The ecological approach to visual perception.* Houghton Mifflin.

Gneezy, U., & Rustichini, A. (2000). A fine is a price. *The Journal of Legal Studies, 29*, 1–17.

Goldberg, L. R. (1992). The development of markers for the big-five factor structure. *Psychological Assessment, 4*(1), 26–42.

Goncalves Filho, A. P., & Waterson, P. (2018). Maturity models and safety culture: A critical review. *Safety Science, 105*, 192–211. https://doi.org/10.1016/j.ssci.2018.02.017

Gotcheva, N., & Ylönen, M. (2021). *Regulatory lessons from accidents due to institutional failures: Boeing 737 MAX and deepwater horizon.* VTT Technical Research Centre of Finland. https://cris.vtt.fi/en/publications/regulatory-lessons-from-accidents-due-to-institutional-failures-b

Graeber, D. (2018). *Bullshit jobs. A theory.* Simon & Schuster.

Graen, G. B., & Uhl-Bien, M. (1995). Relationship-based approach to leadership: Development of Leader-Member Exchange (LMX) theory of leadership over 25 years: Applying a multi-level multi-domain perspective. *The Leadership Quarterly, 6*(2), 219–247. https://doi.org/10.1016/1048-9843(95)90036-5

Greene, J. (2013). *Moral Tribes. Emotion, reason and the gap between us and them.* Atlantic Books.

Grill, M., & Nielsen, K. (2019). Promoting and impeding safety – A qualitative study into direct and indirect safety leadership practices of constructions site managers. *Safety Science, 114*, 148–159. https://doi.org/10.1016/j.ssci.2019.01.008

Grinerud, K., Aarseth, W. K., & Robertsen, R. (2021). Leadership strategies, management decisions and safety culture in road transport organizations. *Research in Transportation Business & Management, 41*, 100670. https://doi.org/10.1016/j.rtbm.2021.100670

Grote, G. (2012). Safety management in different high-risk domains – All the same? *Safety Science, 50*(10), 1983–1992. https://doi.org/10.1016/j.ssci.2011.07.017

Guldenmund, F. W. (2000). The nature of safety culture: A review of theory and research. *Safety Science, 34*(1–3), 215–257. https://doi.org/10.1016/S0925-7535(00)00014-X

Guldenmund, F. W. (2010). (Mis)understanding safety culture and its relationship to safety management. *Risk Analysis, 30*(10), 1466–1480. https://doi.org/10.1111/j.1539-6924.2010.01452.x

Guo, L., Ryan, B., Leditschke, I. A., Haines, K. J., Cook, K., Eriksson, L., Olusanya, O., Selak, T., Shekar, K., & Ramanan, M. (2022). Impact of unacceptable behaviour between healthcare workers on clinical performance and patient outcomes: A systematic review. *BMJ Quality & Safety, 31*(9), 679–687. https://doi.org/10.1136/bmjqs-2021-013955

Hackitt, J. (2017). *Building a safer future. Independent review of building regulations and fire safety: Interim report.* HMSO.

Hackitt, J. (2018). *Building a safer future. Independent review of building regulations and fire safety: Final report.* HMSO.

Hackman, J. R., & Oldham, G. R. (1980). *Work redesign.* Addison-Wesley.

Hale, A. R. (1995). Occupational health and safety professionals and management: Identity, marriage, servitude or supervision? *Safety Science, 20*(2–3), 233–245. https://doi.org/10.1016/0925-7535(95)00026-D

Hale, A. R., & Hovden, J. (1998). Management and culture: The third age of safety. A review of approaches to organizational aspects of safety, health and environment. In A.-M. Feyer & A. Williamson (Eds.), *Occupational injury. Risk, prevention and intervention.* Taylor & Francis.

Hale, A. R., & Swuste, P. (1998). Safety rules: Procedural freedom or action constraint? *Safety Science, 29*(3), 163–177. https://doi.org/10.1016/S0925-7535(98)00020-4

Hallowell, M. R., Quashne, M., Salas, R., Jones, M., MacLean, B., & Quinn, E. (2021). The statistical invalidity of TRIR as a measure of safety performance. *Professional Safety, April 2021*, 28–34.

Harvey, J. B. (1974). The Abilene paradox: The management of agreement. *Organizational Dynamics, 3*, 63–80.

Harvey, J., Bolam, H., Gregory, D., & Erdos, G. (2001). The effectiveness of training to change safety culture and attitudes within a highly regulated environment. *Personnel Review, 30*, 615–636.

Haslam, S. A. (2004). *Psychology in organizations. The social identity approach* (2nd ed.). SAGE Publications Ltd.
Haslam, S. A., Alvesson, M., & Reicher, S. D. (2024). Zombie leadership: Dead ideas that still walk among us. *The Leadership Quarterly*, *35*(3), 101770. https://doi.org/10.1016/j.leaqua.2023.101770
Haslam, S. A., Jetten, J., Maskor, M., McMillan, B., Bentley, S. V., Steffens, N. K., & Johnston, S. (2022). Developing high-reliability organisations: A social identity model. *Safety Science*, *153*, 105814. https://doi.org/10.1016/j.ssci.2022.105814
Haslam, S. A., Reicher, S. D., & Platow, M. (2020). *The new psychology of leadership: Identity, influence and power* (2nd ed.). Routledge.
Hatch, M. J., & Schultz, M. (2002). The dynamics of organizational identity. *Human Relations*, *55*(8), 989–1018. https://doi.org/10.1177/0018726702055008181
Hayes, J., & Tillement, S. (Eds.). (2022). *Contracting and safety: Exploring outsourcing practices in high-hazard industries.* Springer International Publishing. https://doi.org/10.1007/978-3-030-89792-5
van Herk, H., Poortinga, Y. H., & Verhallen, T. M. M. (2004). Response styles in rating scales: Evidence of method bias in data from six EU countries. Journal of Cross-Cultural Psychology, 35(3), 346–360.
Hildreth, J. A., Gino, F., & Bazerman, M. (2016). Blind loyalty? When group loyalty makes us see evil or engage in it. *Organizational Behavior and Human Decision Processes* 132, 16-36.
Hidden, A. (1989). *Investigation into the Clapham junction railway accident* (The department of transport). HMSO.
Higginbotham, A. (2019). *Midnight in Chernobyl. The untold story of the world's greatest nuclear disaster.* Bantam Press.
Higginbotham, A. (2024). *Challenger. A true story of heroism and disaster on the edge of space.* Viking.
Hillmann, J. (2021). Disciplines of organizational resilience: Contributions, critiques, and future research avenues. *Review of Managerial Science*, *15*(4), 879–936. https://doi.org/10.1007/s11846-020-00384-2
Hinkin, T. R., & Schriesheim, C. A. (2008). An examination of 'nonleadership': From laissez-faire leadership to leader reward omission and punishment omission. *Journal of Applied Psychology*, *93*(6), 1234–1248. https://doi.org/10.1037/a0012875
Hoffmeister, K., Gibbons, A. M., Johnson, S. K., Cigularov, K. P., Chen, P. Y., & Rosecrance, J. C. (2014). The differential effects of transformational leadership facets on employee safety. *Safety Science*, *62*, 68–78. https://doi.org/10.1016/j.ssci.2013.07.004
Hofstede, G. H. (2011). Dimensionalizing cultures: The hofstede model in context. *Online Readings in Psychology and Culture*, *2*(1), 3-26. https://doi.org/10.9707/2307-0919.1014
Hofstede, G. H., Hofstede, G. J., & Minkov, M. (2010). *Cultures and organizations. Software of the mind. Intercultural cooperation and its importance for survival* (3rd ed.). McGraw-Hill.
Hofstede, G. H., Neuijen, B., Ohayv, D. D., & Sanders, G. (1990). Measuring organizational cultures: A qualitative and quantitative study across twenty cases. *Administrative Science Quarterly*, *35*, 286–316.
Hogg, M. A., & Abrams, D. (1988). *Social identifications. A social psychology of intergroup relations and group processes.* Routledge.
Hollnagel, E. (2004). *Barriers and accident prevention.* Ashgate.
Hollnagel, E. (2006). Challenge of the unstable. In E. Hollnagel, D. D. Woods, & N. Leveson (Eds.), *Resilience engineering: Concepts and precepts.* Ashgate.
Hollnagel, E. (2008). Safety management—Looking back or looking forward. In E. Hollnagel, C. P. Nemeth, & S. Dekker W. A. (Eds.), *Resilience engineering perspectives, volume 1. Remaining sensitive to the possibility of failure.* Ashgate.
Hollnagel, E. (2009a). *The ETTO principle: Efficiency-thoroughness trade-off. Why things that go right sometimes go wrong.* CRC Press.
Hollnagel, E. (2009b). The four cornerstones of resilience engineering. In C. P. Nemeth, E. Hollnagel, & S. Dekker W. A. (Eds.), *Resilience engineering perspectives, volume 2. Preparation and restoration.* CRC Press.
Hollnagel, E. (2012). *FRAM: The Functional Resonance Analysis Method: Modelling complex sociotechnical systems.* CRC Press, Taylor & Francis Group.
Hollnagel, E. (2014). *Safety-I and safety-II: The past and future of safety management.* Ashgate.
Hollnagel, E. (2015). Why is work-as-imagined different from work-as-done? In R. L. Wears, E. Hollnagel, & J. Braithwaite (Eds.), *Resilient health care, volume 2: The resilience of everyday clinical work.* Ashgate.
Hollnagel, E. (2017). Prologue: Why do our expectations of how work should be done never correspond exactly to how work is done? In J. Braithwaite, R. L. Wears, & E. Hollnagel (Eds.), *Resilient health care, volume 3. Reconciling work-as-imagined and workas-done.* CRC Press.

Hollnagel, E., Woods, D. D., & Leveson, N. (Eds.). (2006). *Resilience engineering: Concepts and precepts.* Ashgate.

Hopkins, A. (1999). The limits of normal accident theory. *Safety Science, 32*(2–3), 93–102. https://doi.org/10.1016/S0925-7535(99)00015-6

Hopkins, A. (2000). *Lessons from longford: The esso gas plant explosion.* CCH.

Hopkins, A. (2005). *Safety, culture and risk. The organisational causes of disasters.* CCH.

Hopkins, A. (2006). What are we to make of safe behaviour programs? *Safety Science, 44*(7), 583–597. https://doi.org/10.1016/j.ssci.2006.01.001

Hopkins, A. (2008). *Failure to learn: The BP Texas city refinery disaster.* CCH.

Hopkins, A. (2011). Management walk-arounds: Lessons from the Gulf of Mexico oil well blowout. *Safety Science, 49*(10), 1421–1425. https://doi.org/10.1016/j.ssci.2011.06.002

Hopkins, A. (2012). *Disastrous decisions: Human & organisational causes of the gulf of Mexico blowout.* CCH.

Hopkins, A. (2014). Issues in safety science. *Safety Science, 67,* 6–14. https://doi.org/10.1016/j.ssci.2013.01.007

Hopkins, A. (2018). The use and abuse of "culture". In C. Gilbert, B. Journé, H. Laroche, & C. Bieder (Eds.), *Safety cultures, safety models* (pp. 35–45). Springer International Publishing. https://doi.org/10.1007/978-3-319-95129-4_4

Hopkins, A. (2019). *Organising for safety: How structure creates culture.* CCH Australia.

Hopkins, A., & Maslen, S. (2015). *Risky rewards: How company bonuses affect safety.* CRC Press.

Horvath, P., & Zuckerman, M. (1993). Sensation seeking, risk appraisal, and risky behavior. *Personality and Individual Differences, 14*(1), 41–52.

Hudson, P. (2007). Implementing a safety culture in a major multi-national. *Safety Science, 45*(6), 697–722. https://doi.org/10.1016/j.ssci.2007.04.005

Hutchins, E. (1995). *Cognition in the wild.* MIT Press.

Hutchinson, B., Dekker, S. W. A., & Rae, A. (2024a). Audit masquerade: How audits provide comfort rather than treatment for serious safety problems. *Safety Science, 169,* 106348. https://doi.org/10.1016/j.ssci.2023.106348

Hutchinson, B., Dekker, S. W. A., & Rae, A. (2024b). How audits fail according to accident investigations: A counterfactual logic analysis. *Process Safety Progress, 43*(3), 441–454. https://doi.org/10.1002/prs.12579

Hynes, P., Kissoon, N., Hamielec, C. M., Greene, A. M., & Simone, C. (2006). Dealing with aggressive behavior within the health care team: A leadership challenge. *Journal of Critical Care, 21*(2), 224–227.

IAEA. (1986). *Summary report on the post-accident review meeting on the Chernobyl accident* (International Nuclear Safety Advisory Group Insag Series No. 1). International Atomic Energy Agency (IAEA).

IAEA. (1991). *Safety culture.* International Atomic Energy Agency.

IAEA. (2012). *IAEA mission to Onagawa Nuclear Power Station To Examine The Performance Of Systems, Structures And Components Following the Great East Japanese earthquake and tsunami* (IAEA mission report IAEA-2012). International Atomic Energy Agency (IAEA).

IAEA. (2016a). *Leadership and management for safety: General safety requirements.* International Atomic Energy Agency.

IAEA. (2016b). *Performing safety culture self-assessments.* International Atomic Energy Agency.

IAEA. (2020). *A harmonized safety culture model. IAEA working document.* International Atomic Energy Agency (IAEA).

IAEA. (2021a). *Format and content of the safety analysis report for nuclear power plants* (Specific Safety Guide No. SSG-61). International Atomic Energy Agency (IAEA).

IAEA. (2021b). *Systematic approach to training for nuclear facility personnel: Processes, methodology and practices* (Nuclear Energy Series No. NG-T-2.8). International Atomic Energy Agency (IAEA).

Ibrion, M., Paltrinieri, N., & Nejad, A. R. (2020). Learning from non-failure of Onagawa nuclear power station: An accident investigation over its life cycle. *Results in Engineering, 8,* 100185. https://doi.org/10.1016/j.rineng.2020.100185

INPO. (2013). *Traits of a healthy nuclear safety culture* (INPO 12–012 Revision 1). Institute of Nuclear Power Operations (INPO).

Jackson, M. C. (2019). *Critical systems thinking and the management of complexity.* John Wiley & Sons.

Jackson, M. C. (2024). *Critical systems thinking. A practitioners guide.* John Wiley & Sons.

Janis, I. L. (1982). *Groupthink: Psychological studies of policy decisions and fiascoes* (2nd ed.). Houghton Mifflin Company.

Javalagi, A. A., Newman, D. A., & Li, M. (2024). Personality and leadership: Meta-analytic review of cross-cultural moderation, behavioral mediation, and honesty-humility. *Journal of Applied Psychology, 109*(9), 1489–1511. https://doi.org/10.1037/apl0001182

Jensen, C. (1996). *No downlink. A dramatic narrative about the challenger.* Farrar, Straus & Giroux.

Jeschke, K. N. (2022). Understanding how managers balance the paradoxical nature of occupational safety through a practice-driven institutional lens. *Safety Science, 147,* 105627. https://doi.org/10.1016/j.ssci.2021.105627

Johnson, T., Kulesa, P., Cho, Y. I., & Shavitt, S. (2005). The relation between culture and response styles: Evidence from 19 countries. *Journal of Cross-Cultural Psychology, 36*(2), 264–277. https://doi.org/10.1177/0022022104272905

Kahneman, D. (2011). *Thinking, fast and slow.* Farrar, Straus and Giroux.

Kahneman, D., & Klein, G. A. (2009). Conditions for intuitive expertise: A failure to disagree. *American Psychologist, 64*(6), 515–526. https://doi.org/10.1037/a0016755

Kasperson, R. E., Renn, O., Slovic, P., Brown, H. S., Emel, J., Goble, R., Kasperson, J. X., & Ratick, S. (1988). The social amplification of risk: A conceptual framework. *Risk Analysis, 8*(2), 177–187. https://doi.org/10.1111/j.1539-6924.1988.tb01168.x

Katz, D., & Kahn, R. L. (1966). *The social psychology of organizations.* John Wiley & Sons.

Kemeny, J. G. (1979). *Report of the president's commission on the accident at three mile Island—A need for change.* The Legacy of TMI.

Kim, T., & Gausdal, A. H. (2020). Leaders' influence tactics for safety: An exploratory study in the maritime context. *Safety, 6*(1), 8. https://doi.org/10.3390/safety6010008

Kinnersley, S., & Roelen, A. (2007). The contribution of design to accidents. *Safety Science* 45, 31-60.

Kirk, K. (2024). Time for a rebalance: Psychological and emotional wellbeing in the healthcare workforce as the foundation for patient safety. *BMJ Quality & Safety, 33*(8), 483–486. https://doi.org/10.1136/bmjqs-2024-017236

Kirkpatrick, D. L. (1994). *Evaluating training programs: The four levels.* McGraw-Hill Ryerson.

Kirkpatrick, D. L., & Kirkpatrick, J. D. (2005). *Transferring learning to behavior: Using the four levels to improve performance.* Berrett-Koehler Publishers.

Klein, G. A. (2008). Naturalistic decision making. *Human Factors: The Journal of the Human Factors and Ergonomics Society, 50*(3), 456–460. https://doi.org/10.1518/001872008X288385

Klein, G. A. (2003). *The power of intuition. How to use your gut feelings to make better decisions at work.* Doubleday.

Klein, G. A., Orasanu, J., Calderwood, R., & Zsambok, C. E. (Eds.). (1993). *Decision making in action: Models and methods.* Ablex Publishing Corporation.

Klein, K. J., Ziegert, J. C., Knight, A. P., & Xiao, Y. (2006). Dynamic delegation: Shared, hierarchical, and deindividualized leadership in extreme action teams. *Administrative Science Quarterly, 51*(4), 590–621.

Klein, R. L., Bigley, G. A., & Roberts, K. H. (1995). Organizational culture in high reliability organizations: An extension. *Human Relations, 48*(7), 771–793.

Kletz, T., & Amyotte, P. (2019). *What went wrong? Case histories of process plant disasters and how they could have been avoided* (6th ed.). Butterworth-Heinemann. https://aiche.onlinelibrary.wiley.com/doi/10.1002/prs.12245

Kluckhohn, C. (1962). *Culture and behavior: Collected essays of Clyde Kluckhohn.* Free Press of Glencoe.

Knight, C., & Haslam, S. A. (2010). Your place or mine? Organizational identification and comfort as mediators of relationships between the managerial control of workspace and employees' satisfaction and well-being. *British Journal of Management, 21*(3), 717–735. https://doi.org/10.1111/j.1467-8551.2009.00683.x

Knowles, R. N. (2002). *The leadership dance: Pathways to extraordinary organizational effectiveness.* The Center for Self-Organizing Leadership.

Kotter, J. P. (1996). *Leading change. Why transformation efforts fail.* Harvard Business School Press.

Kruger, J., & Dunning, D. (1999). Unskilled and unaware of it: How difficulties in recognizing one's own incompetence lead to inflated self-assessments. *Journal of Personality and Social Psychology, 77*(6), 1121–1134.

Künzle, B., Kolbe, M., & Grote, G. (2010a). Ensuring patient safety through effective leadership behaviour: A literature review. *Safety Science, 48*(1), 1–17. https://doi.org/10.1016/j.ssci.2009.06.004

Künzle, B., Zala-Mezö, E., Kolbe, M., Wacker, J., & Grote, G. (2010b). Substitutes for leadership in anaesthesia teams and their impact on leadership effectiveness. *European Journal of Work and Organizational Psychology, 19*(5), 505–531. https://doi.org/10.1080/13594320902986170

Kurtz, C. F., & Snowden, D. J. (2003). The new dynamics of strategy: Sense-making in a complex and complicated world. *IBM Systems Journal, 42*(3), 462–483.

La Duke, P. (2020). *Blood in my pockets is blood on your hands. Special edition.* Marriah Publishing.
La Porte, T. R. (1996). High reliability organizations: Unlikely, demanding and at risk. *Journal of Contingencies and Crisis Management, 4*(2), 60–71. https://doi.org/10.1111/j.1468-5973.1996.tb00078.x
La Porte, T. R., & Consolini, P. M. (1991). Working in practice but not in theory: Theoretical challenges of "High-reliability organizations". *Journal of Public Administration Research and Theory, 1*(1), 19–48. https://doi.org/10.1093/oxfordjournals.jpart.a037070
Labowitz, S., & Baumann-Pauly, D. (2014). *Business as usual is not an option. Supply chains and sourcing after Rana Plaza.* NYU Stern Center for Business and Human Rights.
Lakoff, G., & Johnson, M. (1980). *Metaphors we live by.* University of Chicago Press.
Le Coze, J.-C. (2019a). How safety culture can make us think. *Safety Science, 118*, 221–229. https://doi.org/10.1016/j.ssci.2019.05.026
Le Coze, J.-C. (2019b). Vive la diversité! High Reliability Organisation (HRO) and Resilience Engineering (RE). *Safety Science, 117*, 469–478. https://doi.org/10.1016/j.ssci.2016.04.006
Le Coze, J.-C. (2021a). Broad (multilevel) safety research and strategy. A sociological study. *Safety Science, 136*, 105132. https://doi.org/10.1016/j.ssci.2020.105132
Le Coze, J.-C. (2021b). In the footsteps of Turner. From grounded theory to conceptual ethnography in safety. In K. Pettersen Gould & C. Macrae (Eds.), *Inside hazardous technological systems: Methodological foundations, challenges and future directions.* CRC Press.
Le Coze, J.-C. (2021c). *Post normal accident: Revisiting Perrow's classic.* CRC Press.
Le Coze, J.-C. (2022). The 'new view' of human error. Origins, ambiguities, successes and critiques. *Safety Science, 154*, 105853. https://doi.org/10.1016/j.ssci.2022.105853
Le Coze, J.-C. (2023). Coupling and complexity at the global scale: Flows, networks, interconnectedness and synchronicity (e.g. Covid-19). *Safety Science, 165*, 106193. https://doi.org/10.1016/j.ssci.2023.106193
Le Coze, J.-C. (2024). Understanding the 'blues of safety professionals'. *International Journal of Occupational Safety and Ergonomics, 30*(2), 351–365. https://doi.org/10.1080/10803548.2023.2298561
Le Coze, J-C. & Reiman, T. (2023, Eds.), *Visualising Safety, an Exploration. Drawings, Pictures, Images, Videos and Movies.* SpringerBriefs in Applied Sciences and Technology.
Lee, F. (1993). Being polite and keeping MUM: How bad news is communicated in organizational hierarchies. *Journal of Applied Social Psychology, 23*(14), 1124–1149. https://doi.org/10.1111/j.1559-1816.1993.tb01025.x
Lee, K., & Ashton, M. C. (2005). Psychopathy, machiavellianism, and narcissism in the five-factor model and the HEXACO model of personality structure. *Personality and Individual Differences, 38*(7), 1571–1582. https://doi.org/10.1016/j.paid.2004.09.016
Lei, Z. (2018). Fast, slow, and pause: Understanding error management via a temporal lens. In J. U. Hagen (Ed.), *How could this happen? Managing errors in organizations.* Palgrave Macmillan.
Leisti, T., & Poskiparta, H. (2022). *Päätöksenteon illuusiot. Ymmärrä paremmin, miten teet päätöksiä.* Finnish. Tuuma.
engineering. MIT Press.
Leveson, N. G., Dulac, N., Marais, K., & Carroll, J. (2009). Moving beyond normal accidents and high reliability organizations: A systems approach to safety in complex systems. *Organization Studies, 30*(2–3), 227–249. https://doi.org/10.1177/0170840608101478
Lloyd, C. (2020). *Next generation safety leadership. From compliance to care.* CRC Press.
Loftus, E. F., & Palmer, J. C. (1974). Reconstruction of automobile destruction: An example of the interaction between language and memory. *Journal of Verbal Learning & Verbal Behavior, 13*(5), 585–589.
Lukes, S. (2005). *Power. A radical view* (2nd ed.). Palgrave Macmillan UK.
Lyubykh, Z., Turner, N., Hershcovis, M. S., & Deng, C. (2022). A meta-analysis of leadership and workplace safety: Examining relative importance, contextual contingencies, and methodological moderators. *Journal of Applied Psychology, 107*(12), 2149–2175. https://doi.org/10.1037/apl0000557
Madigan, C., Johnstone, K., Way, K. A., & Capra, M. (2021). How do safety professionals' influence managers within organizations? – A critical incident approach. *Safety Science, 144*, 105478. https://doi.org/10.1016/j.ssci.2021.105478
Madsen, P. M. (2013). Perils and profits: A reexamination of the link between profitability and safety in U.S. Aviation. *Journal of Management, 39*(3), 763–791. https://doi.org/10.1177/0149206310396374
Maidment, D. (1997). The effect of the accident inquiry process within the railway industry. In A. R. Hale, B. Wilbert, & M. Freitag (Eds.), *After the event. From accident to organisational learning.* Pergamon.
Maier, T. R., Aymans, S. C., & Mühlfelder, M. (2024). Product safety culture as a concept for the development of safety–critical products in regulated industries. *Safety Science, 169*, 106336. https://doi.org/10.1016/j.ssci.2023.106336

Maldonado, J. A. (2019). *A practical guide to the safety profession. The relentless pursuit.* CRC Press.
Manuele, F. A. (2013). *On the practice of safety* (4th ed.). Wiley. https://doi.org/10.1002/9781118574683
Manuele, F. A. (2014). *Advanced safety management. Focusing on Z10 and serious injury prevention.* Wiley.
Manuele, F. A. (2015). Culture change agent. The overarching role of OSH professionals. *Professional Safety, 60*(12), 38–44.
March, J., & Simon, H. A. (1958). *Organizations.* John Wiley & Sons.
Marques, J. M., & Paez, D. (1994). The 'black sheep effect': Social categorization, rejection of ingroup deviates, and perception of group variability. *European Review of Social Psychology, 5*(1), 37–68. https://doi.org/10.1080/14792779543000011
Marquet, L. D. (2020). Leadership Is Language. The Hidden Power of What You Say – and What You Don't. Penguin.
Marriott, C. (2018). *Challenging the Safety Quo.* Routledge.
Martin, J. (2002). *Organizational culture: Mapping the terrain.* SAGE Publications.
Mattson Molnar, M., Von Thiele Schwarz, U., Hellgren, J., Hasson, H., & Tafvelin, S. (2019). Leading for safety: A question of leadership focus. *Safety and Health at Work, 10*(2), 180–187. https://doi.org/10.1016/j.shaw.2018.12.001
Mayer, R. C., Davis, J. H., & Schoorman, F. D. (1995). An integrative model of organizational trust. *The Academy of Management Review, 20*(3), 709–734.
Mazmanian, M., Orlikowski, W. J., & Yates, J. (2013). The autonomy paradox: The implications of mobile email devices for knowledge professionals. *Organization Science, 24*(5), 1337–1357. https://doi.org/10.1287/orsc.1120.0806
McDonald, A. J., & Hansen, J. R. (2009). *Truth, lies, and O-rings: Inside the space shuttle challenger disaster.* University Press of Florida.
McDougall, S. J. (2014). *Managing the gaps in a regimen of control – Operational decision making at a nuclear power plant* (Master's thesis). Lund University.
McEvily, B., Perrone, V., & Zaheer, A. (2003). Trust as an organizing principle. *Organization Science, 14*(1), 91–103.
McMillan, E. (2008). *Complexity, management and the dynamics of change.* Routledge. https://doi.org/10.4324/9780203885659
Meadows, D. H. (2008). *Thinking in systems. A primer.* Chelsea Green Publishing.
Mercier, H., & Sperber, D. (2011). Why do humans reason? Arguments for an argumentative theory. *Behavioral and Brain Sciences, 34*(2), 57–74. https://doi.org/10.1017/S0140525X10000968
Meyer, J.P. & Allen, N. J. (1991). A three-component conceptualization of organizational commitment. *Human Resource Management Review*, 1, 61-89.
Mills, J. H. (2003). *Making Sense of Organizational Change.* Routledge.
Monteiro, G. P., Hopkins, A., & Melo, P. F. F. E. (2020a). How do organizational structures impact operational safety? Part 1 – Understanding the dangers of decentralization. *Safety Science, 123*, 104568.
Monteiro, G. P., Hopkins, A., & Melo, P. F. F. E. (2020b). How do organizational structures impact operational safety? Part 2 – Designing structures that strengthen safety. *Safety Science, 123*, 104534.
Morel, G., Amalberti, R., & Chauvin, C. (2009). How good micro/macro ergonomics may improve resilience, but not necessarily safety. *Safety Science, 47*(2), 285–294. https://doi.org/10.1016/j.ssci.2008.03.002
Morrow, S. L., Koves, K. G., & Barnes, V. E. (2014). Exploring the relationship between safety culture and safety performance in U.S. nuclear power operations. *Safety Science, 69*, 37–47.
Morgan, G. (2006). *Images of organization. Updated edition.* Sage.
Mosey, D. (2006). *Reactor accidents. Institutional failure in the nuclear industry* (2nd ed.). Nuclear Engineering International.
Moskowitz, G. B. (2024). *Introduction to social cognition. The essential questions and ideas.* The Guilford Press.
Moura, R., Beer, M., Patelli, E., Lewis, J., & Knoll, F. (2016). Learning from major accidents to improve system design. *Safety Science, 84*, 37–45. https://doi.org/10.1016/j.ssci.2015.11.022
Mullen, J. E., & Kelloway, E. K. (2009). Safety leadership: A longitudinal study of the effects of transformational leadership on safety outcomes. *Journal of Occupational and Organizational Psychology, 82*(2), 253–272. https://doi.org/10.1348/096317908X325313
Mundy, A. (2001). *Dispensing with the truth. The victims, the drug companies, and the dramatic story behind the battle over Fen-Phen.* St. Martin's Press.
Myers, C. G., Sutcliffe, K. M., & Ferrari, B. T. (2019). Treating the "not-invented-here syndrome" in medical leadership: Learning from the insights of outside disciplines. *Academic Medicine, 94*(10), 1416–1418. https://doi.org/10.1097/ACM.0000000000002860

Naderpajouh, N., Zhang, R. P., & Hayes, J. (2022). Outsourced enforcement: Improving the public accountability of building inspectors. In J. Hayes & S. Tillement (Eds.), *Contracting and safety: Exploring outsourcing practices in high-hazard industries*. Springer International Publishing. https://doi.org/10.1007/978-3-030-89792-5

Nævestad, T.-O. (2008). Safety understandings among crane operators and process operators on a Norwegian offshore platform. *Safety Science*, *46*(3), 520–534. https://doi.org/10.1016/j.ssci.2007.05.013

NAS. (2014). *Lessons learned from the Fukushima nuclear accident for improving safety of the U.S. Nuclear Plants*. The National Academy of Sciences (NAS).

Nguyen, V. Q., Turner, N., Barling, J., Axtell, C. M., & Davies, S. (2023). Reconciling general transformational leadership and safety-specific transformational leadership: A paradox perspective. *Journal of Safety Research*, *84*, 435–447. https://doi.org/10.1016/j.jsr.2022.12.006

Nielsen, M. B., Eid, J., Mearns, K., & Larsson, G. (2013). Authentic leadership and its relationship with risk perception and safety climate. *Leadership & Organization Development Journal*, *34*(4), 308–325. https://doi.org/10.1108/LODJ-07-2011-0065

Noort, M. C., Reader, T. W., & Gillespie, A. (2021). Safety voice and safety listening during aviation accidents: Cockpit voice recordings reveal that speaking-up to power is not enough. *Safety Science*, *139*, 105260. https://doi.org/10.1016/j.ssci.2021.105260

Norros, L. (2004). *Acting under uncertainty: The core-task analysis in ecological study of work*. VTT Publications 546. VTT Technical Research Centre of Finland.

NTSB. (1990). *Marine accident report. Grounding of the U.S. tankship EXXON VALDEZ on Bligh Reef, Prince William Sound near Valdez, Alaska, March 24, 1989* (NTSB/MAR-90/04). National Transportation Safety Board (NTSB).

Obonai, A., Watanabe, T., & Hirata, K. (2014). Successful cold shutdown of Onagawa: The closest nuclear power station to the March 11, 2011, epicenter. *Nuclear Technology*, *186*(2), 280–294. https://doi.org/10.13182/NT13-61

O'Connor, P., O'Dea, A., Flin, R., & Belton, S. (2008). Identifying the team skills required by nuclear power plant operations personnel. *International Journal of Industrial Ergonomics*, *38*(11–12), 1028–1037. https://doi.org/10.1016/j.ergon.2008.01.014

Oedewald, P., Pietikäinen, E., & Reiman, T. (2011). *A guidebook for evaluating organizations in the nuclear industry – An example of safety culture evaluation* (2011:20). Swedish Radiation Safety Authority (SSM).

Oedewald, P., & Reiman, T. (2003). Core task modelling in cultural assessment: A case study in nuclear power plant maintenance. *Cognition, Technology & Work*, *5*(4), 283–293. https://doi.org/10.1007/s10111-003-0132-0

Oedewald, P., & Reiman, T. (2007). *Special characteristics of safety critical organizations. Work psychological perspective*. VTT Publications 633. VTT Technical Research Centre of Finland.

Oedewald, P., Skjerve, A. B., Axelsson, C., Viitanen, K., Pietikäinen, E., & Reiman, T. (2014). *The expected and experienced benefits of Human performance tools in nuclear power plant maintenance activities – Intermediate report of HUMAX project* (NKS-300). Nordic Nuclear Safety Research (NKS).

Office of Inspector General. (2015). *FAA lacks an effective staffing model and risk-based oversight process for organization designation authorization* (Audit Report AV-2016-001). U.S. Department of Transportation.

Ouchi, W. G. (1980). Markets, bureaucracies, and clans. *Administrative Science Quarterly*, *25*(1), 129. https://doi.org/10.2307/2392231

Packer, D. J. (2009). Avoiding groupthink: Whereas weakly identified members remain silent, strongly identified members dissent about collective problems. *Psychological Science*, *20*(5), 546–548. https://doi.org/10.1111/j.1467-9280.2009.02333.x

Palmer, D. (2012). *Normal organizational wrongdoing. A critical analysis of theories of misconduct in and by organizations*. Oxford University Press.

Palmer, I., Dunford, R. W., & Buchanan, D. A. (2022). *Managing organizational change: A multiple perspectives approach* (4th ed.). McGraw Hill.

Pariès, J. (2021). *Another look at rule-based and managed safety… Some further thoughts in response to René Amalberti*. TRIBUNES DE LA SÉCURITÉ INDUSTRIELLE n°09.

Pariès, J., Macchi, L., Valot, C., & Deharvengt, S. (2019). Comparing HROs and RE in the light of safety management systems. *Safety Science*, *117*, 501–511. https://doi.org/10.1016/j.ssci.2018.02.026

Parker, D., Lawrie, M., & Hudson, P. (2006). A framework for understanding the development of organisational safety culture. *Safety Science*, *44*(6), 551–562. https://doi.org/10.1016/j.ssci.2005.10.004

Patankar, M. S., & Taylor, J. C. (2004). *Applied human factors in aviation maintenance: A practical guide to improving safety*. Ashgate.

Paulhus, D. L., & Williams, K. M. (2002). The dark triad of personality: Narcissism, machiavellianism, and psychopathy. *Journal of Research in Personality, 36*(6), 556–563. https://doi.org/10.1016/S0092-6566(02)00505-6

Peadon, R., Hurley, J., & Hutchinson, M. (2020). Hierarchy and medical error: Speaking up when witnessing an error. *Safety Science, 125*, 104648. https://doi.org/10.1016/j.ssci.2020.104648

Perin, C. (2005). *Shouldering risks. The culture of control in the nuclear power industry*. Princeton University Press.

Perrow, C. (1984). *Normal accidents. Living with high-risk technologies*. Basic Books.

Perrow, C. (1999). *Normal accidents: Living with high-risk technologies* (2nd ed.). Princeton University Press.

Perrow, C. (2007). *The next catastrophe: Reducing our vulnerabilities to natural, industrial, and terrorist disasters*. Princeton University Press.

Peters, T., & Waterman, R. H. (1982). *In search of excellence. Lessons from America's best-run companies*. Harper & Row.

Peterson, R. A., Rhi-Perez, P., & Albaum, G. (2014). A cross-national comparison of extreme response style measures. *International Journal of Market Research, 56*(1), 89–110. https://doi.org/10.2501/IJMR-2014-005

Pettersen, K. A., & Schulman, P. R. (2019). Drift, adaptation, resilience and reliability: Toward an empirical clarification. *Safety Science, 117*, 460–468. https://doi.org/10.1016/j.ssci.2016.03.004

Pettersen Gould, K., & Macrae, C. (2021). Hazardous technological systems from the inside out. An introduction. In K. Pettersen Gould & C. Macrae (Eds.), *Inside hazardous technological systems: Methodological foundations, challenges and future directions*. CRC Press.

Pettigrew, A. M. (1979). On studying organizational cultures. *Administrative Science Quarterly, 24*(4), 570–581.

Pfeffer, J. (1981). *Power in organizations*. Pitman Publishing.

Pidgeon, N. (1991). Safety culture and risk management in organizations. *Journal of Cross-Cultural Psychology, 22*(1), 129–140. https://doi.org/10.1177/0022022191221009

Pidgeon, N. (1998). Safety culture: Key theoretical issues. *Work & Stress, 12*(3), 202–216. https://doi.org/10.1080/02678379808256862

Pidgeon, N., Hood, C., Jones, D., Turner, B., & Gibson, R. (1992). Risk perception. In The Royal Society (Ed.), *Risk: Analysis, perception and management*. The Royal Society.

Pidgeon, N., Kasperson, R. E., & Slovic, P. (Eds.). (2003). *The social amplification of risk*. Cambridge University Press.

Pilbeam, C., Denyer, D., & Sutliff, M. (2025). Guided by principles or rules: A Delphi study on how safety professionals frame safety practices. *Safety Science, 184*, 106772. https://doi.org/10.1016/j.ssci.2024.106772

Pilbeam, C., Doherty, N., Davidson, R., & Denyer, D. (2016). Safety leadership practices for organizational safety compliance: Developing a research agenda from a review of the literature. *Safety Science, 86*, 110–121. https://doi.org/10.1016/j.ssci.2016.02.015

Plokhy, S. (2018). *Chernobyl. History of a tragedy*. Basic Books.

Poole, G., & Dekker, S. W. A. (2024). *Random noise. Measuring your company's safety performance*. Routledge.

Pratama, G. B., & Caponecchia, C. (2025). Examining the influence of national culture on aviation safety: A systematic review. *Journal of Safety Research, 92*, 317–330. https://doi.org/10.1016/j.jsr.2024.11.020

Presidential Commission. (1986). *Report to the president by the presidential commission on the space shuttle challenger accident*.

Prigogine, I., & Stengers, I. (1984). *Order out of chaos: Man's new dialogue with nature*. Bantam Books.

Probst, T. M., & Estrada, A. X. (2010). Accident under-reporting among employees: Testing the moderating influence of psychological safety climate and supervisor enforcement of safety practices. *Accident Analysis & Prevention, 42*(5), 1438–1444. https://doi.org/10.1016/j.aap.2009.06.027

Provan, D. J. (2021). *A field guide to safety professional practice*. Safety Futures.

Provan, D. J., Dekker, S. W. A., & Rae, A. J. (2017). Bureaucracy, influence and beliefs: A literature review of the factors shaping the role of a safety professional. *Safety Science, 98*, 98–112. https://doi.org/10.1016/j.ssci.2017.06.006

Provan, D. J., Woods, D. D., Dekker, S. W. A., & Rae, A. J. (2020). Safety II professionals: How resilience engineering can transform safety practice. *Reliability Engineering & System Safety, 195*, 106740. https://doi.org/10.1016/j.ress.2019.106740

Pupulidy, I., & Vesel, C. (2023). *Human & organization potential*. Dynamic Inquiry LLC.

Quinlan, M., Hampson, I., & Gregson, S. (2013). Outsourcing and offshoring aircraft maintenance in the US: Implications for safety. *Safety Science*, *57*, 283–292. https://doi.org/10.1016/j.ssci.2013.02.011

Quinn, R. E. (1988). *Beyond rational management: Mastering the paradoxes and competing demands of high performance*. Jossey-Bass.

Quinn, R. E. (2015). *The positive organization. Breaking free from conventional cultures, constraints, and beliefs*. Berrett-Koehler.

Quinn, R. E., & Rohrbaugh, J. (1983). A spatial model of effectiveness criteria: Towards a competing values approach to organisational analysis. *Management Science*, *29*(3), 363–377.

Rae, A. J., Provan, D. J., Weber, D. E., & Dekker, S. W. A. (2018). Safety clutter: The accumulation and persistence of 'safety' work that does not contribute to operational safety. *Policy and Practice in Health and Safety*, *16*(2), 194–211. https://doi.org/10.1080/14773996.2018.1491147

Ramanujam, R. (2003). The effects of discontinuous change on latent errors in organizations: The moderating role of risk. *Academy of Management Journal*, *46*(5), 608–617. https://doi.org/10.2307/30040652

Rasmussen, J. (1982). Human errors. A taxonomy for describing human malfunction in industrial installations. *Journal of Occupational Accidents*, *4*, 311–333.

Rasmussen, J. (1997). Risk management in a dynamic society: A modelling problem. *Safety Science*, *27*(2–3), 183–213. https://doi.org/10.1016/S0925-7535(97)00052-0

Rasmussen, J., & Jensen, A. (1974). Mental procedures in real-life tasks: A case study of electronic trouble shooting. *Ergonomics*, *17*(3), 293–307. https://doi.org/10.1080/00140137408931355

Rasmussen, J., & Svedung, I. (2000). *Proactive risk management in a dynamic society*. Swedish Rescue Services Agency.

Reader, T. W., Noort, M. C., Shorrock, S., & Kirwan, B. (2015). Safety sans frontières: An international safety culture model. *Risk Analysis*, *35*(5), 770–789. https://doi.org/10.1111/risa.12327

Reason, J. T. (1990). *Human error*. Cambridge University Press.

Reason, J. T. (1997). *Managing the risks of organizational accidents*. Ashgate.

Reason, J. T. (2000). Human error: Models and management. *BMJ*, *320*, 768–770.

Reason, J. T. (2016). *Organizational accidents revisited*. CRC Press.

Reason, J. T., & Hobbs, A. (2003). *Managing maintenance error: A practical guide*. Ashgate.

Reason, J. T., Parker, D., & Lawton, R. (1998). Organizational controls and safety: The varieties of rule-related behaviour. *Journal of Occupational and Organizational Psychology*, *71*, 289–304.

Reiman, T., & Oedewald, P. (2007). Assessment of complex sociotechnical systems – Theoretical issues concerning the use of organizational culture and organizational core task concepts. *Safety Science*, *45*(7), 745–768. https://doi.org/10.1016/j.ssci.2006.07.010

Reiman, T., & Oedewald, P. (2009). *Evaluating safety-critical organizations – Emphasis on the nuclear industry* (2009:12). Swedish Radiation Safety Authority (SSM).

Reiman, T., & Pietikäinen, E. (2012). Leading indicators of system safety – Monitoring and driving the organizational safety potential. *Safety Science*, *50*(10), 1993–2000. https://doi.org/10.1016/j.ssci.2011.07.015

Reiman, T., Pietikäinen, E., & Oedewald, P. (2008). *Turvallisuuskulttuuri: Teoria ja arviointi*. Finnish. VTT Publications 700. VTT Technical Research Centre of Finland.

Reiman, T., & Rollenhagen, C. (2011). Human and organizational biases affecting the management of safety. *Reliability Engineering & System Safety*, *96*(10), 1263–1274. https://doi.org/10.1016/j.ress.2011.05.010

Reiman, T., & Rollenhagen, C. (2014). Does the concept of safety culture help or hinder systems thinking in safety? *Accident Analysis & Prevention*, *68*, 5–15. https://doi.org/10.1016/j.aap.2013.10.033

Reiman, T., & Rollenhagen, C. (2018). Safety culture. In N. Möller, S. O. Hansson, J. E. Holmberg, & C. Rollenhagen (Eds.), *Handbook of safety principles*. John Wiley & Sons.

Reiman, T., Rollenhagen, C., Pietikäinen, E., & Heikkilä, J. (2015). Principles of adaptive management in complex safety–critical organizations. *Safety Science*, *71*, 80–92. https://doi.org/10.1016/j.ssci.2014.07.021

Reiman, T., & Viitanen, K. (2019a). Modelling the influence of safety management tools on resilience. In S. Wiig & B. Fahlbruch (Eds.), *Exploring resilience: A scientific journey from practice to theory*. Springer.

Reiman, T., & Viitanen, K. (2019b). Towards actionable safety science. In J.-C. Le Coze (Ed.), *Safety science research: Evolution, challenges and new directions* (1st ed.). CRC Press. https://doi.org/10.4324/9781351190237

Reiman, T., Viitanen, K., Hakala, J., & Wrona, K. (2023). Safety culture in nuclear power plant construction. In N. Balfe, R. Charles, & D. Golightly (Eds.), *Contemporary ergonomics & human factors*. Chartered Institute of Ergonomics & Human Factors (CIEHF).

Renn, O., Laubichler, M., Lucas, K., Kröger, W., Schanze, J., Scholz, R. W., & Schweizer, P. (2022). Systemic risks from different perspectives. *Risk Analysis*, *42*(9), 1902–1920. https://doi.org/10.1111/risa.13657

Rijpma, J. A. (1997). Complexity, tight–coupling and reliability: Connecting normal accidents theory and high reliability theory. *Journal of Contingencies and Crisis Management*, *5*(1), 15–23. https://doi.org/10.1111/1468-5973.00033

Rittel, H. W. J., & Webber, M. M. (1973). Dilemmas in a general theory of planning. *Policy Sciences*, *4*(2), 155–169.

Roberts, K. H. (1989). New challenges in organizational research: High reliability organizations. *Industrial Crisis Quarterly*, *3*(2), 111–125. https://doi.org/10.1177/108602668900300202

Roberts, K. H. (Ed.). (1993). *New challenges to understanding organizations*. Macmillan.

Roberts, K. H., Rousseau, D. M., & La Porte, T. R. (1994). The culture of high reliability: Quantitative and qualitative assessment aboard nuclear-powered aircraft carriers. *The Journal of High Technology Management Research*, *5*(1), 141–161.

Robison, P. (2021). *Flying -blind: The 737 MAX tragedy and the fall of boeing*. Doubleday.

Roe, E., & Schulman, P. R. (2008). *High reliability management: Operating on the edge*. Stanford University Press. https://doi.org/10.1515/9781503627123

Rollenhagen, C. (2010). Can focus on safety culture become an excuse for not rethinking design of technology? *Safety Science*, *48*(2), 268–278. https://doi.org/10.1016/j.ssci.2009.07.008

Rollenhagen, C. (2021). From classical human factors toward a system view—Experiences from the human factors nuclear field in Sweden. In A.-M. Teperi & N. Gotcheva (Eds.), *Human factors in the nuclear industry* (pp. 55–71). Elsevier. https://doi.org/10.1016/B978-0-08-102845-2.00003-X

Rotteau, L., Goldman, J., Shojania, K. G., Vogus, T. J., Christianson, M., Baker, G. R., Rowland, P., & Coffey, M. (2022). Striving for high reliability in healthcare: A qualitative study of the implementation of a hospital safety programme. *BMJ Quality & Safety*, *31*(12), 867–877. https://doi.org/10.1136/bmjqs-2021-013938

Roughton, J. E., Crutchfield, N., & Waite, M. (2019). *Safety culture: An innovative leadership approach* (2nd ed.). Butterworth-Heinemann.

Ryu, A., & Meshkati, N. (2014). *Why you haven't heard about Onagawa nuclear power station after the earthquake and tsunami of March 11, 2011*.

Sagan, S. D. (1993). *The limits of safety: Organizations, accidents, and nuclear weapons*. Princeton University Press.

Salas, E., Cooke, N. J., & Rosen, M. A. (2008). On Teams, teamwork, and team performance: Discoveries and developments. *Human Factors*, *50*(3), 540–547.

Salas, E., Wilson, K. A., Burke, C. S., & Wightman, D. C. (2006). Does crew resource management training work? An update, an extension, and some critical needs. *Human Factors: The Journal of the Human Factors and Ergonomics Society*, *48*(2), 392–412. https://doi.org/10.1518/001872006777724444

Saleem, F., Malik, M. I., & Malik, M. K. (2021). Toxic leadership and safety performance: Does organizational commitment act as stress moderator? *Cogent Business & Management*, *8*(1), 1960246. https://doi.org/10.1080/23311975.2021.1960246

Salgado, J. F. (2002). The big five personality dimensions and counterproductive behaviors. *International Journal of Selection and Assessment*, *10*(1–2), 117–125. https://doi.org/10.1111/1468-2389.00198

Salgado, J. F., Anderson, N., & Moscoso, S. (2020). Personality at Work. In P. J. Corr & G. Matthews (Eds.), *The Cambridge handbook of personality psychology* (2nd ed.). Cambridge University Press.

Samuelson, W., & Zeckhauser, R. (1988). Status quo bias in decision making. *Journal of Risk and Uncertainty*, 1, 7-59.

Sandberg, R., & Lowstedt, M. (2025). 'The happily overworked professional': Unpacking the autonomy paradox in excessive work regimes amongst construction site managers. *Safety Science*, *184*, 106760. https://doi.org/10.1016/j.ssci.2024.106760

Sanne, J. M. (2008). Framing risks in a safety-critical and hazardous job: Risk-taking as responsibility in railway maintenance. *Journal of Risk Research*, *11*(5), 645–658. https://doi.org/10.1080/13669870701715550

Sapolsky, R. (2017). *Behave. The biology of humans at our best and worst*. Vintage.

Saunders, F. C., Gale, A. W., & Sherry, A. H. (2016). Responding to project uncertainty: Evidence for high reliability practices in large-scale safety–critical projects. *International Journal of Project Management*, *34*(7), 1252–1265. https://doi.org/10.1016/j.ijproman.2016.06.008

Savage, S. (2021, April 27). Eight years on from factory collapse garment workers are again at risk, unions warn. *The Daily Telegraph*.

Schein, E. H. (1985). *Organizational culture and leadership. A dynamic view*. Jossey-Bass.

Schein, E. H. (2010). *Organizational culture and leadership* (4th ed.). Jossey-Bass.
Schein, E. H. (2013). *Humble inquiry. The gentle art of asking instead of telling.* Berrett-Koehler.
Schein, E. H. (2017). *Organizational culture and leadership* (5th ed.). Wiley.
Schein, E. H., & Schein, P. A. (2018). *Humble leadership. The power of relationships, openness, and trust.* Berrett-Koehler.
Schein, E. H., & Schein, P. A. (2021). *Humble inquiry. The gentle art of asking instead of telling* (2nd ed.). Berrett-Koehler Publishers.
Schraagen, J. M., Chipman, S. F., & Shalin, V. L. (Eds.). (2014). *Cognitive task analysis* (Reprinted). Psychology Press.
Schulman, P. R. (1993). The negotiated order of organizational reliability. *Administration & Society, 25*(3), 353–372.
Schulman, P. R. (1996). Heroes, organizations and high reliability. *Journal of Contingencies and Crisis Management, 4*(2), 72–82. https://doi.org/10.1111/j.1468-5973.1996.tb00079.x
Schulman, P. R. (2020). Organizational structure and safety culture: Conceptual and practical challenges. *Safety Science, 126*, 104669. https://doi.org/10.1016/j.ssci.2020.104669
Senge, P. M. (1990). *The fifth discipline. The art and practice of the learning organization.* Century Business.
Sennett, R. (1998). *The corrosion of character. The personal consequences of work in the new capitalism.* W.W. Norton.
Serling, R. J. (1992). *Legend and legacy. The story of boeing and its people.* St. Martin's Press.
Sexton, J. B., Adair, K. C., Profit, J., Bae, J., Rehder, K. J., Gosselin, T., Milne, J., Leonard, M., & Frankel, A. (2021). Safety culture and workforce wellbeing associations with positive leadership walkrounds. *The Joint Commission Journal on Quality and Patient Safety, 47*(7), 403–411. https://doi.org/10.1016/j.jcjq.2021.04.001
Sexton, J. B., Helmreich, R. L., Neilands, T. B., Rowan, K., Vella, K., Boyden, J., Roberts, P. R., & Thomas, E. J. (2006). The safety attitudes questionnaire: Psychometric properties, benchmarking data, and emerging research. *BMC Health Services Research, 6*(1), 44. https://doi.org/10.1186/1472-6963-6-44
Sgobba, T. (2019). B-737 MAX and the crash of the regulatory system. *Journal of Space Safety Engineering, 6*(4), 299–303. https://doi.org/10.1016/j.jsse.2019.09.006
Sharman, A. (2019). *Naked safety: Exploring the dynamics of safety in a fast-changing world.* Routledge.
Silbey, S. S. (2009). Taming prometheus: Talk about safety and culture. *Annual Review of Sociology, 3*, 341–369.
Simon, H. A. (1957). *Models of man, social and rational: Mathematical essays on rational human behavior in a social setting.* John Wiley & Sons.
Singer, J. (2022). *There are no accidents. The deadly rise of injury and disaster—who profits and who pays the price.* Simon & Schuster.
Skogstad, A., Einarsen, S., Torsheim, T., Aasland, M. S., & Hetland, H. (2007). The destructiveness of laissez-faire leadership behavior. *Journal of Occupational Health Psychology, 12*(1), 80–92. https://doi.org/10.1037/1076-8998.12.1.80
Skogstad, A., Hetland, J., Glasø, L., & Einarsen, S. (2014). Is avoidant leadership a root cause of subordinate stress? Longitudinal relationships between laissez-faire leadership and role ambiguity. *Work & Stress, 28*(4), 323–341. https://doi.org/10.1080/02678373.2014.957362
Sleiman, A. A., Sigurjonsdottir, S., Elnes, A., Gage, N. A., & Gravina, N. E. (2020). A quantitative review of performance feedback in organizational settings (1998–2018). *Journal of Organizational Behavior Management, 40*(3–4), 303–332. https://doi.org/10.1080/01608061.2020.1823300
Slovic, P., Peters, E., Finucane, M. L., & MacGregor, D. G. (2005). Affect, risk, and decision making. *Health Psychology, 24*(4, Suppl), S35–S40. https://doi.org/10.1037/0278-6133.24.4.S35
Smaggus, A. (2019). Safety-I, Safety-II and burnout: how complexity science can help clinician wellness. *BMJ Quality & Safety*, 28, 667-671.
Smallman, C., & Weir, D. (1999). Communication and cultural distortion during crises. *Disaster Prevention and Management: An International Journal, 8*(1), 33–41. https://doi.org/10.1108/09653569910258219
Smircich, L. (1983). Concepts of culture and organizational analysis. *Administrative Science Quarterly, 28*(3), 339. https://doi.org/10.2307/2392246
Smircich, L., & Morgan, G. (1982). Leadership: The management of meaning. *Journal of Applied Behavioral Science, 18*(3), 257–273.
Smith, G. (2024). *Proving safety. Wicked problems, legal risk management and the tyranny of metrics.* Wayland Legal.
Snook, S. A. (2000). *Friendly fire. The accidental shootdown of U.S. Black Hawks over Northern Iraq.* Princeton University Press.

Starbuck, W. H., & Milliken, F. J. (1988). Challenger: Fine-tuning the odds until something breaks. *Journal of Management Studies*, *25*(4), 319–340. https://doi.org/10.1111/j.1467-6486.1988.tb00040.x

Steele, F. I. (1972). Organizational overlearning. *Journal of Management Studies*, *9*(3), 303–313. https://doi.org/10.1111/j.1467-6486.1972.tb00558.x

Steffens, N. K., Haslam, S. A., & Reicher, S. D. (2014). Up close and personal: Evidence that shared social identity is a basis for the 'special' relationship that binds followers to leaders. *The Leadership Quarterly*, *25*(2), 296–313. https://doi.org/10.1016/j.leaqua.2013.08.008

Steffens, N. K., Haslam, S. A., Schuh, S. C., Jetten, J., & van Dick, R. (2017). A meta-analytic review of social identification and health in organizational contexts. *Personality and Social Psychology Review*, *21*(4), 303–335.

Stiles, S., Ryan, B., & Golightly, D. (2018). Evaluating attitudes to safety leadership within rail construction projects. *Safety Science*, *110*, 134–144. https://doi.org/10.1016/j.ssci.2017.12.030

Strauss, C., & Quinn, N. (1997). *A cognitive theory of cultural meaning*. Cambridge University Press.

Svedung, I., & Rasmussen, J. (2002). Graphic representation of accident scenarios: Mapping system structure and the causation of accidents. *Safety Science*, *40*(5), 397–417. https://doi.org/10.1016/S0925-7535(00)00036-9

Swuste, P., Groeneweg, J., Guldenmund, F. W., Gulijk, C. V., Lemkowitz, S., Oostendorp, Y., & Zwaard, W. (2021). *From safety to safety science: The evolution of thinking and practice*. Routledge. https://doi.org/10.4324/9781003001379

Tajfel, H. (1978). *Differentiation between social groups: Studies in the social psychology of intergroup relations*. London: Academic Press.

Taleb, N. N. (2012). *Antifragile: Things that gain from disorder*. Random House.

Tawfik, D. S., Adair, K. C., Palassof, S., Sexton, J. B., Levoy, E., Frankel, A., Leonard, M., Proulx, J., & Profit, J. (2023a). Leadership behavior associations with domains of safety culture, engagement, and health care worker wellbeing. *The Joint Commission Journal on Quality and Patient Safety*, *49*, 156–165. https://doi.org/10.1016/j.jcjq.2022.12.006

Tawfik, D. S., Adair, K. C., Palassof, S., Sexton, J. B., Levoy, E., Frankel, A., Leonard, M., Proulx, J., & Profit, J. (2023b). Leadership behavior associations with domains of safety culture, engagement, and health care worker wellbeing. *The Joint Commission Journal on Quality and Patient Safety*, *49*(3), 156–165. https://doi.org/10.1016/j.jcjq.2022.12.006

Taylor, C., Maben, J., Jagosh, J., Carrieri, D., Briscoe, S., Klepacz, N., & Mattick, K. (2024). Care under pressure 2: A realist synthesis of causes and interventions to mitigate psychological ill health in nurses, midwives and paramedics. *BMJ Quality & Safety*33: 523–538. https://doi.org/10.1136/bmjqs-2023-016468

Taylor, J. R. (2007). Statistics of design error in the process industries. *Safety Science*, *45*(1–2), 61–73. https://doi.org/10.1016/j.ssci.2006.08.013

Tett, G. (2015). *The Silo effect. The peril of expertise and the promise of breaking down barriers*. Simon & Schuster.

The Culture Factor. (2025). Global report 2025. Culture and safety. https://www.theculturefactor.com/resources/report/global-report-2025

Thompson, L. L. (2018). *Making the team: A guide for managers* (6th ed.). Pearson.

Tillement, S., & Leuridan, G. (2022). Contracting and safety: Lessons from observing an outsourcing process "in the making". In J. Hayes & S. Tillement (Eds.), *Contracting and safety: Exploring outsourcing practices in high-hazard industries*. Springer. https://doi.org/10.1007/978-3-030-89792-5

Trist, E. L., & Bamforth, K. W. (1951). Some social and psychological consequences of the Longwall method of coal getting. *Human Relations*, *4*, 3–38.

TUKES. (2004). *Onnettomuustutkintaraportti. Sähköasentajan kuolemaan johtanut sähkötapaturma Fortum Power and Heat Oy:n Loviisan ydinvoimalaitoksella* (Dnro 3634/06/2004). Finnish. Turvatekniikan Keskus (TUKES).

Turner, B. A. (1978). *Man-made disasters*. Wykeham Publications.

Turner, B. A., & Pidgeon, N. (1997). *Man-made disasters* (2nd ed.). Butterworth-Heinemann.

Turner, J. C. (1991). *Social influence*. Open University Press.

Turner, J. C., Brown, R. J., & Tajfel, H. (1979). Social comparison and group interest in ingroup favouritism. *European Journal of Social Psychology*, *9*(2), 187–204. https://doi.org/10.1002/ejsp.2420090207

Tversky, A., & Kahneman, D. (1974). Judgment under uncertainty: Heuristics and biases. *Science*, *185*(4157), 1124–1131.

Tversky, A., & Kahneman, D. (1991). Loss aversion in riskless choice: A reference-dependent model. *The Quarterly Journal of Economics*, *106*(4), 1039–1061.

United States Congress. (2011). *Final report of the President's National Commission on the BP deepwater horizon oil spill and offshore drilling: Oversight hearing before the Committee on Natural Resources* (House. Committee On Natural Resources U.S. House of Representatives, One Hundred Twelfth Congress, first session, Wednesday). U.S. G.P.O.

Väänänen, A. (2024). *The rise of mental vulnerability at work. A socio-historical and cultural analysis.* Policy Press.

Vaish, A., Grossmann, T., & Woodward, A. (2008). Not all emotions are created equal: The negativity bias in social-emotional development. *Psychological Bulletin, 134*(3), 383–403. https://doi.org/10.1037/0033-2909.134.3.383

Vartiainen, M., & Vanharanta, O. (2024). True nature of hybrid work. *Frontiers in Organizational Psychology, 2*, 1448894. https://doi.org/10.3389/forgp.2024.1448894

Vaughan, D. (1996). *The challenger launch decision: Risky technology, culture, and deviance at NASA.* University of Chicago Press.

Vaughan, D. (1999). The dark side of organizations: Mistake, misconduct, and disaster. *Annual Review of Sociology, 25*(1), 271–305. https://doi.org/10.1146/annurev.soc.25.1.271

Viitanen, K. (2021). Human performance tools as a part of programmatic human performance improvement. In A.-M. Teperi & N. Gotcheva (Eds.), *Human factors in the nuclear industry* (pp. 107–126). Elsevier. https://doi.org/10.1016/B978-0-08-102845-2.00006-5

Viitanen, K., Airola, M., & Gotcheva, N. (2022). *Effective improvement of leadership and safety culture. Intermediate report.* VTT Research Report No. VTT-R-00545-22. VTT Technical Research Centre of Finland.

Viitanen, K., Airola, M., & Nadezhda, G. (2023). *Approaches to assessing effective improvement of safety culture.* VTT Research Report No. VTT-R-00112-23. VTT Technical Research Centre of Finland.

Viitanen, K., Gotcheva, N., Rollenhagen, C., & Reiman, T. (2018). *Safety culture assurance and improvement methods in complex projects. Final report from the NKS-R SC_AIM* (NKS-405). Nordic Nuclear Safety Research (NKS).

Viitanen, K., Reiman, T., Jakobsson, F., Sylvander, C., & Karadeniz, S. (2025). *Internal nuclear safety oversight as part of organizational defence-in-depth – Lessons learned for the Nordic nuclear industry. Final report from the NKS-R INSOLE activity* (NKS-494). Nordic Nuclear Safety Research (NKS).

Viitanen, K., Reiman, T., Karadeniz, S., Airola, M., Jakobsson, F., Sylvander, C., & Lind, S. (2024). *Internal nuclear safety oversight as part of organizational defence-in-depth. Lessons learned for the nordic nuclear industry. Intermediate report from the NKS-R INSOLE activity* (NKS-482). Nordic Nuclear Safety Research (NKS).

Vogus, T. J., & Sutcliffe, K. M. (2007). The impact of safety organizing, trusted leadership, and care pathways on reported medication errors in hospital nursing units. *Medical Care, 45*(10), 997–1002.

Vogus, T. J., Sutcliffe, K. M., & Weick, K. E. (2010). Doing no harm: Enabling, enacting, and elaborating a culture of safety in health care. *Academy of Management Perspectives, 24*, 60–77.

Vogus, T. J., & Welbourne, T. M. (2003). Structuring for high reliability: HR practices and mindful processes in reliability-seeking organizations. *Journal of Organizational Behavior, 24*(7), 877–903. https://doi.org/10.1002/job.221

Walumbwa, F. O., Avolio, B. J., Gardner, W. L., Wernsing, T. S., & Peterson, S. J. (2008). Authentic leadership: Development and validation of a theory-based measure. *Journal of Management, 34*(1), 89–126. https://doi.org/10.1177/0149206307308913

WANO. (2013). *Traits of a healthy nuclear safety culture* (WANO Principles PL 2013-1). World Association of Nuclear Operators (WANO).

WANO & IAEA. (2018). *Independent oversight* (No. GL 2018-01). World Association of Nuclear Operators (WANO).

Weeks, J. (2004). *Unpopular culture. The ritual of complaint in a British Bank.* University of Chicago Press.

Weick, K. E. (1987). Organizational culture as a source of high reliability. *California Management Review, 29*(2), 112–127. https://doi.org/10.2307/41165243

Weick, K. E. (1988). Enacted sensemaking in crisis situations. *Journal of Management Studies, 25*(4), 305–317. https://doi.org/10.1111/j.1467-6486.1988.tb00039.x

Weick, K. E. (1990). The vulnerable system: An analysis of the Tenerife air disaster. *Journal of Management, 16*(3), 571–593.

Weick, K. E. (1993). The collapse of sensemaking in organizations: The Mann Gulch disaster. *Administrative Science Quarterly, 38*(4), 628. https://doi.org/10.2307/2393339

Weick, K. E. (1995). *Sensemaking in organizations.* SAGE Publications.

Weick, K. E. (1996). Drop your tools: An allegory for organizational studies. *Administrative Science Quarterly, 41*(2), 301–313.
Weick, K. E. (1998). Foresights of failure: An appreciation of Barry Turner. *Journal of Contingencies and Crisis Management, 6*(2), 72–75.
Weick, K. E. (2005). Organizing and failures of imagination. *International Public Management Journal, 8*(3), 425–438. https://doi.org/10.1080/10967490500439883
Weick, K. E. (2006). The role of imagination in the organizing of knowledge. *European Journal of Information Systems, 15*(5), 446–452. https://doi.org/10.1057/palgrave.ejis.3000634
Weick, K. E. (2009). *Making sense of the organization. Volume 2. The impermanent organization.* John Wiley & Sons.
Weick, K. E., & Quinn, R. E. (1999). Organizational change and development. *Annual Review of Psychology, 50*, 361–386.
Weick, K. E., & Roberts, K. H. (1993). Collective mind in organizations: Heedful interrelating on flight decks. *Administrative Science Quarterly, 38*(3), 357–381.
Weick, K. E., & Sutcliffe, K. M. (2001). *Managing the unexpected. Assuring high performance in an age of complexity.* Jossey-Bass.
Weick, K. E., & Sutcliffe, K. M. (2007). *Managing the unexpected: Resilient performance in the age of uncertainty* (2nd ed.). Jossey-Bass.
Weick, K. E., & Sutcliffe, K. M. (2015). *Managing the unexpected. Sustained performance in a complex world* (3rd ed.). Wiley.
Weick, K. E., Sutcliffe, K. M., & Obstfeld, D. (1999). Organizing for high reliability: Processes of collective mindfulness. In R. S. Sutton & B. M. Staw (Eds.), *Research in organizational behavior* (Vol. 1, pp. 81–123). Jai Press.
Weick, K. E., Sutcliffe, K. M., & Obstfeld, D. (2005). Organizing and the process of sensemaking. *Organization Science, 16*(4), 409–421.
Weiss, M., Kolbe, M., Grote, G., Spahn, D. R., & Grande, B. (2018). We can do it! Inclusive leader language promotes voice behavior in multi-professional teams. *The Leadership Quarterly, 29*(3), 389–402. https://doi.org/10.1016/j.leaqua.2017.09.002
Westrum, R. (1993). Cultures with requisite imagination. In J. A. Wise, V. D. Hopkin, & P. Stager (Eds.), *Verification and validation of complex systems: Human factors issues* (pp. 401–416). Springer Berlin Heidelberg. https://doi.org/10.1007/978-3-662-02933-6_25
Westrum, R. (2004). A typology of organisational cultures. *Quality and Safety in Health Care, 13*(suppl_2), ii22–ii27. https://doi.org/10.1136/qshc.2003.009522
Wetzel, C. M., Kneebone, R. L., Woloshynowych, M., Nestel, D., Moorthy, K., Kidd, J., & Darzi, A. (2006). The effects of stress on surgical performance. *The American Journal of Surgery, 191*(1), 5–10. https://doi.org/10.1016/j.amjsurg.2005.08.034
White, D. G. J. (2017). *Rethinking culture. Embodied cognition and the origin of culture in organizations.* Routledge.
White, D. G. J. (2021). *Disrupting corporate culture. How cognitive science alters accepted beliefs about culture and culture change and its impacts on leaders and change agents.* Routledge.
Whittingham, R. B. (2008). *Preventing corporate accidents: An ethical approach.* Butterworth-Heinemann.
WHO. (2009). *Implementation manual. WHO surgical safety checklist 2009* (p. 16). World Health Organization (WHO). https://iris.who.int/handle/10665/44186
Wiggins, M. W. (2022). *Introduction to human factors for organisational psychologists.* CRC Press. https://doi.org/10.1201/9781003229858
Wilson, J. R. (2014). Fundamentals of systems ergonomics/human factors. *Applied Ergonomics, 45*(1), 5–13. https://doi.org/10.1016/j.apergo.2013.03.021
Woods, D. D. (2006). How to design a safety organization: Test case for resilience engineering. In E. Hollnagel, D. D. Woods, & N. Leveson (Eds.), *Resilience engineering. Concepts and precepts* (1st ed., pp. 315–325). Ashgate. https://doi.org/10.1201/9781315605685-26
Woods, D. D. (2009). Escaping failures of foresight. *Safety Science, 47*(4), 498–501. https://doi.org/10.1016/j.ssci.2008.07.030
Woods, D. D. (2018a). Resilience is a verb. In B. D. Trump, M.-V. Florin, & I. Linkov (Eds.), *IRGC resource guide on resilience: Domains of resilience for complex interconnected systems* (Vol. 2). EPFL International Risk Governance Center (IRGC). http://infoscience.epfl.ch/record/262527
Woods, D. D. (2018b). The theory of graceful extensibility: Basic rules that govern adaptive systems. *Environment Systems and Decisions, 38*(4), 433–457. https://doi.org/10.1007/s10669-018-9708-3

Woods, D. D. (2019). Essentials of resilience, revisited. In M. Ruth & S. Goessling-Reisemann (Eds.), *Handbook on resilience of sociotechnical systems*. Edward Elgar Publishing. https://doi.org/10.4337/9781786439376.00009

Woods, D. D., Dekker, S. W. A., Cook, R., Johannesen, L., & Sarter, N. (2010). *Behind human error* (2nd ed.). CRC Press. https://doi.org/10.1201/9781315568935

Woods, D. D., & Hollnagel, E. (2006). Prologue: Resilience engineering concepts. In E. Hollnagel, D. D. Woods, & N. Leveson (Eds.), *Resilience engineering: Concepts and precepts* (Repr). Ashgate.

Wreathall, J. (2009). Leading? Lagging? Whatever! *Safety Science*, *47*(4), 493–494. https://doi.org/10.1016/j.ssci.2008.07.031

Yankelovich, D. (1971). Interpreting the New Life Styles. *Sales Management, the Marketing Magazine,* 15 November 1971, *pp. 26-27.*

Yeung, S. K., Yay, T., & Feldman, G. (2022). Action and inaction in moral judgments and decisions: Meta-analysis of omission bias omission-commission asymmetries. *Personality and Social Psychology Bulletin,* 48, 1499-1515.

Yukl, G. A., & Gardner, W. L. (2020). *Leadership in organizations* (9th ed.). Pearson Education.

Yun, S., Faraj, S., & Sims, H. P. (2005). Contingent leadership and effectiveness of trauma resuscitation teams. *Journal of Applied Psychology*, *90*(6), 1288–1296. https://doi.org/10.1037/0021-9010.90.6.1288

Zohar, D. (1980). Safety climate in industrial organizations: Theoretical and applied implications. *Journal of Applied Psychology*, *65*(1), 96–102.

Zohar, D. (2002). The effects of leadership dimensions, safety climate, and assigned priorities on minor injuries in work groups. *Journal of Organizational Behavior*, *23*(1), 75–92.

Zohar, D., & Luria, G. (2005). A multilevel model of safety climate: Cross-level relationships between organization and group-level climates. *Journal of Applied Psychology*, *90*(4), 616–628. https://doi.org/10.1037/0021-9010.90.4.616

Index

Note: The page numbers provided in **bold** refer to figures.

D

E

T

U

V

W

For Product Safety Concerns and Information please contact our EU
representative GPSR@taylorandfrancis.com
Taylor & Francis Verlag GmbH, Kaufingerstraße 24, 80331 München, Germany

www.ingramcontent.com/pod-product-compliance
Ingram Content Group UK Ltd.
Pitfield, Milton Keynes, MK11 3LW, UK
UKHW052219190926
475702UK00006B/1102

* 9 7 8 1 0 3 2 7 8 9 1 6 3 *